Sustainable Fertilization Management Consequences to Horticultural Crops

Sustainable Fertilization Management Consequences to Horticultural Crops

Guest Editors

Wacław Jarecki
Maciej Balawejder
Natalia Matłok

MDPI

Basel • Beijing • Wuhan • Barcelona • Belgrade • Novi Sad • Cluj • Manchester

Guest Editors
Wacław Jarecki
Department of Crop Production
University of Rzeszów
Rzeszów
Poland

Maciej Balawejder
Department of Chemistry and Food Toxicology
University of Rzeszów
Rzeszów
Poland

Natalia Matłok
Department of Food and Agriculture Production Engineering
University of Rzeszów
Rzeszów
Poland

Editorial Office
MDPI AG
Grosspeteranlage 5
4052 Basel, Switzerland

This is a reprint of the Special Issue, published open access by the journal *Horticulturae* (ISSN 2311-7524), freely accessible at: https://www.mdpi.com/journal/horticulturae/special_issues/Fertilization_Horticultural_Crops.

For citation purposes, cite each article independently as indicated on the article page online and as indicated below:

Lastname, A.A.; Lastname, B.B. Article Title. *Journal Name* **Year**, *Volume Number*, Page Range.

ISBN 978-3-7258-5351-9 (Hbk)
ISBN 978-3-7258-5352-6 (PDF)
https://doi.org/10.3390/books978-3-7258-5352-6

Cover image courtesy of Natalia Matłok

Contents

About the Editors

Wacław Jarecki

Wacław Jarecki is Fixed-term Assistant Professor of the Faculty of Technology and Life Sciences of the University of Rzeszów. His scientific activity concerns research topics related to the optimization of plant production in diverse habitat conditions.

His research includes improving the production technology of crop plants, especially cereals, legumes and vegetables, the influence of agrotechnical and environmental factors on the differentiation of the quality parameters of seeds of various species of cultivar plants, and the influence of stress factors on the course of physiological processes in crop plants.

He has authored and co-authored over 100 scientific works and is an experienced researcher working on two research projects funded by the Ministry of Science and Higher Education and the National Centre for Research and Development. He has been a scholarship holder at the following universities: Wrocław University of Environmental and Life Sciences, the University of Natural Resources and Life Sciences in Vienna, the University of Oradea and the Slovak University of Agriculture in Nitra. He is also a member of various organizations within the industry, including the Polish Agronomic Society.

Maciej Balawejder

Maciej Balawejder is a Full Professor of the Faculty of Technology and Life Sciences of the University of Rzeszów.

The scientific interests of Professor Maciej Balawejder concern the following issues: the modification of the phytochemical composition of agricultural and horticultural plants, methods of extending the shelf life of plant raw materials, chromatographic analysis of volatile compounds and methods of separating stereoisomers by crystallization.

Prof. Balawejder has established significant cooperation with the ago-food industry. This cooperation concerns the implementation of R&D projects and innovative technological and technical solutions for agricultural practice. He has been and is involved as a participant and/or has scientific responsiblity in several research projects concerning different aspects of gardening. He has been a member of the organizing committees of several national and international scientific conferences. He has published more than 100 items indexed on SCOPUS.

Natalia Matłok

Natalia Matłok is Fixed-term Assistant Professor of the Faculty of Technology and Life Sciences of the University of Rzeszów. Her scientific activity concerns research topics related to the cultivation and modification of the phytochemical composition of agricultural and horticultural plants, and methods of extending the shelf life of plant materials. She is a member of international and national societies of horticulture and fruitculture. He has published more than 50 items indexed on SCOPUS. She undertakes numerous activities aimed at establishing cooperation with the socio-economic environment. This collaboration involves the implementation of research and the development of projects aimed at developing and implementing innovative technological and technical solutions for agricultural and horticultural practices. Furthermore, she is the author of numerous technical studies, including patents and utility models, some of which have been commercialized and implemented in industrial practice. Selected inventions have been presented and awarded at international invention exhibitions.

Preface

Horticultural crops require many essential nutrients for optimum growth, yield and quality (N, P, K, Mg, S, Ca and microelements). Hence, these are applied through fertilizers. Fertilizer application must be planned to accommodate any variability. The main factors for consideration are crop type, anticipated yield, the naturally available nutrients in the soil and the changes in nutrients required by a particular crop during its growth cycle. Calculations are usually based on a combination of scientific research and local experience. A crop's nutrient uptake often varies, both by field and by year. Field variability is primarily due to different soil conditions, while annual variability is often a result of changing weather patterns. Application programmes that do not sufficiently account for these variations in uptake can lead to incorrect fertilization.

Wacław Jarecki, Maciej Balawejder, and Natalia Matłok
Guest Editors

horticulturae

Editorial

Sustainable Fertilization Management Consequences to Horticultural Crops

Wacław Jarecki [1,*], Maciej Balawejder [2] and Natalia Matłok [3]

1 Department of Crop Production, University of Rzeszow, Zelwerowicza 4 St., 35-601 Rzeszow, Poland
2 Department of Chemistry and Food Toxicology, University of Rzeszow, St. Ćwiklińskiej 2D, 35-601 Rzeszow, Poland; mbalawejder@ur.edu.pl
3 Department of Food and Agriculture Production Engineering, University of Rzeszow, Zelwerowicza 4 St., 35-601 Rzeszow, Poland; nmatlok@ur.edu.pl
* Correspondence: wjarecki@ur.edu.pl

1. Introduction

Agricultural systems must identify fertilization strategies in line with the principles of sustainable development and the circular economy to achieve environmentally friendly food production and to meet the fertilizer-reduction targets set by some regions in the world, e.g., the EU (European Union) (Contribution 2). The promotion of sustainable plant fertilization is currently very important in agriculture, as it helps reduce greenhouse gas emissions and carbon footprint, and is economically justified [1]. The improper use of fertilizers in agriculture has led to environmental pollution and adverse effects on human health. Therefore, in line with the requirements of sustainable development, it has become necessary to reduce the use of chemical fertilizers and to seek alternative methods of crop fertilization [2,3]. Current achievements in increasing agricultural production, particularly in developed countries, are attributed to three main factors: the development of irrigation, the advancement of improved, more productive, and disease-resistant crop varieties, and the use of chemical fertilizers [4]. However, agricultural activity is widely recognized as a main source of environmental pollution due to the introduction of heavy metals through fertilizers, pesticides, sewage sludge, and irrigation water [5]. Therefore, it is necessary to find a compromise between ensuring global food security and protecting the natural environment [6–8].

In agriculture, nitrogen is a primary macronutrient which is essential for the growth and development of crops. Its widespread use increases yields and profitability; however, it should be noted that plant productivity is influenced by many other factors (Contributions 3, 4). Avoiding the excessive use of nitrogen (N) in agriculture without negatively impacting yields has long been a priority for both scientific research and agricultural practices [9,10].

Previous research has indicated the need for region-specific and sustainable fertilization strategies to optimize agricultural production. Future research should focus on improving fertilization methods through interdisciplinary approaches, considering soil–plant–climate interactions and adapting these practices to diverse agroecosystems. By aligning agricultural practices with ecological principles, farmers can protect soil health, the foundation of environmental sustainability and human well-being, while ensuring food security for present and future generations [11].

https://doi.org/10.3390/horticulturae11091049

2. The Impact of Sustainable Fertilization Management on Soil Properties

Venig and Peticilă (Contribution 11) report that the NPK fertilization rate for a cultivated field should be determined after assessing the soil properties and the nutrient requirements of individual species. They recommend conducting regular soil chemical analyses to determine the optimal fertilization regime. Šimansky et al. (Contribution 6) demonstrate that, in addition to NPK fertilization, sowing grass strips between vineyard rows exerts beneficial effects, improving both soil chemical composition and crop yield. Zapałowska et al. [12] argue that fertilization with compost derived from organic waste is a promising approach that can reduce the environmental impact of chemical fertilizers, improve waste management, and soil quality, while promoting sustainable agricultural practices.

3. The Impact of Sustainable Fertilization Management on the Yield and Quality of Horticultural Crops

Venig and Peticilă (Contribution 11) observe that achieving optimal growth of grafted trees requires a combination of irrigation and NPK fertilization. Therefore, it is necessary to monitor soil moisture levels and irrigate during periods of water deficit, when trees have increased water demand. Similar results are obtained by Bello et al. (Contribution 3) in cucumber cultivation. In dry regions, irrigation and nitrogen fertilization are considered particularly important agronomic practices. Tallarita et al. (Contribution 5) conclude that the ratio of nitrogen to potassium applied is important in cauliflower cultivation. Chatzistathis et al. (Contribution 4) demonstrate that the effects of organic or mineral fertilization could modify the performance of applied Rhizobacteria and arbuscular mycorrhizal fungi. Arshad et al. (Contribution 7) consider improving fruit quality and reducing pre-harvest fruit drop as key objectives in *Citrus reticulata* cultivation. For this reason, appropriate fertilization strategies for this crop should be developed. The latter authors also demonstrate that the application of K + Zn fertilization can be a practical approach for citrus growers aiming to increase fruit yield. Tallarita et al. (Contribution 10) indicate that the biofortification of leafy vegetables with selenium (Se) and iodine (I) provides an important basis for optimizing food composition for human health.

4. Conventional and Alternative Fertilization Methods in Horticultural Crops

Recently, numerous new and alternative fertilizers and their application methods have been developed, yet many farmers still use inorganic fertilizers in a traditional manner. Many commonly used chemical fertilizers are supplied in simple forms, which are typically characterized by low efficiency. Therefore, they are applied in large quantities to achieve high yields [13]. Zhang et al. (Contribution 8) demonstrate that traditional calcium fertilizers often have low solubility, poor uptake, and limited mobility. In contrast, calcium-chelated fertilizers provide improved efficiency and uptake. Ruamrungsri et al. (Contribution 1) confirm that plasma nitrate could be an alternative source of nitrate N, offering a safer option for the environment and human health regarding nitrate accumulation. However, the economic feasibility of this approach must be assessed before implementing the technology on a commercial scale. Castillo-Díaz et al. (Contribution 2) argue that biodecontamination is an effective fertilization strategy that allows the use of inorganic fertilizers to be reduced in long-term, intensive horticultural cycles. This is particularly important for farmers who require alternative solutions to adapt their production systems to sustainable agriculture principles. Wang et al. (Contribution 9) find that bioorganic material from the safe disposal of cattle and poultry carcasses can be used as fertilizer, with potential to partially or fully replace chemical fertilizers, contributing to their reduction.

5. Conclusions

The challenge of modern agriculture is to ensure food security for a constantly growing global population without excessive burden on the natural environment. For example, overuse of fertilizers negatively affects soil and generates additional risks. Therefore, sustainable nutrient management provides a pathway to addressing these issues, encompassing both changes in fertilizer production and their application in agricultural practice. Horticultural crops require a range of essential macro- and micronutrients for optimal growth, development, yield, and quality. Sustainable fertilization management involves the optimal supply of all nutrients to plants throughout their growing period. A deficiency of one nutrient cannot be compensated for by an excess of others. It is also important to consider the synergistic and antagonistic interactions between macro- and micronutrients. When planning fertilization, factors to consider include the species and variety, expected yield, soil chemical composition, plant tests, applied agricultural practices, and the selection of appropriate fertilizers, their doses, and application timing. The effects of fertilization often depend on local site conditions and variable weather patterns over the years. Fertilization is undergoing rapid development, directly linked to advances in satellite navigation, variable-rate application, soil and yield mapping, and precision machinery operation. Consequently, research in this area must be continuously advanced and is essential for both horticultural science and practice.

Author Contributions: The authors (W.J., M.B. and N.M.) contributed equally to this work. All authors have read and agreed to the published version of the manuscript.

Funding: This research received no external funding.

Acknowledgments: The Guest Editors of the Special Issue "Sustainable Fertilization Management Consequences to Horticultural Crops" wish to thank all the authors whose valuable manuscripts was published under this title and thus contributed to the success of the collection. We also want to thank all who assisted in developing this Special Issue.

Conflicts of Interest: The authors declare no conflicts of interest.

List of Contributions:

1. Ruamrungsri, S.; Sawangrat, C.; Panjama, K.; Sojithamporn, P.; Jaipinta, S.; Srisuwan, W.; Intanoo, M.; Inkham, C.; Thanapornpoonpong, S.-n. Effects of Using Plasma-Activated Water as a Nitrate Source on the Growth and Nutritional Quality of Hydroponically Grown Green Oak Lettuces. *Horticulturae* **2023**, *9*, 248. https://doi.org/10.3390/horticulturae9020248.
2. Castillo-Díaz, F.J.; Belmonte-Ureña, L.J.; Batlles-delaFuente, A.; Camacho-Ferre, F. Impact of Soil Biodisinfection Techniques in Horticultural Crops on Profitability within the Framework of the Circular Economy. *Horticulturae* **2023**, *9*, 859. https://doi.org/10.3390/horticulturae9080859.
3. Bello, A.S.; Huda, S.; Chen, Z.-H.; Khalid, M.F.; Alsafran, M.; Ahmed, T. Evaluation of Nitrogen andWater Management Strategies to Optimize Yield in Open Field Cucumber (*Cucumis sativus* L.) Production. *Horticulturae* **2023**, *9*, 1336. https://doi.org/10.3390/horticulturae9121336.
4. Chatzistathis, T.; Zoukidis, K.; Vasilikiotis, C.; Apostolidis, A.; Giannakoula, A.E.; Bountla, A.; Chatziathanasiadis, A. Plant-Growth-Promoting Rhizobacteria and Arbuscular Mycorrhizal Fungi May Improve Soil Fertility and the Growth, Nutrient Uptake, and Physiological Performance of Batavia Lettuce (*Lactuca sativa* L. var. longifolia) Plants. *Horticulturae* **2024**, *10*, 449. https://doi.org/10.3390/horticulturae10050449.
5. Tallarita, A.V.; Cozzolino, E.; Salluzzo, A.; Sekara, A.; Pokluda, R.; Murariu, O.C.; Vecchietti, L.; del Piano, L.; Lombardi, P.; Cuciniello, A.; et al. Effect of Transplanting Time and Nitrogen–Potassium Ratio on Yield, Growth, and Quality of Cauliflower Landrace Gigante di Napoli in Southern Italy. *Horticulturae* **2024**, *10*, 518. https://doi.org/10.3390/horticulturae10050518.
6. Šimansky, V.; Kačániová, M.; Juriga, M.; Čmiková, N.; Borotová, P.; Aydın, E.;Wójcik-Gront, E. Impact of Soil Management Practices on Soil Culturable Bacteriota and Species Diversity in

Central European a Productive Vineyard under Warm and Dry Conditions. *Horticulturae* **2024**, *10*, 753. https://doi.org/10.3390/horticulturae10070753.
7. Arshad, I.; Saleem, M.; Akhtar, M.; Shani, M.Y.; Farid, G.; Jarecki, W.; Ashraf, M.Y. Enhancing Fruit Retention and Juice Quality in 'Kinnow' (*Citrus reticulata*) Through the Combined Foliar Application of Potassium, Zinc, and Plant Growth Regulators. *Horticulturae* **2024**, *10*, 1245. https://doi.org/10.3390/horticulturae10121245.
8. Zhang, J.; Du, M.; Liu, G.;Ma, F.; Bao, Z. Lignin Sulfonate-Chelated Calcium Improves Tomato Plant Development and Fruit Quality by Promoting Ca^{2+} Uptake and Transport. *Horticulturae* **2024**, *10*, 1328. https://doi.org/10.3390/horticulturae10121328.
9. Wang, J.; Li, X.; Chen, A.; Li, Y.; Xue, M.; Feng, S. Effects of Exogenous Organic Matter on Soil Nutrient Dynamics and Its Role in Replacing Chemical Fertilizers for Vegetable Yield and Quality. *Horticulturae* **2024**, *10*, 1355. https://doi.org/10.3390/horticulturae10121355.
10. Tallarita, A.V.; Golubkina, N.; De Pascale, S.; Sękara, A.; Pokluda, R.; Murariu, O.C.; Cozzolino, E.; Cenvinzo, V.; Caruso, G. Effects of Selenium/Iodine Foliar Application and Seasonal Conditions on Yield and Quality of Perennial Wall Rocket. *Horticulturae* **2025**, *11*, 211. https://doi.org/10.3390/horticulturae11020211.
11. Venig, A.; Peticilă, A. Plum Trees' Leaf Area Response to Fertilization and Irrigation in the Nursery. *Horticulturae* **2025**, *11*, 737. https://doi.org/10.3390/horticulturae11070737.

References

1. Ding, Z.; Zeng, J.; He, Z.; Zhu, B.; Nie, J.; Zhou, Y.; Jiang, M.; Liu, Z. Optimizing fertilization strategies to reduce carbon footprints and enhance net ecosystem economic benefits in ratoon rice systems. *Agriculture* **2025**, *15*, 1715. [CrossRef]
2. Chen, Y.; Hu, S.; Guo, Z.; Cui, T.; Zhang, L.; Lu, C.; Yu, Y.; Luo, Z.; Fu, H.; Jin, Y. Effect of balanced nutrient fertilizer: A case study in Pinggu District, Beijing, China. *Sci. Total Environ.* **2021**, *754*, 142069. [CrossRef] [PubMed]
3. Gaballah, M.S.; Mansour, H.A.; Nofal, O.A. Balanced fertilization of major crops in Egypt: A review. *Plant Arch.* **2020**, *20*, 2453–2458.
4. Ryan, J. A perspective on balanced fertilization in the Mediterranean region. *Turk. J. Agric. For.* **2008**, *32*, 79–89.
5. Madjar, R.M.; Vasile Scăeteanu, G. An overview of heavy metal contamination in water from agriculture: Origins, monitoring, risks, and control measures. *Sustainability* **2025**, *17*, 7368. [CrossRef]
6. Wang, J.L.; Liu, K.L.; Zhao, X.Q.; Zhang, H.Q.; Li, D.; Li, J.J.; Shen, R.F. Balanced fertilization over four decades has sustained soil microbial communities and improved soil fertility and rice productivity in red paddy soil. *Sci. Total Environ.* **2021**, *793*, 148664. [CrossRef] [PubMed]
7. Montoya, D.; Gaba, S.; de Mazancourt, C.; Bretagnolle, V.; Loreau, M. Reconciling biodiversity conservation, food production and farmers, demand in agricultural landscapes. *Ecol. Model.* **2020**, *416*, 108889. [CrossRef] [PubMed]
8. Huang, W.; Lin, M.; Liao, J.; Li, A.; Tsewang, W.; Chen, X.; Sun, B.; Liu, S.; Zheng, P. Effects of potassium deficiency on the growth of tea (*Camelia sinensis*) and strategies for optimizing potassium levels in soil: A critical review. *Horticulturae* **2022**, *8*, 660. [CrossRef]
9. Cai, S.; Zhao, X.; Pittelkow, C.M.; Fan, M.; Zhang, X.; Yan, X. Optimal nitrogen rate strategy for sustainable rice production in China. *Nature* **2023**, *615*, 73–79. [CrossRef] [PubMed]
10. Asadkhani, E.; Ramroudi, M.; Asgharipour, M.R.; Shahhosseini, H.R. Challenges of sustainability of rice agrosystem: Insights from energy use, ecological footprint, and greenhouse gas emissions (case study: Golestan Province, Iran). *Agrosyst. Geosci. Environ.* **2025**, *8*, e70061. [CrossRef]
11. Xing, Y.; Xie, Y.; Wang, X. Enhancing soil health through balanced fertilization: A pathway to sustainable agriculture and food security. *Front. Microbiol.* **2025**, *16*, 1536524. [CrossRef] [PubMed]
12. Zapałowska, A.; Jarecki, W.; Skwiercz, A.; Malewski, T. Optimization of compost and peat mixture ratios for production of pepper seedlings. *Int. J. Mol. Sci.* **2025**, *26*, 442. [CrossRef]
13. Souri, M.K. Aminochelate fertilizers: The new approach to the old problem; a review. *Open Agric.* **2016**, *1*, 118–123. [CrossRef]

horticulturae

MDPI

Article

Plum Trees' Leaf Area Response to Fertilization and Irrigation in the Nursery

Adelina Venig [1,*] and Adrian Peticilă [2]

1 Department of Agriculture-Horticulture, Faculty of Environmental Protection, University of Oradea, 26 Magheru Street, 410048 Oradea, Romania
2 Department of Bioengineering of Horti-Viticultural Systems, Faculty of Horticulture, University of Agronomic Sciences and Veterinary Medicine of Bucharest, 59 Mărăști Boulevard, 011464 Bucharest, Romania; adrian.peticila@horticultura-bucuresti.ro
* Correspondence: adelina.venig@uoradea.ro

Abstract: This study addressed a significant and relevant issue, specifically the production of high-quality fruit planting material linked to an economically viable nursery operation. The process considered both the pedo-climatic conditions of the region where the fruit planting material was cultivated and the technological elements utilized. The objective of this research was to gather information regarding the necessity and effectiveness of implementing localized irrigation for plum trees in the nursery in the context of various fertilization treatments. It also aimed to investigate the variations in leaf area among Cacanska Lepotica and Stanley plum cultivars subjected to various irrigation (non-irrigated control, 10 mm, 20 mm, and 30 mm) and fertilization (unfertilized control, $N_8P_8K_8$, $N_{16}P_{16}K_{16}$, and $N_{24}P_{24}K_{24}$) methods. The study was conducted within a private nursery situated in the northwest region of Romania using a 4 × 2 × 4 split-split-plot design with five replications. This research took place in the summer of 2024, in the second field of the nursery during the growth stage of grafted trees. The implementation of various NPK fertilization methods (8%, 16%, and 24%) led to enhancements in leaf surface developments (increased by 6.53–16.14% compared to the control). The application of fertilization ranging from 8 to 16% and subsequently from 16 to 24% was effectively absorbed by the plum trees, resulting in a substantial growth of 180–226 cm^2. Irrigation with 30 mm generated significant increases in the leaf area of 4.42–14.27% compared to the control. To obtain optimal yields of grafted trees, it is advisable to utilize a combination of irrigation and NPK fertilization. To promote the appropriate growth and development of the trees, it is essential to monitor the soil moisture levels and to implement irrigation during times of water shortage when the trees exhibit heightened water usage. The research findings indicated that both cultivars experienced similar advantages from 24% NPK fertilization and 30 mm of irrigation; therefore, the implementation of the aforementioned technological elements is strongly recommended.

Keywords: macronutrients; water; leaf area; planting material; Cacanska Lepotica; Stanley

1. Introduction

The development of fruit trees in nurseries, like all cultivated species, is greatly impacted by the existing climatic and soil conditions. Water, in conjunction with heat, light, air, and minerals, is one of the key factors that greatly contributes to this development. According to Smith [1], water constitutes approximately 75–85% of the total weight of various tree organs, and in some cases, this proportion may exceed that range. Water not only facilitates the transport of nutrients from the soil to the plant but also serves as a

 https://doi.org/10.3390/horticulturae11070737

fundamental component in the synthesis of all organic substances that form the tissues of both rootstocks and trees. Therefore, it is essential for plants to have a consistent and adequate supply of water to ensure that growth processes occur at the optimal intensity [2].

The management of water and nutrients applied through fertilizers constitutes two primary factors that influence growth and productivity within tree nurseries. The emphasis on water and sustainable fertilizers in these nurseries has become increasingly important, leading to the implementation of management strategies that ensure satisfactory yields and thereby enhancing the efficiency of both fertilizer and water usage. In recent years, there has been a significant market demand for fruit trees, with the harvested fruit serving as a vital source of vitamins. Additionally, the planting materials produced in the nursery have substantial water requirements, necessitating irrigation throughout the growing seasons. Nurseries have faced challenging production periods due to weather-related factors in recent years. Consequently, it is deemed essential to gather information regarding the effective combination of irrigation practices and irrigation volumes. This research topic was selected due to the limited availability of data regarding the impacts of localized irrigation when combined with additional technological elements on fruit tree seedlings in nurseries. The current body of literature and prior studies have mainly focused on irrigation and fertilization practices in fruit tree orchards, rather than in nursery settings. This research holds considerable scientific importance as it may provide essential insights for fruit tree specialists engaged in seedling production. Practically, it aims to implement various irrigation and fertilization standards for maiden plum trees in nurseries, thereby enhancing the horticultural sector and elevating productivity levels. The anticipated outcomes resulting from the application of differentiated irrigation and fertilization practices include achieving greater economic efficiency compared to nurseries that do not utilize the aforementioned methods. Prior findings indicate that, when fertilization is applied, the following factors must be considered: soil conditions, species, cultivar, rootstock, planting density, and expected yield.

With regard to the need for water in the nursery, the present research started from the premise that it is necessary for the trees to have sufficient water available at all times so that the growth processes can take place with the greatest possible intensity. While a substantial quantity of water from the soil is absorbed by the roots for the synthesis of organic compounds, only a small fraction is utilized by the rootstock, with the remainder being lost through transpiration. Rootstocks cultivated in saturated, nutrient-deficient soils characterized by lower concentrations of fertilizing elements in the soil solution generally exhibit higher transpiration rates compared to the rootstocks planted in more fertile soils. In Romania, their cultivation is primarily concentrated in regions with abundant rainfall.

The addition of fertilizers and the provision of adequate amounts of water in the soil decrease the transpiration coefficient, thereby enhancing the efficiency of water usage by the rootstock; the transpiration coefficient represents the amount of water eliminated in the process of perspiration in the period in which one gram of dry substance is biosynthesized [3].

In horticulture, the administration of irrigation systems involves multiple components concerning water management and distribution, plant care, and the maintenance of irrigation infrastructure. The effectiveness of irrigation is significantly dependent on the correct implementation of such strategies. Research indicates that neglecting the aforementioned aspects may decrease the effectiveness of investments allocated to irrigation initiatives. Consequently, administering water to plants without considering their specific needs can adversely affect not only the current year's yield but also the overall quality of the produce [4].

Experiences in horticulture so far have demonstrated that providing a controlled supply of supplementary water to soil and plants beyond what is naturally available supports consistent agricultural output at high levels [5]. This approach ultimately aims to foster conditions for efficient and sustainable agricultural practices while safeguarding environmental elements. Among these factors, soil plays a crucial role, necessitating the maintenance and enhancement of its fertility. It is advisable to integrate irrigation, a vital method for combating drought, with other technical and agricultural practices, such as drainage (when necessary), fertilization, and the selection of superior plant cultivars [6–8].

Sullivan et al. [9] associated irrigation with a crucial technological intervention that has the potential to greatly improve agricultural productivity, irrespective of natural precipitation levels. Irrigation can effectively mitigate water shortages by providing additional water in line with the requirements of plants, which in turn enhances the efficient utilization of fertilizers due to the increased solubility of nutrients. Furthermore, it has the potential to improve product quality, prolong the growing season, and enable more efficient production planning.

In the climatic conditions in Romania, irrigation serves to supplement the naturally available water from rainfall during times when it is insufficient compared to crop requirements. In fact, it is used to effectively mitigate the variability in harvests from year to year, which is caused by natural factors. The aim of implementing irrigation is to achieve stable yields that align closely with the productive potential of crops under specific soil and climate conditions. Research conducted in Romania has revealed that there are years when inadequate rainfall leads to significantly diminished harvests, sometimes resulting in total crop failure [10].

Furthermore, studies in horticulture indicate that irrigating only along tree rows does not adversely affect root system development. Under such conditions, roots tend to cluster in a more confined area, without significantly shortening their overall length, thus maintaining their capacity for water and nutrient absorption [11].

In Romania, the circumstances are unique and influenced by its geographical location, which results in areas defined by a temperate continental climate with average annual rainfall ranging between 400 and 550 in irrigated agricultural regions. Thus, precipitation levels within the aforementioned limits prevent the agglomeration of the root system, especially along the rows of the plants. The roots grow both horizontally and vertically, similarly to the patterns observed with irrigation techniques such as sprinkling and surface runoff, without significant differences [12]. In fruit nurseries, the water content in the soil should be maintained within the limits of the active moisture range at a level higher than the minimum threshold. Thus, the soil water content should be kept above the wilting coefficient, near which plants undergo a series of changes during the growing process. Soil moisture in the range between the wilting coefficient (Co) and the field soil water capacity (C) represents the water that plants can utilize. The greatest quantity of water available to plants is characterized by the accessible water capacity (Caa) or the useful water capacity (Cau) of the soil. The moisture interval between these two constants is referred to as the active moisture range (AWR).

In modern fruit growing, fertilization, along with irrigation, stand out as some of the most crucial technological components. As fruit plants tend to occupy the same plot of land for extended periods, they develop extensive root systems and, due to their high yields, draw significant amounts of nutrients from the soil during the harvest time. In such circumstances, as pointed out by Jäger [13], yearly fertilization interventions in several rounds might ensure, on the one hand, the achievement of a certain level of production and, on the other hand, a certain level and ratio of nutrients. In horticulture, the fertilizer application system is primarily structured to address issues related to the effective replenishment of easily accessible nutrients in the soil and to restore these nutrients in a balanced manner, according to the requirements of rootstocks and seedlings. This is a crucial factor, as every harvest depletes the soil of specific mineral and organic compounds. Additionally, it is important to provide organic matter to the microflora of the soil; otherwise, these microorganisms may compete with higher plants for essential mineral nutrients [14]. Most microorganisms in the arable layer are heterotrophic, which means that, to obtain the energy needed for nutrition, they rely on organic matter previously synthesized by other organisms. Fertilization plays crucial roles in regulating the ionic composition of the soil and in adjusting the pH levels to meet the biological needs of rootstocks and seedlings.

Shibiao et al. [15] demonstrated that the application of fertilizers in fruit nurseries is intended to enhance nutritional conditions and boost organic matter synthesis while also facilitating substantial and cost-effective production with high-quality standards. This approach ensures that plant resistance to diseases and pests remains intact and minimizes environmental pollution. The effectiveness of fertilizers in increasing production is maximized only when they are integrated into a comprehensive system of technological practices, with the application rates tailored to the specific plant, soil and climatic conditions, and cultivation techniques.

Furthermore, replenishing soil nutrients depleted during harvest and providing organic matter to support the microflora is most effectively achieved through the application of both chemical and organic fertilizers.

The suitable dosages for each situation in the field should be determined based on a thorough analysis of various soil characteristics, alongside a good understanding of crops' requirements and the need to maintain specific qualitative and quantitative production standards. Given the wide range of conditions and circumstances, fertilization practices in orchards should offer tailored solutions for each unique case. Research on fertilizers, whether conducted in field settings or in greenhouses, seeks to clarify the mechanisms governing the mineral nutrition of trees and assesses the nutrient balance within both the plant and the soil. This research might also explore ways to improve fertilization in an efficient way [16].

Concerning the specific needs of plum trees in the nursery, fertilizers are applied during the active season, typically 2–3 times throughout the summer or more frequently, albeit in smaller amounts. Fertilizers with a high nitrogen content are utilized to promote the vegetative growth of plum trees. Nitrogen plays a crucial role in encouraging the vegetative growth of maiden plum trees, aiding in the development of leaves and green branches. A sufficient supply of nitrogen is vital for healthy tree growth and enhances the production of leaves, which are essential for photosynthesis and energy generation. Phosphorus is critical for the establishment of the root system in trees, as it aids in the formation of robust roots and facilitates the effective uptake of water and other nutrients from the soil. Potassium improves the trees' resilience to environmental stress and bolsters their defence mechanisms against diseases. However, the application of excessive fertilizers to maiden plum trees in the nursery may lead to rampant vegetative growth, reduced drought resistance, or complications in the absorption of other nutrients.

In the climatic context of recent years, Romania is encountering the issue of complex agricultural drought, a climatic hazard that results in the most severe repercussions for agriculture. In particular, within the nursery sector, irrigation provides the opportunity to produce vigorous, viable, high-quality, healthy, and damage-free planting material that is resistant to diseases and pests. Besides water, another crucial element in the production of quality fruit planting material is the fulfilment of fertilizer requirements, which underpins the rationale for this research. The objective of this research has been to establish the influences of irrigation, fertilization, and the plant cultivar on a physiological character in the plum species in the nursery. Based on this research, it can be concluded that the efficient and optimal utilization of fertilizers within a rational fertilization system is possible only by adhering to all the necessary requirements dictated by the cultivation technology. Effective agricultural techniques, even when employing reduced amounts of fertilizers, can result in higher crop yields as a consequence of the improved fertilizer utilization coefficient. The efficiency of the fertilization system is higher when the physical–chemical properties of the fertilizer are considered and a differentiated application is made, corresponding to the soil properties and the biological particularities of the species. This research holds practical significance through the implementation of various irrigation and fertilization techniques on maiden plum trees within the nursery aimed at advancing the horticultural sector and enhancing productivity. The anticipated outcomes from the application of these diverse irrigation and fertilization methods include the production of robust and healthy fruit planting material, an increase in nursery output, and the attainment of improved economic efficiency compared to nurseries that do not utilize irrigation or fertilization.

2. Materials and Methods

2.1. Characterization of the Climatic Conditions of the Experimental Field

This research was carried out in a private nursery in field II. It is important to note that although the rootstocks are located in field I of the nursery, they transition to field II after grafting in the subsequent spring. In this field, the scion is thus present as a bud that has been grafted onto the rootstock. Regardless of the establishment method used, the efforts undertaken in field II of the coppice are focused on maximizing the number of grafted shoots that are established in vegetation. In this particular instance, the rootstocks were planted in the spring of 2023, grafted during the summer of that same year, and by the spring of 2024, they had transitioned to field II. The nursery is located in the northwestern part of Romania on a plain unit, being the result of a long process of accumulation and erosion through the diversion of the hydrographic network, which descends from the mountainous and hilly areas.

Table 1 displays the average monthly temperatures and average monthly precipitation for that specific year. The warmest month was August with 23.2 °C, and the coldest month was January.

Table 1. The monthly average temperatures and monthly average rainfall in the research field in 2024.

Month	I	II	III	IV	V	VI	VII	VIII	IX	X	XI	XII
monthly average temperatures (°C)	−1.5	4.4	6.9	11.0	14.5	20.2	21.6	23.2	18.9	12.7	5.1	4.9
monthly average rainfall (L/mp)	13.5	45.2	43.8	15.4	60.0	101.5	48	45	48	48	35	70

I—January, II—February, III—March, IV—April, V—May, VI—June, VII—July, VIII—August, IX—September, X—October, XI—November, XII—December.

In 2024, recognized as the driest year, three periods could be identified based on the fluctuations in monthly precipitation relative to the multi-year average. The initial period, spanning from January to May, exhibited negative deviations ranging from 2 to 16 mm, followed by the months of June and July, with positive deviations of 39–85 mm and the July–December period, characterized by negative variations of 16–25 mm. The distribution of monthly precipitation throughout these three periods in 2024 also reflected negative variations of 2–48 mm in the period between January and May and of 9–21 mm in the last part of the year. Rainfall exceeding the multiannual average by 39–85 mm during

June and July significantly contributed to the overall precipitation for the year, positioning it as intermediary between the other two years in terms of rainfall amounts (Table 1). Information regarding climate conditions was sourced from the meteorological station, which monitors the temperature and precipitation patterns for each area. The rainfall patterns were assessed using a rain gauge placed in the experimental field.

2.2. Soil Characteristics and the Organization of the Experimental Field

To identify, delimit, and characterize the soil units within the land where the study was conducted, a pedological profile was developed. From it, samples were gathered on 4 horizons (Table 2). The assessments conducted and the methodologies employed were as follows: potentiometric pH measurement using a glass electrode in a 1:2 aqueous soil/water suspension; humus percentage determination via the Walkley–Black method; a mobile phosphorus assessment using the Eqner–Riehm–Domingo method; and a mobile potassium analysis also following the Eqner–Riehm–Domingo method. The concentration of potassium in the analysed soil extract was determined using a potassium interference filter and a photoelectric cell, and the mechanical analysis was conducted via the Kacinschi method after pre-treatment with hexametaphosphate for dispersion. Following the analyses, the identified soil taxonomic was arable, weakly glaciated, loamy clay on fluvial deposits. The soil texture was fine throughout the depth of the studied profile. The soil reaction was slightly alkaline at 0–30 cm and 50–80 cm, and neutral at 30–50 cm and 80–120 cm.

Table 2. Physical and hydrophysical indices of alluvial gley soil from the nursery.

Soil Horizon	Sampling Depth (cm)	AD (t/m^3)	HC (%)	WC (%)	FC (%)	AHR (%)	Tmin (%)
Ap	0–30	1.26	9.04	13.56	26.85	13.29	20.21
A0	30–50	1.27	8.96	13.44	26.51	13.07	19.98
AC	50–80	1.33	8.87	13.31	26.39	13.09	19.85
Cg	80–120	1.34	8.83	13.25	25.81	12.57	19.53

AD—apparent density; HC—hygroscopicity coefficient; WC—wilting coefficient; FC—field capacity; AHR—active humidity range; Tmin—minimum threshold of soil humidity.

As a rule, fruit trees thrive in soil that has a neutral pH, as well as in slightly acidic or slightly alkaline conditions. The base saturation level for our experiment was notably high, ranging from 84% to 90%. The humus content was moderate, while the levels of phosphorus and potassium were exceptionally elevated.

The observed density within the examined soil profile ranged from 1.26 and 1.34 t/m^3, which can be classified as average. The hygroscopicity coefficient ranged from 8.83% to 9.04%, which correlated with elevated wilting coefficient values of 13.25% to 13.56% and field capacity values of 25.81% to 26.85%. The minimum threshold recorded values ranged between 19.53% and 20.21%, corresponding to relative values of 75% to 76% of the field capacity.

The study was conducted utilizing a 4 × 2 × 4 split-split-plot design with five replications, featuring plots of four two-year-old trees planted at a spacing of 0.7 × 0.25 m^2. The primary factor was irrigation, which included four levels (non-irrigated control, 10 mm, 20 mm, and 30 mm), while the second factor was the cultivar (Stanley and Cacanska Lepotica), and the third factor was fertilization at four levels (unfertilized control, $N_8P_8K_8$, $N_{16}P_{16}K_{16}$, and $N_{24}P_{24}K_{24}$). To achieve the three levels of fertilization ($N_8P_8K_8$, $N_{16}P_{16}K_{16}$, and $N_{24}P_{24}K_{24}$), a complex fertilizer with a ratio of 16:16:16 was applied in three increments (50, 100, and 150 kg/ha). The irrigation schedule was based on variations in soil moisture, with applications occurring on 17 August, 22 August, and 27 August. Fertilizers were incorporated into the soil at a sampling depth of 15 cm on 19 August, 25 August, and 29 August. The rootstocks were planted in the spring of 2023 and grafted during the summer of the same year. Leaf area measurements were conducted in August 2024, following leaf drop after the first frost, which typically occurs in October. Considering that the trees are removed from the second field of the nursery and prepared for sale, they are not affected by frost, since when frost occurs, the trees are already planted in the orchard. The leaf area was measured on one tree in each plot. One typical tree was chosen in each plot, and was normally developed, healthy, and without defects. Different leaves of an average size located on the average annual growth area were chosen. Measurements were made on 10 leaves from each tree, and the arithmetic mean of the results obtained was calculated. A leaf area meter, AM 350, produced by ADC BioScientific Ltd. (Hoddesdon, UK), bought from a Romanian distributor from Bucharest, was employed for these measurements, featuring a scanning speed of up to 20 mm/s, a maximum width of 103 mm, a maximum length of 2 m, linear precision/repeatability of 1%, area accuracy of ±2%, perimeter accuracy of ±5%, and a resolution of 0.065 mm^2.

2.3. Calculations

To assess the direct water usage, the soil water balance was evaluated by measuring the soil water reserves at the beginning and the end of each month from April to September. The soil moisture content was determined using a gravimetric method, where soil samples were collected from the field at the beginning and midpoint of each month, and weighed before and after being dried in an oven. The calculation of the soil moisture content was based on the difference between the two weight measurements using the following formula [17]:

$$W = 100 \times (B - A)/A; \tag{1}$$

where W represents soil moisture (%); A is the mass of the wet soil sample (g); and B is the mass of dry soil sample.

When the soil moisture fell below 10%, it was deemed low, potentially signalling a water deficit, which could necessitate the application of irrigation. The irrigation rate was calculated by applying the following formula [17]:

$$M = 110 \times H \times Gv\ (C - P\ min.) \tag{2}$$

where m is the watering rate (m^3/ha); H is the thickness of the soil layer intended to be supplied with water in meters; Gv is the volumetric weight of the soil (t/m^3); C is the field capacity for water, in weight percentages relative to dry soil; and P is the minimum threshold or momentary supply, as a percentage of dry soil weight.

The calculation of water consumption from the soil, which includes the volume of water absorbed by the trees in the nursery during transpiration along with the water lost through evaporation from the land surface, was performed using the following formula [17]:

$$ET = \sum (t + e) \tag{3}$$

where ET = evapotranspiration (m^3/ha), t = the amount of water lost from the soil through plant transpiration (m^3/ha); and e = water evaporated on the soil surface (m^3/ha).

The application of the initial irrigation, defined as the period between plant emergence and the first watering, was determined by the initial water reserves in the soil at the start of the growing season, the specific water requirements of the particular crop, and the rainfall that occurred during that timeframe [17]:

$$T = \frac{Ri - p\,min}{(e+t) - P} \tag{4}$$

and the interval between two waterings was calculated according to the following formula [17]:

$$T = \frac{m}{(e+t) - P} \tag{5}$$

In this context, T represents the interval measured in days; e + t denotes the total daily water consumption for the specific crop, expressed in m^3/ha/day; Ri refers to the initial water reserve present in the soil at the depth of saturation at the onset of the growing season, which can be determined using a formula akin to that used for calculating the final reserve, but substituting the withering coefficient with the field water capacity of the soil (m^3/ha); and P indicates the immediate moisture content (water level at the time of irrigation) of the soil layer H, represented as a percentage of the dry soil's weight.

Beneficial precipitation is determined by the disparity between total precipitation and the losses incurred through evaporation and excessive infiltration [17]:

$$P_e = (P_0 - E) - (I + R) \tag{6}$$

where P_0 is the total precipitation, E is evaporation, I is infiltration, and R is surface runoff.

The irrigation standard for the growing season, in the context of a closed-loop balance (a hydrological irrigation regime that does not involve groundwater recharge), was determined as follows [17]:

$$\sum m = \sum (E + T) + Rf + Ri - Pv\ (m^3/ha) \tag{7}$$

- $\sum$ (E + T) the total consumption of water from the soil during the vegetation period;
- The soil water reserve Rf, at the conclusion of the growing season is established through a relationship involving the volumetric weight of the soil at the specified depth, the thickness of the soil layer designated for irrigation, and the wilting coefficient at that same depth (m^3/ha);
- The initial soil water reserve at the wetting depth at the onset of the growing season is similarly calculated using a relationship akin to that of the final reserve, with the soil water capacity in the soil field (m^3/ha) substituted for the wilting coefficient;
- The sum of summer precipitation is defined as being equal to or exceeding 5 mm.

The data regarding the leaf area were statistically processed using ANOVA associated with the 4 × 2 × 4 split-split-plot design, and an LSD test at $p < 0.05$ [18]. The results were expressed as means ± standard errors.

2.4. *Analysis of the Water Balance and Consumption Under Different Irrigation Conditions*

In April 2024, the recorded soil water reserve ranged from 2405 to 2463 m^3/ha, representing about 86.5% of the field water capacity, indicating a favourable soil supply. However, due to a negative water balance in April, the soil water reserve at the start of May fell to between 2210 and 2228 m^3/ha, which was marginally above the minimum threshold (Table 3).

Table 3. The elements of the soil water balance for maiden plum trees in the research field.

Irrigation Norm		Month					
	Elements	April	May	June	July	August	September
Non-irrigated 0 mm	Ri (m^3/ha)	2405	2210	2223	2291	2296	1864
	Pe (m^3/ha)	150	542	850	1048	277	295
	ET (m^3/ha)	345	529	782	1044	708	464
	Rf_1 (m^3/ha)	2210	2223	2291	2296	1864	1695
	M (m^3/ha)	-	-	-	-	-	-
	Rf_2 (m^3/ha)	-	-	-	-	-	-
Irrigated with 10 mm	Ri (m^3/ha)	2431	2214	2251	2305	2294	2106
	Pe (m^3/ha)	150	542	850	1048	277	295
	ET (m^3/ha)	367	505	796	1059	765	528
	Rf_1 (m^3/ha)	2214	2251	2305	2294	1806	1873
	M (m^3/ha)	-	-	-	-	300	-
	Rf_2 (m^3/ha)	-	-	-	-	2106	-
Irrigated with 20 mm	Ri (m^3/ha)	2463	2228	2222	2313	2247	2275
	Pe (m^3/ha)	150	542	850	1048	277	295
	ET (m^3/ha)	385	548	759	1114	849	562
	Rf_1 (m^3/ha)	2228	2222	2313	2247	1675	2008
	M (m^3/ha)	-	-	-	-	600	-
	Rf_2 (m^3/ha)	-	-	-	-	2275	-
Irrigated 30 mm	Ri (m^3/ha)	2419	2210	2218	2294	2258	2519
	Pe (m^3/ha)	150	542	850	1048	277	295
	ET (m^3/ha)	359	534	774	1084	916	584
	Rf_1 (m^3/ha)	2210	2218	2294	2258	1619	2230
	M (m^3/ha)	-	-	-	-	900	-
	Rf_2 (m^3/ha)	-	-	-	-	2519	-

Ri—initial reserve; Pe—useful precipitation; ET—evapotranspiration; Rf_1—final reserve without irrigation; M—irrigation rate; Rf_2—final reserve with irrigation.

In June and July, the abundant rainfall, associated with a relatively stable soil water balance, facilitated the maintenance of soil moisture above the minimum threshold. In August, the low level of rainfall and high temperatures generated a significant loss of soil water that reached values between 1619 and 1864 m^3/ha, well below the minimum threshold. To compensate for the water deficit, irrigation was applied in accordance with the norms of 300, 600, and 900 m^3/ha so that the soil water reserve exceeded the minimum threshold, reaching levels of 2106–2519 m^3/ha. In September, due to a low amount of rainfall, the water stress experienced in the previous month persisted, resulting in the soil water reserves dropping below the minimum threshold for both the non-irrigated variant (1695 m^3/ha) and the irrigated variant with the norm of 10 mm (1873 m^3/ha). Throughout the 2024 vegetation period, the average daily consumption exhibited a steady rise until July (refer to Figure 1). Thus, against the background of values ranging from 11.5 to 12.83 $m^3/ha/day$ observed in April, the climatic conditions in May generated an increase in water consumption with rates ranging between 33.2% and 48.35%.

The peak water consumption level, recorded at 33.67–35.94 m^3/ha in July, was primarily influenced by the substantial precipitation levels in conjunction with the irrigation applied, rather than the temperature increase of 1.4 °C compared to June. Water consumption declined during the April to June period compared to the previous year, whereas an increase was noted in July and August, and fairly close values in September. At the beginning of April 2024, the soil moisture content in the non-irrigated variant had a value of 22.8%, close to that recorded in the previous year. During the May–June period, the soil moisture maintained its values above the minimum threshold (Figure 2).

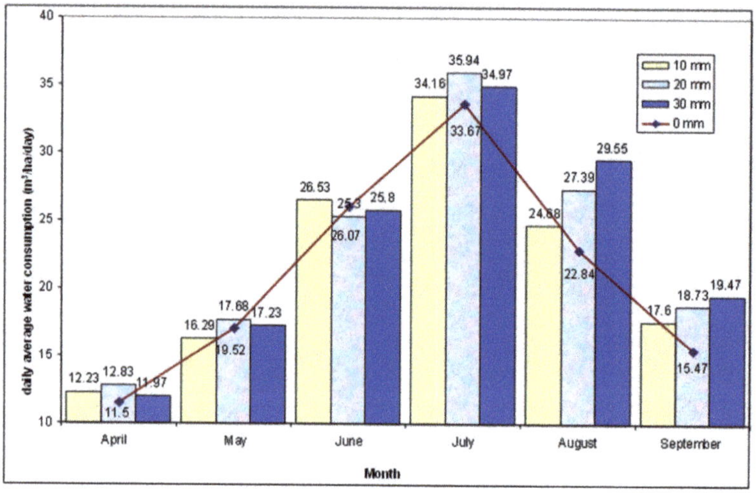

Figure 1. Daily average water consumption for plum planting material in the nursery.

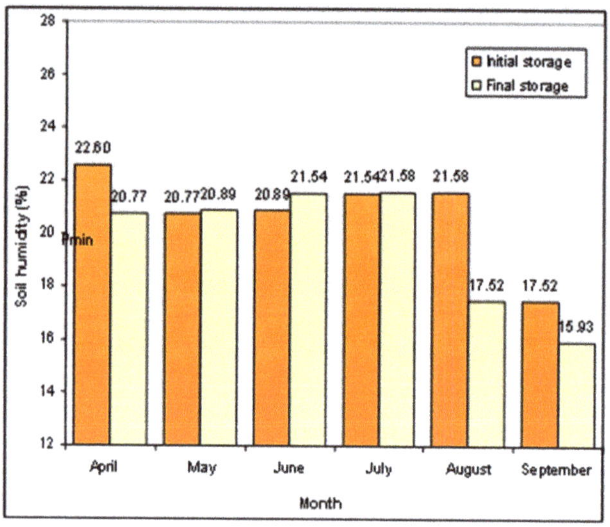

Figure 2. Dynamics of soil humidity for the non-irrigated variant.

Against the background of high temperatures in association with the lower level of precipitation in August, the soil moisture level decreased to 17.52%, a value that was significantly lower than the minimum threshold. In September, the climatic conditions led to a heightened water deficit, resulting in a reduction in the soil moisture content to 15.93%. For the variant involving a 10 mm irrigation standard, the water deficit experienced in August was mitigated by the application of an irrigation volume of 300 m^3/ha, which facilitated an increase in the soil moisture content of approximately 2.8%, bringing it close to the minimum threshold, of 16.97% (Figure 3).

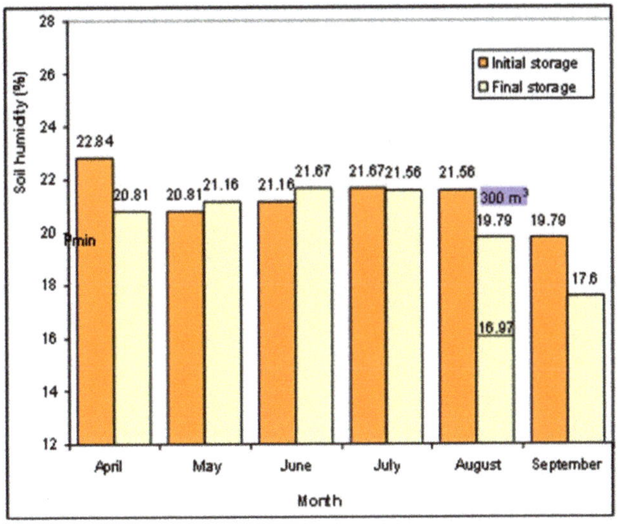

Figure 3. Dynamics of soil humidity for the variant irrigated with 10 mm.

The drought experienced in August necessitated, based on the variant associated with the 20 mm standard, the implementation of three irrigation sessions, amounting to an irrigation norm of 600 m^3/ha, which maintained a soil moisture level of 21.38%, approximately 1.5% above the minimum threshold, of 15.74% (Figure 4).

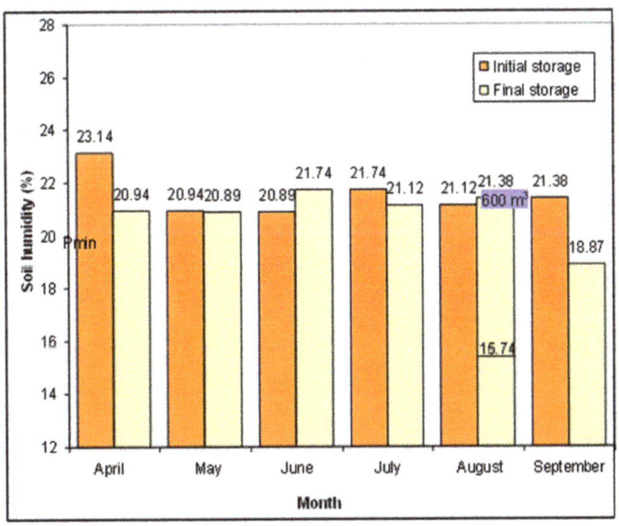

Figure 4. Dynamics of soil humidity for the variant irrigated with 20 mm.

The water deficit in September resulted in a reduction in the soil moisture levels by as much as 18.87%, which was 2.9% higher compared to the non-irrigated variant. For the irrigated variant with the norm of 30 mm, three irrigations totalling 900 m^3/ha were administered in response to the water deficit caused by the drought in August. This resulted in a soil moisture level of 23.87% by the end of the month, exceeding the minimum threshold, of 15.22% (Figure 5).

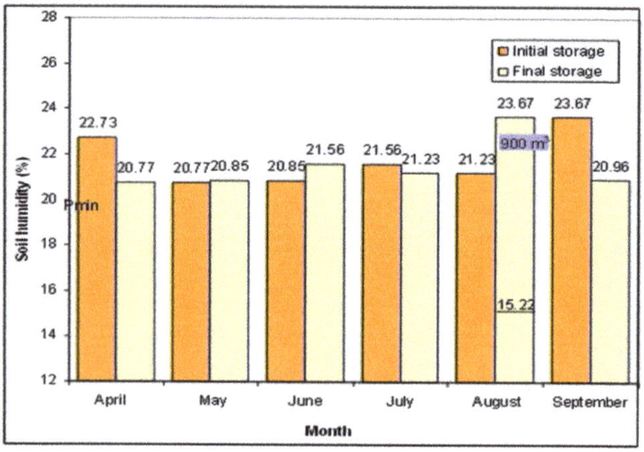

Figure 5. Dynamics of soil humidity for the variant irrigated with 30 mm.

The water shortage in September led to a decrease in the soil moisture content of up to 20.96%. However, that level remained above the minimum threshold and was notably around 5% higher than that of the non-irrigated variant.

3. Results

3.1. Analysis of the Variance Components

The findings from the variance analysis presented in Table 4 indicate that only irrigation and fertilization had significant impacts on the leaf area, with irrigation demonstrating a more pronounced effect compared to the cultivar, which had a negligible influence. Furthermore, while the combined effects of these factors were significant as regards the leaf area, they were notably less impactful than their individual effects. The interaction between the cultivar and irrigation was particularly emphasized due to its substantial effect, followed by the interaction between fertilization and the cultivar.

Table 4. Analysis of variance regarding the effects of irrigation, the cultivar, and fertilization on the leaf area.

Source of Variation	SS	DF	MS	F Test
Total	43,868,124	159		
Replication	216,761	4	54,190	0.59
Irrigation (A)	8,493,021	3	2,831,007	30.75 **
Error A	1,104,856	12	92,071	
Cultivar (B)	9986	1	9986	0.12
A × B	970,711	3	323,570	3.84 *
Error B	1,349,748	16	84,359	
Fertilization (C)	9,982,917	3	3,327,639	22.07 **
A × C	3,044,252	9	338,250	2.24 *
B × C	1,398,818	3	466,273	3.09 *
A × B × C	2,819,710	9	313,301	2.08 *
Error C	14,477,344	96	150,806	

* Significant at $p < 0.05$; ** significant at $p < 0.01$.

3.2. Effects of Irrigation and Cultivar on the Leaf Area

An analysis of the interaction between irrigation and the cultivar on the leaf area of plum trees revealed that for the Cacanska Lepotica cultivar, the three irrigation levels resulted in significant differences when compared to the control, amidst variations among the irrigation levels. Conversely, the Stanley cultivar demonstrated effective utilization of only the 20 and 30 mm irrigation levels, leading to significant enhancements in this characteristic relative to the control, whereas the 10 mm level had a non-significant effect and was inferior to that observed in the Cacanska Lepotica cultivar (Table 5).

Table 5. The effects of irrigation and the cultivar on the leaf area.

Cultivar	Irrigation				Cultivar Mean
	Control	10 mm	20 mm	30 mm	
Stanley	4316 ± 59 a ^{z}	4498 ± 71 a ^{yz}	4651 ± 65 a ^{xy}	4821 ± 86 a ^{x}	4572 ± 41 A
Cacanska Lepotica	4234± 65 a ^{z}	4429 ± 60 a ^{z}	4738 ± 88 a ^{y}	4948 ± 90 a ^{x}	4587 ± 49 A

Cultivar $LSD_{5\%}$ = 97 cm^2; cultivar–irrigation $LSD_{5\%}$ = 193 cm^2; irrigation–cultivar $LSD_{5\%}$ = 195 cm^2; the same letter (a) in the column indicates non-significant differences ($p > 0.05$) between cultivars; different superscript letters (x, y, z) in the row indicate significant differences ($p < 0.05$) between watering norms. The same letter for the cultivar means (A) indicates a non-significant difference ($p > 0.05$).

Given the interaction between irrigation and the leaf area in Stanley cultivar trees, it was observed that only irrigation with 20–30 mm allowed a significant variation of 7.76–11.7%, while the effect of the 10 mm watering norm was smaller and non-significant. Under the effect of different watering norms, trees of the Cacanska Lepotica cultivar recorded a leaf area ranging between 4234 cm^2 in the case of the control and 4821 cm^2 for the 30 mm watering norm, against a background of a 9.5% variability between watering norms. Compared to the control, in the trees of this cultivar, the three watering norms generated significant increases of 4.61–16.86%. The gradual adjustment of the watering standard by 10 mm resulted in changes in this character, ranging from 210 to 309 cm^2.

3.3. *Effects of Irrigation and Fertilization on the Leaf Area*

The leaf surface exhibited a variation of 682 cm^2 due to different fertilization treatments, with measurements ranging from 4226 cm^2 in the unfertilized agricultural setting to 4908 cm^2 when a dose of 24% NPK was applied, reflecting a variability of 6.3% across the treatments (Table 6).

Table 6. The effects of irrigation and fertilization on the leaf area.

Irrigation	Fertilization				Irrigation Mean
	Control	$N_8P_8K_8$	$N_{16}P_{16}K_{16}$	$N_{24}P_{24}K_{24}$	
Control	4006 ± 63 b ^{y}	4196 ± 45 c ^{y}	4343 ± 76 c ^{xy}	4557 ± 57 c ^{x}	4275 ± 43 D
10 mm	4216 ± 91 ab ^{y}	4364 ± 64 bc ^{y}	4520 ± 64 bc ^{xy}	4753 ± 62 bc ^{x}	4464 ± 46 C
20 mm	4267 ± 58 ab ^{z}	4608 ± 55 ab ^{yz}	4856 ± 43 ab ^{xy}	5048 ± 71 ab ^{x}	4695 ± 54 B
30 mm	4415 ± 47 a ^{z}	4838 ± 98 a ^{y}	5008 ± 86 a ^{xy}	5277 ± 77 a ^{x}	4885 ± 62 A

Irrigation $LSD_{5\%}$ = 148 cm^2; irrigation–fertilization $LSD_{5\%}$ = 327 cm^2; fertilization–irrigation $LSD_{5\%}$ = 345 cm^2; different letters (a, b, c) in the column indicate significant differences ($p < 0.05$) between watering norms; different superscript letters (x, y, z) in the row indicate significant differences ($p < 0.05$) between NPK levels. Capital letters were used for irrigation mean (A, B, C, D) comparisons.

The application of NPK fertilization variants determined the recording of significant increases in the growth of this character by 6.53–16.14% compared to the non-fertilized agricultural background. The addition of fertilization from 8 to 16% and from 16 to 24% was effectively utilized by the seedlings that achieved a significant growth of 180–226 cm^2.

By analysing the impact of fertilization on the growth of the leaf structure at a specific watering rate, it is obvious that the greatest variation was noted with the 30 mm norm, whereas the 10 mm norm exhibited a smaller range of variation between the NPK dosage levels.

Without fertilization, the examined irrigation standards resulted in leaf surface area measurements ranging from 4006 cm^2 under non-irrigated conditions to 4415 cm^2 at the 30 mm irrigation level, indicating a variability of 5.79% among the different watering standards.

Compared to the non-irrigated version, the three irrigation standards facilitated an increase of 5.24 to 10.21% in the leaf apparatus in these conditions; however, only the effect of the 30 mm standard was significant. The modification of watering norms did not generate substantial increases against the background of variations in the leaf surface area of 51–199 cm^2. Under fertilization conditions involving 8% NPK, a variation in this character was evident, ranging from 4196 cm^2 in the non-irrigated case to 4838 cm^2 when applying a watering norm of 30 mm. On that agricultural land, only the watering norms of 20 and 30 mm showed significant effects associated with increases of 11.81–15.31% compared to the non-irrigated version. Furthermore, adjusting the watering rate by 10 mm had a negligible impact; thus, only the increase in the watering rate from 10 mm to 30 mm resulted in a significant variation of 10.86% in the leaf surface area. Under fertilization conditions involving 16% NPK, the leaf surface ranged between 4343 cm^2 and 5008 cm^2. As such, in that case, irrigation with 20–30 mm generated a significant increase of 11.81–15.31% of that character compared to the non-irrigated variant. The adjustment in the irrigation rate from 10 mm to 20 mm resulted in a change of 7.43% in the leaf surface area, whereas a further increase in irrigation from 20 mm to 30 mm produced a lesser and statistically insignificant effect of 4.54%.

In the context of fertilization with 24% NPK, the application of irrigation with 20–30 mm favoured a significant increase in that character by 491–720 cm^2. In the case of this agricultural background, only the increase in the watering rate from 10 to 30 mm showed a high efficiency, which materialized as the significant development of the leaf apparatus with 524 cm^2. According to the data presented in Figure 4, the non-irrigated variant demonstrated that the exponential regression revealed an average foliar apparatus development rate of 22.96 cm^2 for every kilogram of NPK applied. That rate was 18.3 cm^2 at application levels of 8 to 16% and increased to 26.75 cm^2/kg NPK for doses between 16 and 24%. The estimates provided are highly precise, with a confidence level of 99.62%, based on a leaf area of 4009 cm^2 in the absence of fertilization.

For the norm of 10 mm, the effect of fertilization on that character was expressed by a regression that was based on a precision of 99.08% and indicated an average growth with a relatively constant rate of 18–19.5 cm^2/kg NPK up to the dose of 16% and a higher rate of 26.75 cm^2/kg NPK between the last levels. Under irrigation conditions of 20 mm, the leaf surface exhibited a variation corresponding to an average rate of 32.52 cm^2/kg NPK, associated with a coefficient of determination of 97.67%, with the initial value of that characteristic being 4310 cm^2 in the non-irrigated agricultural context. In the context of irrigation at 30 mm, the average growth rate of the leaf structure was 35.92 cm^2/kg NPK applied, with different values from one dose to another (21.25–52.87 cm^2/kg NPK) under conditions that allowed for an estimation of the accuracy of that character at 96% (Figure 6).

Compared to the non-fertilized agricultural background, the treatments with 8–16% resulted in a positive impact, reflected by minor increases of 4.74–8.41% in that character. Conversely, the treatment with 24% NPK had a significant effect of 13.75%. Notably, the change in the NPK dosage from 8 to 24% was associated with a significant effect of an 8.6% increase in the leaf area.

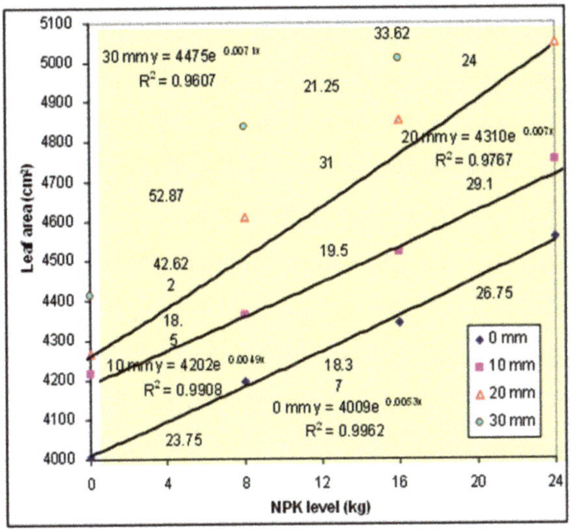

Figure 6. Variation in the leaf area under the effects of different watering norms and fertilizations.

In the context of irrigation set at a standard of 10 mm, the fertilization methods employed resulted in values ranging from 4216 cm^2 for the unfertilized agricultural setting to 4753 cm^2 for the application of 24% NPK, with an amplitude of variation of 537 cm^2 and a variability of 6.52% between treatments. Under such soil moisture conditions, only the treatment with 24% NPK generated a significant increase of 12.74% compared to the unfertilized version. Furthermore, the application of 24% NPK demonstrated a markedly greater impact on that characteristic when compared to the 8% NPK dosage. Under the effect of irrigation with 20 mm, fertilization with 16–24% NPK showed a significant influence on the increase in the leaf surface, associated with increases of 589–781 cm^2. Only the addition of fertilization from 8 to 24% generated a significant variation of 9.55% in that character. An examination of the impact of fertilization on the leaf surface area of seedlings at varying watering rates reveals that in the absence of irrigation, the seedlings recorded values ranging from 4006 cm^2 in the case of the unfertilized variant up to 4557 cm^2 in the 24% NPK variant, with a variability between treatments of 6.46%.

In the context of applying a watering norm of 30 mm, the seedlings exhibited leaf surface area values ranging from 4415 cm^2 in the unfertilized variant to 5277 cm^2 in the variant supplemented with 24% NPK, with a variability of 8.06% observed between the treatments. Compared to the non-fertilized agricultural background, the NPK treatments had a significantly higher efficiency, which materialized as increases of 9.58–19.52%. Changing the dose of NPK from 8 to 24% was associated with a significant increase in the development of the leaf apparatus of 9.07%.

3.4. *Effects of the Cultivar and Fertilization on the Leaf Area*

Upon examining the interaction between the cultivar and fertilization on leaf surface area characteristics (Table 7), it was observed that in the Stanley cultivar, only the treatments utilizing 16–24% NPK produced noteworthy positive effects, amidst relatively minor differences across the treatments. Conversely, for the seedlings of the Cacanska Lepotica cultivar, fertilization had a significant impact on this type.

Table 7. The effects of the cultivar and fertilization on the leaf area.

Cultivar	Fertilization			
	Control	$N_8P_8K_8$	$N_{16}P_{16}K_{16}$	$N_{24}P_{24}K_{24}$
Stanley	4293 ± 63 a ^{z}	4500 ± 82 a ^{yz}	4657 ± 88 a ^{xy}	4837 ± 79 a ^{x}
Cacanska Lepotica	4160 ± 65 a ^{z}	4503 ± 85 a ^{y}	4707 ± 82 a ^{xy}	4980 ± 94 a ^{x}
Fertilization mean	4226 ± 69 W	4502 ± 71 Z	4682 ± 76 Y	4908 ± 85 X

Fertilization $LSD_{5\%}$ = 172 cm^2; cultivar–fertilization $LSD_{5\%}$ = 15 cm^2; fertilization–cultivar $LSD_{5\%}$ = 7 cm^2; the same letter (a) in the column indicates non-significant differences ($p > 0.05$) between cultivars; different superscript letters (x, y, z) in the row indicate significant differences ($p < 0.05$) between NPK levels. Capital letters were used for fertilization mean (X, Y, Z, W) comparisons.

Concerning the impact of fertilization on the leaf surface area of each cultivar (Table 8), it could be noticed that for the Stanley cultivar, the values ranged between 4293 cm^2 for the unfertilized variant and 4837 cm^2 in the case of applying the dose of 24% NPK. Compared to the unfertilized version, only the treatments with 16–24 kg had significant effects and were associated with increases of 364–544 kg. It was also found that only the addition of fertilization from 8 to 24% NPK generated a significant increase in that character of 7.5%.

Table 8. The combined impacts of irrigation and fertilization on the leaf area of the two cultivars.

Specification	0 mm			
	Control			
	Fertilization			
Cultivar	Control	$N_8P_8K_8$	$N_{16}P_{16}K_{16}$	$N_{24}P_{24}K_{24}$
Stanley	4124 ± 80 a ^{x}	4261 ± 73 a ^{x}	4363 ± 150 a ^{x}	4518 ± 100 a ^{x}
Cacanska Lepotica	3888 ± 68 a ^{y}	4131 ± 41 a ^{xy}	4323 a ± 55 ^{x}	4595 a ± 63 ^{x}
	10 mm			
	Fertilization			
Cultivar	Control	$N_8P_8K_8$	$N_{16}P_{16}K_{16}$	$N_{24}P_{24}K_{24}$
Stanley	4308 ± 167 a ^{y}	4406 ± 94 a ^{xy}	4483 ± 106 a ^{xy}	4794 ± 123 a ^{x}
Cacanska Lepotica	4125 ± 74 a ^{y}	4322 ± 94 a ^{xy}	4558 ± 32 a ^{xy}	4713 ± 38 a ^{x}
	20 mm			
	Fertilization			
Cultivar	Control	$N_8P_8K_8$	$N_{16}P_{16}K_{16}$	$N_{24}P_{24}K_{24}$
Stanley	4322 ± 72 a ^{y}	4519 ± 87 a ^{xy}	4873 ± 68 a ^{xy}	4891 ± 70 a ^{x}
Cacanska Lepotica	4212 ± 91 a ^{y}	4697 ± 47 a ^{xy}	4839 ± 59 a ^{x}	5204 ± 77 a ^{x}
	30 mm			
	Fertilization			
Cultivar	Control	$N_8P_8K_8$	$N_{16}P_{16}K_{16}$	$N_{24}P_{24}K_{24}$
Stanley	4417 ± 44 a ^{z}	4814 ± 197 a ^{yz}	4910 ± 143 a ^{xy}	5143 ± 98 a ^{x}
Cacanska Lepotica	4414 ± 37 a ^{z}	4862 ± 66 a ^{yz}	5107 ± 91 a ^{xy}	5410 ± 70 a ^{x}

Cultivar $LSD_{5\%}$ = 459 cm^2; fertilization $LSD_{5\%}$ = 487 cm^2; the same letter (a) in the column indicates non-significant differences ($p > 0.05$) between cultivars; different superscript letters (x, y, z) in the row indicate significant differences ($p < 0.05$) between NPK levels.

As regards the leaf surface area of the seedlings from the Cacanska Lepotica cultivar, the variation in the effects of fertilization was more pronounced, being characterized by an amplitude of 820 cm^2 ranging from 4160 cm^2 on untreated soil to 4980 cm^2 at a 24% NPK application rate. The three treatments resulted in an improvement, with an increase of 8.25% to 19.71% in the leaf apparatus. The addition of fertilization by increasing the NPK levels from 8 to 16 and 24% determined a significant increase in this character of 273–477 cm^2.

Figure 5 provides data based on the exponential regression suggesting that for the Cacanska Lepotica cultivar, the leaf surface area varied by an average of 34.17 cm^2 for each kg of NPK fertilizer applied, with fluctuations between 25.5 cm/kg NPK for levels ranging from 8 to 16% and reaching 42.87 cm^2/kg NPK at the 8% level. These estimates were made with a high degree of accuracy, specifically 98.61%, based on a leaf surface area of 4194 cm^2 when no fertilization was used.

In the case of the Stanley cultivar, the impact of fertilization on the growth of the leaf structure was reduced, showing an average growth rate equivalent to 22.67 cm^2/kg NPK against the background of variations from 19.62 cm^2/kg NPK between treatments with 8–16% NPK and 25.97 cm^2/kg NPK between the levels of 8% and the control.

The predictability of the logarithmic regression between the fertilization dose and the variation of that character for the Stanley cultivar was 99.55%, based on an initial value of 4306 cm^2 on the unfertilized agricultural background (Figure 7).

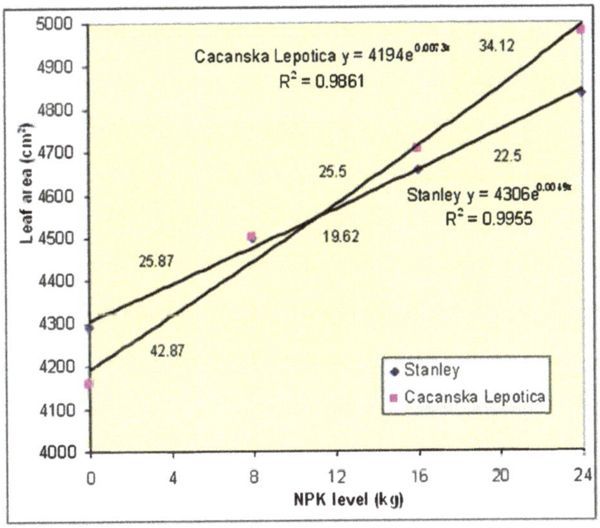

Figure 7. Variations in the leaf area under the effects of different watering norms and fertilizations.

3.5. Combined Effects of Irrigation, Fertilization and the Cultivar on the Leaf Area

Considering the combined effects of the three factors (Table 8), fertilization did not significantly influence the leaf surface area of the Stanley cultivar on the non-irrigated agricultural background under the circumstances of a 394 cm^2 amplitude. For the Cacanska Lepotica cultivar, an enhancement of 435–707 cm^2 in the leaf surface area was observed solely with fertilization using 16–24% NPK.

On the arable land irrigated with 10 mm, only the fertilization with 24% NPK determined a significant increase in the leaf surface area of both cultivars of 486–688 cm^2 under the conditions of small and insignificant variations between the three doses applied.

When irrigated with 20 mm, an increase in the leaf area was recorded, with the Stanley cultivar seedlings showing an increase of 569 cm^2 and the Cacanska Lepotica cultivar seedlings exhibiting an increase of 992 cm^2, both fertilized with 24% NPK, compared to the untreated variant; however, alterations in the NPK dosage resulted in negligible variations in this character.

The application of irrigation with 30 mm revealed that fertilization had a more pronounced effect on the leaf surface area in both cultivars that was associated with differences of 725–996 cm^2 compared to the non-fertilized variant and insignificant variations between treatments. Within agricultural contexts that represented combinations between fertilization and irrigation, there were no significant distinctions between the two cultivars for that character.

4. Discussion

Utilizing irrigation and fertilization in nurseries results in the production of robust and healthy fruit planting material, enhancing overall yield and providing superior economic returns compared to nurseries lacking irrigation and fertilization. Although nurseries are typically situated on fertile land, the application of fertilizers is essential to create optimal conditions for the growth of rootstocks and trees, as well as to support the activity of soil microorganisms. While climatic conditions generally ensure the normal growth of trees without irrigation, in excessively dry years, irrigation becomes a necessity. Prolonged periods of drought hinder the growth of plants within nurseries. Additionally, it is crucial for trees in a nursery to achieve rapid growth within a limited timeframe, particularly when the root systems have been significantly reduced through shaping, necessitating swift restoration to enable them to access a larger volume of soil. Therefore, it is essential to apply fertilizers and take measures to enhance soil fertility in order to achieve high yields of fruit trees in nurseries. Nursery fertilization should encompass a series of actions that contribute to the long-term enhancement of physical and chemical properties while ensuring the availability of necessary nutrients in forms that can be assimilated during each phase of vegetation, according to the requirements of the trees [19]. This research highlights that the climate and soil conditions are favourable to tree cultivation and do not pose any limiting factors.

Additionally, irrigation must maintain optimal water levels, ensuring a supply of 65–75% of the total soil retention capacity, avoiding excessive moisture levels beyond 80%.

In terms of fertilizer application within the nursery, it is essential to note that nitrogen (N), phosphorus (P), and potassium (K) are the key nutrients that should be utilized to ensure optimal plant growth in the field, as shown by Rosen and Swanson [20]. Fertilization influences crop development, nutrient uptake, and overall productivity [21]. Regarding the application of fertilizers, the current understanding indicates that factors such as soil conditions, species, cultivar, rootstock, density, forecasted production, etc., must be considered. Establishing the optimal doses for each situation in the field must be performed after analysing a series of soil properties, knowing the requirements imposed by the crop and those related to ensuring a certain quantitative and qualitative level of production. Although at first glance they seem to be expensive elements of technology, they are actually only rational methods of using water and fertilizers to obtain planting material. The improper use of fertilizers in nurseries, particularly when applied in excessive, unilateral, and unbalanced quantities, as well as the use of fertilizer types that do not align with the biological needs of specific species and soils, can lead to detrimental outcomes. Instead of achieving the anticipated increases in tree production, a decline may be observed. Furthermore, the economic viability and yield in irrigated agriculture are largely influenced by the irrigation practices employed for each crop. A significant consequence of the overuse of chemical fertilizers is the leaching of nutrients from the soil due to irrigation or rainfall, which can infiltrate groundwater and exacerbate the eutrophication of watercourses.

The investigations carried out so far on the use of irrigation and fertilization in the nursery refer to fruit shrubs or woody species, not to fruit tree planting materials. Consequently, it is not possible to compare the various results obtained. But as a general idea, other recent research shows that both irrigation and fertilization have been proven to improve plant growth, accelerate photosynthesis and growth metabolism activities, and eventually achieve higher crop yield and quality [22–24]. Besides obtaining a high yield, another objective of nursery owners is to obtain a profit. According to Fulcher and

Fernandez [25], irrigation can shorten the production period for field nursery crops and increase quality, which has a positive impact on nursery profitability. Drip irrigation and fertilization involves some additional costs in terms of purchasing and installing equipment, but it is compensated by a high and good quality of fruit tree production.

In the same study, the effects of irrigation, fertilization, and the cultivar on different morphological traits of plum, including shoot growth, shoot weight, crown diameter, branching growth, and tree yield, were also determined. Concerning the physiological traits of plum, the leaf area and photosynthesis rate were examined under the influences of irrigation and fertilization.

5. Conclusions

The implementation of various NPK fertilization methods resulted in statistically assured enhancements in leaf surface growth, ranging from 6.53% to 16.14% when compared to the control group that did not receive fertilization. Especially when added in amounts between 8 to 16% and 16 to 24%, fertilizers were effectively absorbed by the grafted trees, leading to substantial growth improvements of 180–226 cm^2. In the Stanley cultivar, only the treatments with 16–24% NPK showed significantly positive influences on the leaf surface area against the background of small variations between the treatments. In the case of the Cacanska Lepotica cultivar, fertilization had a substantially higher influence on the leaf area.

Irrigation led to increases in the leaf area ranging from 4.42% to 14.27% when compared to the control group. For the Cacanska Lepotica trees, the applied watering norms resulted in significant differences in the leaf area relative to the control, highlighting the substantial variations among the watering norms. Conversely, the Stanley cultivar trees effectively utilized only the 20 mm and 30 mm watering norms, which resulted in significant enhancements of this characteristic compared to the control. As future research directions, it is appropriate to analyse the economic efficiency of producing grafted trees under different conditions of irrigation and fertilization.

It is essential for nursery owners and horticulture enthusiasts to consider various factors, such as soil conditions, species, cultivar, rootstock, density, and anticipated production, when implementing irrigation and fertilization practices. Additionally, understanding the rainfall patterns, the low water permeability of the soil, and groundwater levels can be helpful for establishing appropriate irrigation standards. The optimal level of NPK in the field must be set after analysing a series of soil properties, knowing the requirements imposed by the crop and those related to ensuring a certain quantitative and qualitative level of production. It is also recommended to carry out chemical laboratory analyses on the qualitative and quantitative presence of various nutrients in the soil.

Author Contributions: Conceptualization, A.V. and A.P.; methodology, A.V.; software, A.V.; validation, A.V.; formal analysis, A.V.; investigation, A.V. and A.P.; resources, A.V.; data curation, A.V.; writing—original draft preparation, A.V.; writing—review and editing, A.P.; visualization, A.V.; supervision, A.P.; project administration, A.V.; funding acquisition, A.V. All authors have read and agreed to the published version of the manuscript.

Funding: The APC was funded by the University of Oradea.

Data Availability Statement: The data are contained within the article or the references.

Conflicts of Interest: The authors declare no conflicts of interest.

References

1. Smith, P. *Trees: From Root to Leaf*; University of Chicago Press: Chicago, IL, USA, 2022; p. 59.
2. Jie, F.; Fei, L.; Peng, Y.; Li, S.; Ge, Y. Optimal Allocation of Water Resources in Irrigation Areas Considering Irrigation Return Flow and Uncertainty. *Appl. Sci.* **2025**, *15*, 2380. [CrossRef]
3. Soman, P. *Fertigation Scheduling of Field Crops*; New India Publishing Agency: New Delhi, India, 2022; p. 152.
4. Harun, K.; Hamide, G.; Tezcan, A.; Can, M.; Özbek, Ö. Water- yield relationship of greenhouse- grown strawberry under limited irrigation. *Not. Bot. Horti Agrobot. Cluj Napoca* **2023**, *51*, 2.
5. Paschoalinotto, B.H.; Polyzos, N.; Liava, V.; Mandim, F.; Pires, T.C.S.P.; Añibarro-Ortega, M.; Ferreira, I.C.F.R.; Dias, M.I.; Barros, L.; Petropoulos, S.A. The Effect of Cropping System and Irrigation Regime on the Plant Growth and Biochemical Profile of Cichorium spinosum. *Horticulturae* **2025**, *11*, 306. [CrossRef]
6. Saisai, G.; Xue, M.; Xue, M.; Sun, H.; Li, H.; Han, Q.; Li, R. Effects of Different Irrigation Rates and Microbial Fertilizers on Inter-Root Soil Environment and Yield and Quality of *Brassica chinensis* L. *Horticulturae* **2025**, *11*, 321. *Horticulturae* **2025**, *11*, 321.
7. Jaoudé, R.A.; Ficca, A.G.; Luziatelli, F.; Bececco, A.; Brunori, E.; Biasi, R.; Baraccani, I.; Ruzzi, M. Impact of a Synthetic Microbial Community on Salvia officinalis Under Optimal and Reduced Irrigation. *Horticulturae* **2025**, *11*, 374. [CrossRef]
8. Jeet, B.; Thangavel, M.; Prajapati, J. *Irrigation Techniques in Horticulture, Innovations and Emerging Technologies in Horticulture*; Elite Publishing House: New Delhi, India, 2025; p. 24.
9. Sullivan, T.; Yost, M.; Boren, D.; Barker, B.; Creech, E.; Bugbee, B. Impacts of irrigation system, irrigation rate, and cultivar on hemp production in the semiarid Intermountain West. *Agrosyst. Geosci. Environ.* **2025**, *8*, e70083. [CrossRef]
10. Stănică, F. *Experimental Research on Fertilization in Orchards*; Invel Multimedia Publishing House: Bucharest, Romania, 2004; p. 74.
11. Küçükyumuk, C.; Açal, E.K.; Yildiz, H. Effects of Different Deficit Irrigation Strategies on Yield, Fruit Quality and Some Parameters: 'Braeburn' Apple Cultivar. *Not. Bot. Horti Agrobot. Cluj Napoca* **2013**, *41*, 2. [CrossRef]
12. Septar, L.; Moale, C.; Caplan, I. Peach Response to Water Deficit under the Climatic Conditions of South-Eastern Romania. *Sci. Pap. Ser. B Hortic.* **2024**, *LXVIII*, 2.
13. Jäger, H. *The Fruit Trees Nursery*; Salzwasser Publishing House: Paderborn, Germany; p. 39.
14. Bolohan, D.E.; Volf, M.; Chelariu, E.L.; Răus, L. Apple Trees Fertilization and its Influence on the Potassium Content in Soil and Plants in Correlation with Calcium and Magnesium Absorption. *Sci. Pap. Ser. B Hortic.* **2023**, *LXVII*, 2.
15. Cai, S.; Zheng, B.; Zhao, Z.; Zheng, Z.; Yang, N.; Zhai, B. Precision Nitrogen Fertilizer and Irrigation Management for Apple Cultivation Based on a Multilevel Comprehensive Evaluation Method of Yield, Quality, and Profit Indices. *Water* **2023**, *15*, 468. [CrossRef]
16. Vidican, R.; Rusu, M. *Application of Fertilizers—Practical Guide*; Colorama Publishing House: Cluj Napoca, Romania, 2018; p. 84.
17. Bradford, R.M. *The Handbook of Technical Irrigation Information*; Hunter Industries Incorporated: San Marcos, CA, USA, 2002; pp. 46–56.
18. Ciulcă, S. *Experimental Methodologies in Agriculture and Biology*; Agroprint Press: Timisoara, Romania, 2006; pp. 286–294.
19. Hoza, D. *Pomology*; Prahova Press: Ploiești, Romania, 2000; p. 73.
20. Rosen, C.; Swanson, B.T. *Fertilization for Nursery and Landscape Management*; University of Minnesota: Minneapolis, MN, USA, 2010; p. 1.
21. Lyu, X.; Hassan, H.M.; Zan, Y.; Tan, J. Interactive effects of irrigation and fertilization on the growth and physiological characteristics of greenhouse tomatoes, *Solanum lycopersicum* L. *Sci. Rep.* **2025**, *15*, 6007. [CrossRef]
22. Zhang, F.; Chen, M.; Zheng, Y.; Xie, Y.; Xing, Y. Optimizing Irrigation and Fertilization to Simultaneously Improve Potato Tuber Yield, Water and Fertilizer Use Efficiency and Net Income in Northwest China. *Agronomy* **2024**, *14*, 1124. [CrossRef]
23. Liang, C.; Liu, X.; Lv, J.; Zhao, F.; Yu, Q. The Impact of Different Phosphorus Fertilizers Varieties on Yield under Wheat–Maize Rotation Conditions. *Agronomy* **2024**, *14*, 1317. [CrossRef]
24. Barbosa, A.S.; Silva, A.O.d.; Sousa, G.G.d.; Souza, M.V.P.d.; Freire, M.H.d.C.; Goes, G.F.; Pereira, A.P.d.A.; Viana, T.V.d.A.; Costa, R.N.T.; Lacerda, C.F.d.; et al. Brackish Water, Phosphate Fertilization and Trichoderma in the Agronomic Performance of Beet Crops. *Agronomy* **2024**, *14*, 1306. [CrossRef]
25. Fulcher, A.; Fernandez, T. *Sustainable Nursery Irrigation Management Series*; Part 1; Institute of Agriculture, The University of Tennessee: Knoxville, TN, USA, 2002.

horticulturae

MDPI

Article

Effects of Selenium/Iodine Foliar Application and Seasonal Conditions on Yield and Quality of Perennial Wall Rocket

Alessio Vincenzo Tallarita [1], Nadezhda Golubkina [2,*], Stefania De Pascale [1,*], Agnieszka Sękara [3], Robert Pokluda [4], Otilia Cristina Murariu [5], Eugenio Cozzolino [6], Vincenzo Cenvinzo [1] and Gianluca Caruso [1]

1 Department of Agricultural Sciences, University of Naples Federico II, Portici, 80055 Naples, Italy; alessiovincenzo.tallarita@unina.it (A.V.T.); vincenzo.cenvinzo2@unina.it (V.C.); gcaruso@unina.it (G.C.)
2 Analytical Laboratory Department, Federal Scientific Vegetable Center, 143072 Moscow Region, Russia
3 Department of Horticulture, Faculty of Biotechnology and Horticulture, University of Agriculture, 31-120 Krakow, Poland; agnieszka.sekara@urk.edu.pl
4 Department of Vegetable Growing and Floriculture, Faculty of Horticulture, Mendel University, 613 00 Brno, Czech Republic; robert.pokluda@mendelu.cz
5 Department of Food Technologies, 'Ion Ionescu de la Brad' Iasi University of Life Sciences, 700490 Iasi, Romania; otilia.murariu@iuls.ro
6 Research Center for Cereal and Industrial Crops, Council for Agricultural Research and Economics (CREA), 81100 Caserta, Italy; eugenio.cozzolino@crea.gov.it
* Correspondence: segolubkina45@gmail.com (N.G.); depascal@unina.it (S.D.P.)

Abstract: The biofortification of leafy vegetables with selenium (Se) and iodine (I) provides the basis for the Se/I status optimization and preservation of human health. The effect of foliar Se, I, and Se + I supply in three different crop cycles (autumn, autumn–winter, and winter) on yield, quality, and mineral composition of wall rocket leaves was investigated using biochemical and ICP-MS methods of analysis. Joint foliar supply with selenate/iodide increased yield, antioxidant activity, total phenolic, ascorbic acid, and protein levels by 1.63, 1.24, 1.22, 1.25, and 1.50 times, respectively, and the content of Ca, Mg, P, K, Fe, Cu, and Zn by 1.27, 1.24, 1.35, 1.46, 3.67, 2.76, and 1.44 times, respectively. High correlations between Se, antioxidants, P, Mg, and Ca ($r > 0.80$) as well as between yield and K/protein content were recorded. Despite a significant decrease in yield, protein, and K, Fe, Cu, and Mn contents in the third crop cycle, compared to the first one, 50 g of wall rocket biofortified with Se/I may provide up to 100% of the Se adequate consumption level, 34.3% of I, 9% of K, 24% of Fe, and 17.7% Ca. The results of the present research confirm the high efficiency of Se/I supply to produce *D. tenuifolia* leaves as a new functional food.

Keywords: *Diplotaxis tenuifolia* (L.) D.C.; functional food; biofortification; Se and I; antioxidants; mineral elements

1. Introduction

Selenium and iodine are essential microelements for human organism, responsible for antioxidant defence, high immunity, and brain activity, playing an important role in growth and development, protection against chronic diseases including cancer, and heart failure [1]. Selenium and iodine deficiency is widespread in many countries in the world, decreasing life expectation, human education, and labour efficiency, thus causing significant economic losses [2]. The metabolism of these elements is closely related to each other, as Se is important to an active centre of tri-iodothyronine deiodinases participating in the thyroid hormone biosynthesis [3]. In this respect, the optimization of the human Se and I status should be carried out simultaneously [4]. Among different approaches for solving this problem (food supplements and iodized salt), the joint biofortification of agricultural crops with Se and I has become more and more attractive due to the so-called buffer effect of plants preventing Se/I toxicosis in humans and thanks to the low cost of the applied technology [5,6].

Many studies have been carried out regarding the joint Se/I biofortification of agricultural crops, i.e., in carrot and lettuce [7–10], chicory [11], potato [12], Indian mustard [13], chickpea [14], chervil [15], pumpkin seedlings [16], kohlrabi [17], radish [18], buckwheat [19], spinach [20], and pea [21]. The results demonstrated the high variability of yield, antioxidant status, and the mineral composition of biofortified plants, which greatly hampered the development of industrial technology. The reason of the latter situation is connected with several factors including (1) the extremely narrow concentration ranges suitable for safe biofortification with Se and I [1], (2) species and varietal differences in plant tolerance to Se and I [22], (3) a different efficiency of Se and I biofortification which hinders the production of Se/I-enriched plants, thereby allowing them to sufficiently meet the adequate Se/I consumption levels (70 and 150 μg day^{-1}, respectively) [1,22], and (4) the unpredictable Se/I interaction during biofortification, which may result either in synergism or antagonism between the two microelements, or no effects [11,19].

To date, Se and I are essential for mammals, while they are considered only beneficial in plants, showing growth stimulation and stress protection at low concentrations, which opens new opportunities for the yield improvement of agricultural crops [1]. However, iodine essentiality was recorded via plant protein iodination [23], whereas that of Se was previously demonstrated in algae [24]. The investigation by Rakoczy-Lelek et al. [25] demonstrated that the biofortification of plants with iodide resulted in the formation of iodotyrosine, iodobenzoates, iodosalicylates, and plant-derived thyroid hormone analogues.

Among the agricultural crops tested, leafy vegetables are very interesting targets for biofortification due to the possibility of Se/I losses during drying and food processing because of the formation of volatile methyl iodides and methyl selenides [26–28]. In addition, fresh leafy vegetables provide the highest levels of other antioxidants (vitamins, polyphenols, sterols, etc.) known to act synergistically with Se in humans [29,30].

The agronomic biofortification of the Brassicaceae species, which are important dietary sources of Ca, Mg, and Fe [14], is highly desirable due to high levels of sulphur, with the latter being a Se chemical analogue easily substituted with Se in living organisms [31]. Furthermore, selenium is known to form Se derivatives of glucosinolates present in Brassicaceae, with higher anti-carcinogenic activity compared to the unsubstituted compounds [32].

Among the leafy Brassicaceae species, only Indian mustard has been jointly fortified with Se and I [13], and, in this respect, perennial wall rocket (*Diplotaxis tenuifolia* L.-DC) may become an attractive target of Se/I biofortification. Indeed, *Diplotaxis tenuifolia* is a perennial plant mostly cultivated to be commercialized as baby leaves rich in glucosinolates, vitamin C, and polyphenol content [33–36]. The mentioned plant is highly valuable as a food and a spice and is popular in medicine, providing protection against cancer and heart failure due to its high antioxidant activity [37,38]. Its remarkable adaptability and tolerance to salinity, pathogen attack, and diseases [38–41] as well as quick growth provide additional benefits for wall rocket utilization as a target of Se/I biofortification. To date, no research regarding the joint biofortification of perennial wall rocket with Se and I has been carried out. In previous studies, Se biofortification in hydroponic conditions stimulated wall rocket growth and the accumulation of chlorophyll, polyphenols, and Se [42–44], while iodine application via foliar/soil supply did not lead to a significant effect on plant yield but only resulted in tendential chlorophyll accumulation under the foliar treatment [45]. Plant foliar biofortification both with Se [46] and iodine [47] is more efficient than the hydroponic and soil applications, does not cause Se/I interactions with soil components, is easy to perform and, therefore, is preferable to produce functional vegetables enriched with the mentioned essential microelements.

Based on the concepts expressed above, the present work aimed to assess the interactions between foliar joint Se/I biofortification and crop cycle on yield, selenium, iodine, and other antioxidant contents, as well as the mineral composition of *D. tenuifolia* leaves.

2. Materials and Methods

2.1. Experimental Protocol and Growing Conditions

Research was carried out on *Diplotaxis tenuifolia* (L.) D.C. cultivar Marte at the Department of Agricultural Sciences, University of Naples Federico II, Portici (Naples), Southern Italy, in 2022–2024 under an unheated greenhouse covered with a thermal polyethylene film in sandy loam soil (76% sand, 17% silt, 7% clay), with 0.12% N, 95.5 ppm NO_3-N, 2.4 ppm NH_4-N, 96.1 ppm P_2O_5, 1822 ppm K_2O, pH 7.2, 2.86% organic matter, and 689 μS cm^{-1} electrical conductivity. During the cropping seasons, the mean monthly temperatures inside the greenhouse were 18 °C in November, 15 °C in December, 11 °C in January, 14 °C in February, and 18 °C in March.

The experimental protocol was based on the factorial combination of 3 crop cycles (autumn, starting on 2 November; autumn–winter, starting on 9 December; and winter, starting on 24 January) and 3 biofortification treatments (selenium, iodine, and selenium + iodine) plus an untreated control. A randomized complete block design was used for the treatment distribution in the field with three replicates, and each experimental unit had a 1 m^2 (1 × 1 m) surface area.

Sowing was performed on 2 November, both in 2022 and 2023, with a spacing of 2 cm along the rows which were 5 cm apart. Foliar spraying with selenium, iodine, and selenium + iodine solutions was performed 3 times at 5-day intervals, starting when the leaves were 1 cm long on average, applying the same following doses in each of the three crop cycles: sodium selenate at 50 mg Se L^{-1} (0.26 mM), potassium iodide at 100 mg I L^{-1} (0.59 mM), and sodium selenate + potassium iodide at the same mentioned doses. The Se and I doses were chosen based on the tolerable dose of Se in most agricultural crops and the significantly lower ability of plants to accumulate iodine compared to selenium [1,15]. Each solution was applied until runoff at the average volume of 300 mL per plant in each crop cycle, corresponding to the supply of 10 mg of Na and 28 mg of K per cycle from selenite and iodide, respectively. As the latter amounts have an insignificant impact on plants, the untreated control was irrigated with water only at the same volume as the experimental treatments, with no K and Na addition. As can be observed in Table 1, the first cutting was conducted on 9 December which was the starting date of the second cycle, characterized by the same scheme of treatments as the first one. The second cutting was performed on 24 January which was the beginning of the third cycle. The biofortification treatments were conducted according to the same scheme in the three crop cycles.

Table 1. Beginning and end dates of crop cycle, as well as biofortification treatments.

Crop Cycle	Starting Date	First Se/I Treatment	Second Se/I Treatment	Third Se/I Treatment	Date of Leaf Cutting
First	2 November (sowing)	15 November	20 November	25 November	9 December
Second	9 December	25 December	30 December	4 January	24 January
Third	24 January	10 February	15 February	20 February	9 March

Crop management was performed using sustainable management practices, including an organic preplant fertilization with N, P_2O_5, and K_2O (at a rate of 38, 10, and 30 kg ha^{-1}, respectively) and the control of fungal diseases and pests, when necessary, with copper oxychloride and azadirachtin treatments, respectively. During the three growing seasons, drip irrigation was activated when the soil available water at 10 cm depth dropped to 80%, and N, P_2O_5, and K_2O were supplied via fertigation during each crop cycle at a dose of 112, 30, and 90 kg ha^{-1}, respectively.

The harvests were conducted on 9 December, 24 January, and 9 March, when rocket leaves reached 6–7 cm length, cutting at 3–5 cm above the soil surface in order to leave the vegetative apex undamaged, thus allowing for more efficient regrowth.

As the research year did not have a significant effect on the examined variables and the latter were not significantly affected by the interaction between crop cycle and the biofortification treatment, the average values of the two trial years related to the main effects of the two experimental factors are presented.

2.2. *Yield Determination*

At each harvest, the following fresh yield parameters of perennial wall rocket were measured in all the experimental plots: total weight, the number of leaves, and mean leaf weight of 100 g samples.

2.3. *Sample Preparation*

Fresh leaf samples were randomly collected in each experimental plot and stored at −80 °C until the extractions and analyses of the antioxidant compounds (total phenols and the ascorbic acid), and antioxidant activity were performed. Dried plant material was ground to a powder in a grinder; then, a 500 mg sample was used for the extraction and determination of the mineral elements and nitrate.

2.4. *Dry Matter*

The dry matter content was determined gravimetrically after drying baby leaf samples to a constant weight at 70 °C in a forced air oven. The results were expressed as the weight/weight percentage of dry matter (% *w*/*w*).

2.5. *Ascorbic Acid*

The ascorbic acid content was assessed using the manual titration method based on the interaction of the ascorbic acid with sodium 2,6-dichlorophenol indophenolate (Tillman's reagent) [48], with a detected limit of 4 mg L^{-1}.

2.6. *Total Polyphenols (TPs)*

The Folin–Ciocalteu colorimetric method was used for total polyphenol determination by utilizing 70% ethanol extracts of dry leaf homogenates [49]. The concentration of polyphenols was assessed using the absorption value of the Folin reagent–sample ethanol extract reaction mixture at 730 nm (Unico 2804 UV spectrophotometer, Suite E Dayton, NJ, USA). A solution of 0.02% gallic acid was used as an external standard, and all the results were expressed in mg of gallic acid equivalent per g of dry weight (mg GAE g^{-1} d.w.). The detected limit was 1 µg mL^{-1}.

2.7. *Total Antioxidant Activity (AOA)*

To evaluate the total antioxidant activity of perennial wall rocket leaves, a 0.01 N $KMnO_4$ solution was titrated with the ethanol extracts of leaf homogenates until the solution became completely discoloured [49]. Gallic acid (PhytoLab GmbH & Co. KG, Vestenbergsgreuth, Germany) was used as an external standard, and the values obtained were expressed in mg GAE g^{-1} d.w.

2.8. Proteins

The crude protein content was measured via the Kjeldahl methodology, which is based on sample digestion with sulfuric acid, and the quantification of the ammonia content after reaction mixture alkalization was conducted [50]. The detected limit was 0.1 mg.

2.9. Soluble Solids

The determination of wall rocket leaf soluble solids was carried out using a portable digital refractometer, model DBR 35 (Sinergica Soluzioni s.r.l., Pescara, Italy), expressed in °Brix.

2.10. Nitrates

Nitrates were assessed using ion chromatography (ICS-3000, Dionex, Sunnyvale, CA, USA) coupled to a conductivity detector, with an IonPac AG11-HC guard (4 × 50 mm) column and an IonPac AS11-HC analytical column (4 × 250 mm). The flow rate of KOH solution (1 to 50 mM gradient) was 1.5 mL min^{-1}. The detection limit reached 0.08 mg L^{-1}.

2.11. Colour Measurement

Leaf colour was measured with a portable colorimeter (CR-10 Chroma; Minolta, Osaka, Japan) with Cilluminant after calibration with a white standard calibration plate, referring to the CIELab components L, a, and b. The 'L' component represents the relative lightness ranging from 0 (black) to 100 (white); both 'a' and 'b' range from –60 to +60, i.e., 'a' from green to red and 'b' from blue to yellow.

2.12. Mineral Determination

D. tenuifolia's leaf elemental profile, including Ca, K, Mg, P, Cu, Fe, Mn, and Zn, was determined in triplicate on dried, homogenized leaf samples using ICP-MS on the quadruple mass spectrometer Nexion 300D (Perkin Elmer Inc., Shelton, CT, USA), after the digestion of 1 g sample with sub-boiled HNO_3 (Fluka No. 02, 650 Sigma-Aldrich, Co., Saint Louis, MO, USA), 1:150 diluted with distilled deionized water in a Berghof SW-4 DAP-40 microwave digestion system (Berghof Products + Instruments Gmb H, 72, 800 Eningen, Germany). The analysis details were as follows: plasma power and argon flow, 1500 and 18 L min^{-1}, respectively; nebulizer argon flow, 0.98 L min^{-1}; and aux argon flow, 1.6 L min^{-1}. A sample introduction system was represented using the ESI ST PFA concentric nebulizer and the ESI PFA cyclonic spray chamber (Elemental Scientific Inc., Omaha, NE, USA) with the following parameters: injector, ESI Quartz 2.0 mm I.D.; sample flow, 637 L min^{-1}; standard internal flow, 84 L min^{-1}; and dwell time and acquisition mode, 10–100 ms with peak hopping for all the analytes. Rhodium 103 Rh was used as an internal standard to eliminate instability during the measurements. Quantitation was performed using an external standard (Merck IV, a multi-element standard solution); the Perkin–Elmer standard solution for P and all the standard curves were obtained at five different concentrations.

2.13. Selenium Content

The selenium content in rocket leaves was analyzed using the fluorimetric method previously described for tissues and biological fluids [51]. A total of 0.1 g of dried homogenized sample powder was digested via heating with a mixture of nitric and perchloric acids, according to the following sequence: at 120 °C for 1 h, 150 °C for 1 h, and 180 °C for 1 h. To reduce selenates (Se^{+6}) to selenites (Se^{+4}), a solution of 6 N HCl was used. Selenium concentration was obtained based on the fluorescence value of a complex between Se^{+4} and 2,3-diaminonaphtalene (hexane, 376 nm λ excitation; 519 nm λ emission). Each determination was performed in triplicate. The precision of the results was verified using the Mitsuba reference standard of Se-fortified stem powder for each determination, with a Se concentration of 1865 μg kg^{-1} (Federal Scientific Vegetable Center). The detection limit was 0.8 ng.

2.14. Iodine Content

To determine I content, 0.15 g of lyophilized and homogenized leaves was extracted with 10 mL of distilled water and 2 mL of 25% tetramethylammonium hydroxide. The extracts were filtered and analyzed via ICP-MS [52]. The detection limit was 4 μg L^{-1}. An external multi-element standard solution (Merck IV) was used for each determination.

2.15. Statistical Analysis

The presented results are the mean values of three replicates per experimental treatment. The data were statistically processed using analysis of variance (ANOVA), and mean separation was performed through Duncan's multiple range test at $p < 0.05$ using the SPSS software version 29 (IBM, Armonk, NY, USA).

3. Results and Discussion

3.1. Yield and Plant Performance

The wall rocket production is usually targeted to perform at least three cuttings of leaves [39,40]. As no significant interactions arose between the two experimental factors applied, the significant effects of crop cycle on wall rocket yield are presented as the means of the biofortification treatments and the untreated control. The first cycle provided the highest production, 2.12 and 2.4 times higher than those corresponding to the second and third ones, respectively, due to both higher leaf number per m^2 and mean weight (Table 2). Plant dry matter per m^2 showed a similar trend to yield.

Table 2. Effects of crop cycle and the biofortification treatment on yield parameters and plant dry matter of perennial wall rocket (means of the two-year data) (n = 3).

Treatment	Crop Surface (m^2)	Yield ($g\ m^{-2}$)	Number of Leaves per m^2	Mean Leaf Weight (g)	Plant Dry Matter ($g\ m^{-2}$)
Crop cycle (means of 4 values: I, Se, Se + I, control)					
Autumn	12	1536.6 ± 319.8 a	1442.8 ± 300.8 a	1.08 ± 0.13 a	158.7 ± 42.6 a
Autumn–winter	12	724.2 ± 163.7 b	1213.8 ± 208.8 b	0.59 ± 0.05 b	76.4 ± 23.4 b
Winter	12	640.0 ± 98.6 b	1266.9 ± 179.2 b	0.51 ± 0.02 b	66.1 ± 14.5 c
Biofortification treatment (means of the 3 crop cycles)					
Control	9	743.9 ± 347.2 c	987.9 ± 63.3 c	0.75 ± 0.34 ab	67.0 ± 32.1 c
I	9	893.5 ± 357.6 b	1320.1 ± 146.8 b	0.66 ± 0.21 b	89.4 ± 36.2 b
Se	9	1020.7 ± 446.5 a	1391.0 ± 221.0 b	0.71 ± 0.22 ab	111.5 ± 47.5 a
Se + I	9	1209.6 ± 589.0 a	1532.3 ± 147.2 a	0.77 ± 0.31 a	133.7± 62.4 a

Within each column, values followed by different letters are statistically different according to Duncan's test at $p \leq 0.05$.

The significant effects of the biofortification treatment on the yield and growth parameters are reported only as mean values of the three crop cycles, as no significant interactions arose between the two experimental factors applied (Table 2). In this respect, the combined application of Se and I resulted in the highest yield as a consequence of the highest leaf number and mean weight, exceeding the lowest production associated with the untreated control by 62.6%, the single I supply by 20.1%, and the Se treatment by 37.2%. The synergetic effect of Se and I on plant yield has been previously described in chervil plants, with a more pronounced growth stimulation effect in roots than shouts [15]. A similar trend to the yield was recorded for plant dry matter per m^2. The results are in accordance with the growth stimulation effect of Se and I described in different agricultural crops. At present, the beneficial effect of low iodine doses on plant yield is explained by the ability of I to regulate the expression of various genes and to affect the activity and structure of various proteins as a result of iodination [53]. Low doses of Se are known to maintain cell structure and improve water availability and nitrogen assimilation [54], which is consistent with the detected beneficial effect of this element on *D. tenuifolia* in the present research.

Joint Se/I biofortification had a growth stimulation effect on perennial wall rocket, differing from other species which were not significantly affected in terms of plant production. Indeed, carrot [10], potato [12], pumpkin [16], and Indian mustard [13] did not show a yield increase under joint Se/I biofortification. Overall, the results of the present research revealed significant growth stimulation effects of both single and combined applications of Se and I on wall rocket production, suggesting high prospects of this new technology.

3.2. *Quality, Colour, and Antioxidant Parameters*

Overall, the biochemical parameters and mineral element contents of wall rocket leaves, presented using the same pattern as for yield and plant growth variables, were significantly affected by both crop cycle and Se/I biofortification. In this respect, only the dry residue, soluble solids, colour components L, a, and b (Table 3), and vitamin C content (Table 4) were not statistically affected by crop cycle (Table 3); moreover, the colour parameters did not significantly differ between the Se/I biofortification treatments (Table 3).

Table 3. Effects of crop cycle and the biofortification treatment on rocket leaf quality and colour (means of the two-year data) (n = 3).

Treatment	Crop Surface (m^2)	Dry Matter (%)	Soluble Solids (°Brix)	Colour Parameters L	a	b
		Crop cycle (means of 4 values: I, Se, Se + I, control)				
Autumn	12	10.2 ± 0.8	5.7 ± 0.5	38.5 ± 1.9	−13.4 ± 1.1	20.0 ± 1.8
Autumn–winter	12	10.4 ± 1.1	5.7 ± 0.6	39.5 ± 1.0	−12.2 ± 1.0	22.6 ± 1.3
Winter	12	10.2 ± 0.9	5.8 ± 0.6	40.7 ± 1.0	−13.2 ± 1.0	21.8 ± 1.1
		n.s.	n.s.	n.s.	n.s.	n.s.
		Biofortification treatment (means of the 3 crop cycles)				
Control	9	8.97 ± 0.16 b	4.90 ± 0.23 b	37.9 ± 2.6	−12.4 ± 1.0	23.0 ± 1.6
I	9	9.99 ± 0.29 ab	5.95 ± 0.23 a	40.4 ± 1.0	−13.2 ± 1.4	20.9 ± 1.9
Se	9	10.97 ± 0.20 a	6.00 ± 0.20 a	40.1 ± 1.6	−13.4 ± 0.9	21.3 ± 1.4
I + Se	9	11.14 ± 0.42 a	6.08 ± 0.17 a	39.8 ± 1.2	−12.7 ± 0.9	20.7 ± 1.6
				n.s.	n.s.	n.s.

Within each column, values followed by different letters are statistically different according to Duncan's test at $p \leq 0.05$.

Table 4. Effects of crop cycle and the biofortification treatment on rocket leaf antioxidant compounds and activity (means of the two-year data) (n = 3).

Treatment	Crop Surface (m^2)	Total Antioxidant Activity (mg-GAE g^{-1} d.w.)	Total Polyphenols (mg-GAE g^{-1} d.w.)	Vitamin C (g 100 g^{-1} f.w.)
		Crop cycle (means of 4 values: I, Se, Se + I, control)		
Autumn	12	25.5 ± 1.9 b	13.5 ± 1.7 b	20.8 ± 2.2 a
Autumn-winter	12	25.1 ± 2.2 b	14.4 ± 1.3 ab	21.1 ± 2.2 a
Winter	12	27.2 ± 3.4 a	15.3 ± 1.8 a	21.4 ± 2.0 a
		Biofortification treatment (means of the 3 crop cycles)		
Control	9	22.9 ± 1.1 b	12.9 ± 1.0 b	18.5 ± 0.9 a
I	9	25.0 ± 1.6 b	13.5 ± 1.2 b	20.3 ± 0.9 a
Se	9	27.5 ± 1.9 a	15.4 ± 1.3 a	22.4 ± 1.0 a
Se + I	9	28.4 ± 1.8 a	15.8 ± 1.5 a	23.2 ± 1.0 a

Within each column, values followed by different letters are statistically different according to Duncan's test at $p \leq 0.05$.

The Se/I supply resulted in the significant increase in dry residue, with the highest values recorded under the Se + I and Se applications which exceeded the untreated control by 24% and 22%, respectively (Table 3). Furthermore, the soluble solids value was the highest in the Se + I treatment, showing a 1.24-fold increase compared to the control (Table 3).

The protein content decreased from the first to the second and third crop cycle by 19.7% and 23.4%, respectively, displaying a similar trend as yield (Figure 1).

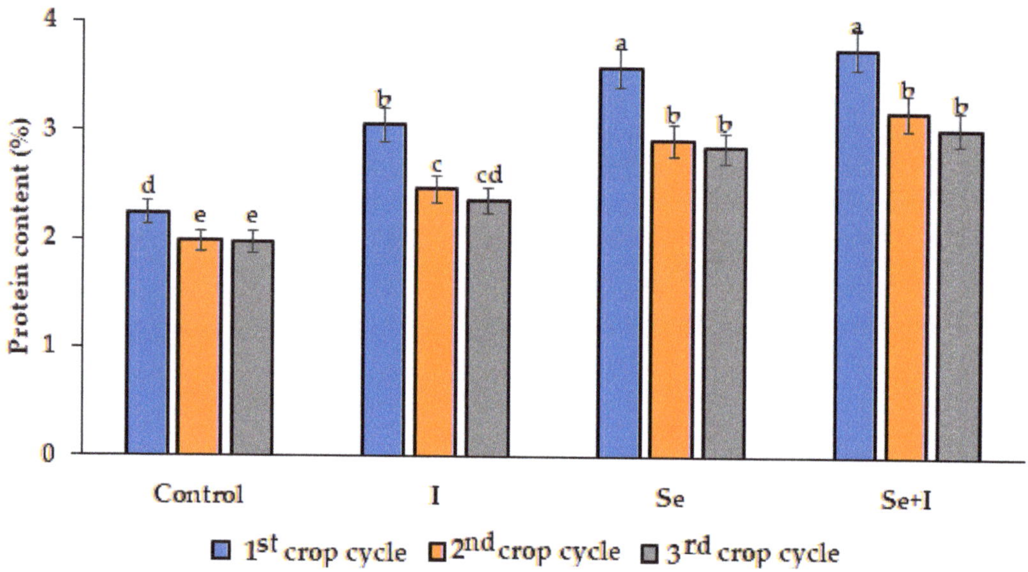

Figure 1. Differences in protein content in wall rocket plants under the Se/I treatment in different crop cycles. Values with the same letters do not differ statistically according to Duncan test at $p < 0.05$ (n = 3).

On the contrary, both Se and I treatments provided similar significant increases in the protein content, by 36–48%, 20–67%, and 54–61% under selenium, iodine, and Se/I joint applications, respectively. The recorded beneficial effect of Se and I on protein accumulation relates to the well-known participation of selenium in nitrogen metabolism and a stimulation of protein biosynthesis by Se [55] and I [23].

The response of plants to environmental stresses is closely connected with the intensity of biosynthesis of secondary metabolites participating in the antioxidant defence [56]. In this respect, the total antioxidant activity and polyphenol content recorded in the third crop cycle were higher than those detected in the two earlier cycles by 2 mg GAE g^{-1} d.w. and 13.3%, respectively, which confirms the enhancement of plant oxidative stress in the latest cycle (Table 4).

According to the obtained data (Table 4), the Se/I supply, especially under joint application, resulted in a significant increase in both the polyphenol content and the total antioxidant activity, suggesting the importance role of the mentioned microelements in the antioxidant defence. In general, wall rocket is known to be a good natural source of polyphenols with the prevalence of quercetin derivatives [38,57,58]. The remarkable antioxidant properties of quercetin and its protection against cardiovascular diseases and cancer [59] suggest the importance of the significant increase in the polyphenol content in this study; under the Se and Se + I supply, it increased by approximately 1.2-fold, compared to both the I treatment and control (Table 4). The total antioxidant activity of wall rocket leaves was higher upon Se + I and Se applications, exceeding the untreated control by 23.6% and 20.2%, respectively (Table 4). The impact of joint biofortification with I and Se recorded in other crops are consistent with the results of the present research. Indeed, the foliar Se/I biofortification of Indian mustard resulted in more than a two-fold increase in flavonoid content but slightly affected the ascorbic acid level [13]. On the contrary, only a 12.5% increase was recorded in the total antioxidant activity and polyphenol content in chickpea due to joint Se/I biofortification [14] and a 12% rise in polyphenols in lettuce [9], confirming the importance of genetic factors for the maintenance of plant antioxidant status.

3.3. Mineral Composition

D. tenuifolia is known to be a good source of Ca, K, Mg, and Fe [60]. In this respect, the sequential cutting of leaves may be considered as an external oxidant stress affecting both the yield and antioxidant status of wall rocket (Tables 2 and 4), as well as causing a mineral imbalance in plant tissues (Tables 5 and 6).

Table 5. Effects of crop cycle and the biofortification treatment on the macroelement content in rocket leaves (means of the two-year data) (n = 3).

Treatment	Ca (g kg^{-1} d.w.)	Mg (g kg^{-1} d.w.)	NO_3^- (mg kg^{-1} f.w.)	P (g kg^{-1} d.w.)
Crop cycle (means of 4 values: I, Se, Se + I, control)				
Autumn	27.4 ± 2.7 b	3.45 ± 0.32 b	5150 ± 408 a	2.84 ± 0.40 b
Autumn-winter	29.0 ± 3.7 ab	3.59 ± 0.37 ab	5079 ± 342 a	2.95 ± 0.42 ab
Winter	31.1 ± 4.2 a	3.81 ± 0.48 a	4206 ± 287 b	3.11 ± 0.45 a
Biofortification treatment (means of the 3 crop cycles)				
Control	25.7 ± 1.6 b	3.24 ± 0.17 b	5260 ± 586 a	2.52 ± 0.16 b
I	26.6 ± 1.8 b	3.33 ± 0.12 b	4763 ± 443 b	2.65 ± 0.15 b
Se	31.6 ± 2.6 a	3.88 ± 0.26 a	4625 ± 478 b	3.29 ± 0.19 a
Se + I	32.7 ± 1.8 a	4.01 ± 0.31 a	4597 ± 494 b	3.39 ± 0.20 a

Within each column, values followed by different letters are statistically different according to Duncan's test at $p \leq 0.05$.

Table 6. Effects of crop cycle and the biofortification treatment on the microelement content in rocket leaves (means of the two-year data) (n = 3).

Treatment	Zn (mg kg^{-1} d.w.)	Fe (mg kg^{-1} d.w.)	Cu (mg kg^{-1} d.w.)	Mn (mg kg^{-1} d.w.)
Crop cycle (means of 4 values: I, Se, Se + I, control)				
Autumn	11.9 ± 1.8 a	239.2 ± 132.7 a	15.9 ± 7.5 a	10.8 ± 4.5 b
Autumn-winter	11.9 ± 1.9 a	233.0 ± 107.6 a	14.9 ± 7.9 a	16.6 ± 2.6 a
Winter	14.1 ± 2.4 a	96.5 ± 21.4 b	6.6 ± 1.8 b	7.0 ± 1.9 c
Biofortification treatment (means of the 3 crop cycles)				
Control	9.7 ± 0.8 b	77.0 ± 20.8 d	5.9 ± 1.3 c	15.6 ± 4.7 a
I	12.9 ± 1.3 a	173.3 ± 65.9 c	19.0 ± 7.8 b	11.8 ± 4.9 b
Se	13.7 ± 2.0 a	225.5 ± 113.7 b	8.4 ± 2.1 a	8.8 ± 4.3 c
Se + I	14.0 ± 1.5 a	282.4 ± 129.9 a	16.4 ± 6.7 c	9.6 ± 4.1 c

Within each column, values followed by different letters are statistically different according to Duncan's test at $p \leq 0.05$.

Among the examined macroelements, Ca, Mg, and P showed significant increased levels in the third crop cycle, exceeding those of the first crop cycle by 13.6%, 10.4%, and 9.5%, respectively, regardless of the Se/I application (Table 5).

In contrast, the decreases in potassium (Figure 2) and nitrate content (Table 5) from the first to the third crop cycle were 40% and 21.5%, respectively.

The changes in potassium content in the wall rocket leaves from the first to the third cycle are consistent with the lowest yield and the highest antioxidant status of plants recorded in the third cycle. The mentioned dynamics suggest the important role of K and nitrogen in plant growth, as well as the oxidant stress enhancement in the leaves harvested in the latest cycle. In this respect, Se/I supply to wall rocket plants was extremely beneficial for improving the potassium accumulation levels.

Considering the negligible amount of K supplied by the potassium iodide treatment, the significant effect of Se has arisen on the increase in potassium level which has not been significantly affected by I application.

A general picture of the Se/I effect on macroelement accumulation in wall rocket, shown in Figure 3, confirms the beneficial effect of Se/I treatments for improving the nutritional value of plants, indicating a much lower effect of iodine compared to Se and Se + I supply.

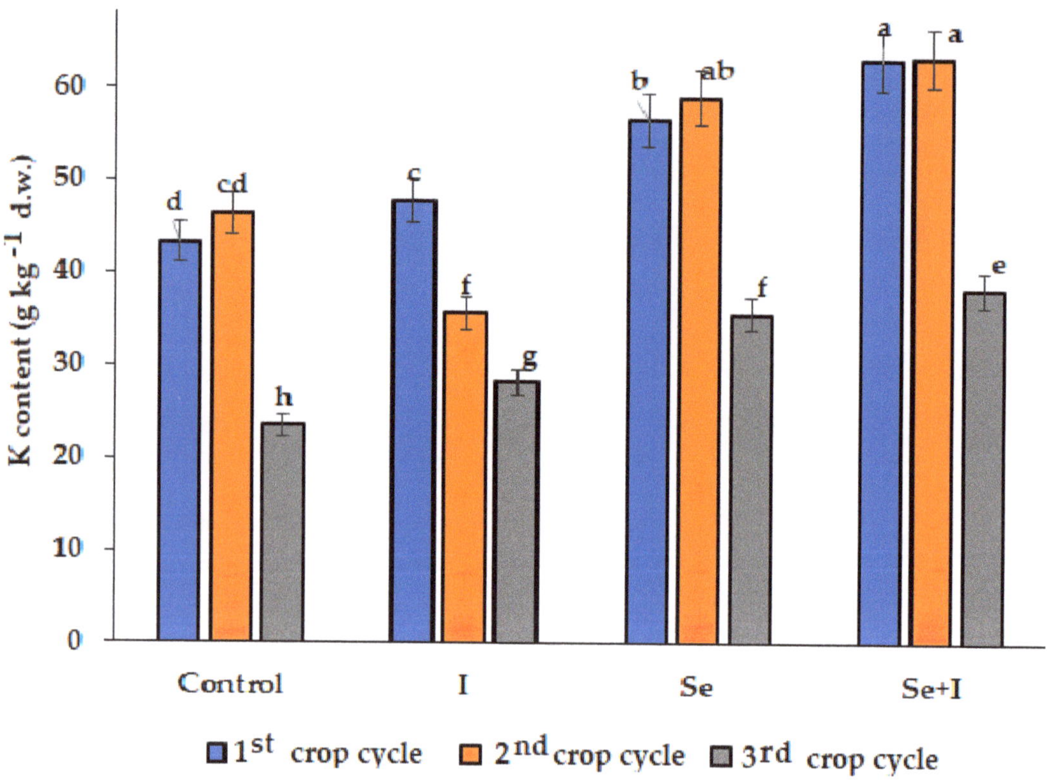

Figure 2. Effect of crop cycle and the Se/I supply on the K content in wall rocket leaves. Values with the same letters do not differ statistically according to Duncan's test at $p < 0.05$ (n = 3).

In this respect, the mean values of calcium content in all three crop cycles were 1.2-fold higher under the Se + I and Se treatments, compared to the I treatment and control (Figure 3). The same pattern was observed for Mg, whose contents under the Se + I and Se treatments exceeded those relevant to I supply and control by 23.8% and 19.8%, respectively. The phosphorus concentration under the Se + I and Se treatments was about 1.3 times higher compared to the I treatment and the untreated control, while the most significant effect was recorded for K.

To date, the effect of Se and I supply on mineral composition of plants has been studied rather sporadically, thus impeding the creation of a univocal frame of changes in the element levels [60,61]. In this respect, the perennial wall rocket mineral profile recorded under the biofortification treatments in the present research is particularly interesting (Table 5).

A different trend was observed for nitrate content which showed the highest value in the wall rocket control leaves, exceeding the average of the biofortification treatments by about 12%. In this respect, it is worth mentioning that iodine treatment to radish did not change the nitrate content [62] but increased the free amino acid concentration. According to the literature reports, Se supply decreases nitrate levels by enhancing the activity of nitrate reductase [13].

Wall rocket is known to be a hyper-accumulator of nitrates capable of accumulating up to 9000 mg kg^{-1} f.w. or even more depending on growth conditions [63,64]. The maximum threshold of nitrate content in wall rocket leaves according to the European Commission Regulation [65] is 6000 mg kg^{-1} f.w. in spring–summer and 7000 mg kg^{-1} in autumn–winter, while the acceptable daily intake (ADI) of nitrates should not exceed 3.7 mg kg^{-1} bw^{-1} day^{-1}. In general, the high consumption of nitrates may cause health risks due to the possibility of carcinogenic nitrosamine formation in the stomach under the presence of secondary amines and methemoglobin synthesis which hamper oxygen accumulation [66]. However, nitrate is also highly valuable for human health, as it prevents hypertension and protects human organism against cardiovascular diseases [67,68] via the regulation of gene expressions of proteins and enzymes involved in nitric oxide synthesis. The obtained results indicate a significant decrease in the nitrate levels both under Se and I supply, which is in accordance with the Se influence on nitrogen metabolism, usually decreasing the content of this ion in plants [13,69], and a competition between iodine and nitrates in plant nutrient uptake [70]. However, the high levels of natural antioxidants present in vegetables and especially in wall rocket guarantee the restriction of nitrosamine formation, thus diminishing health risks and providing conditions for beneficial nitrogen oxide formation [71].

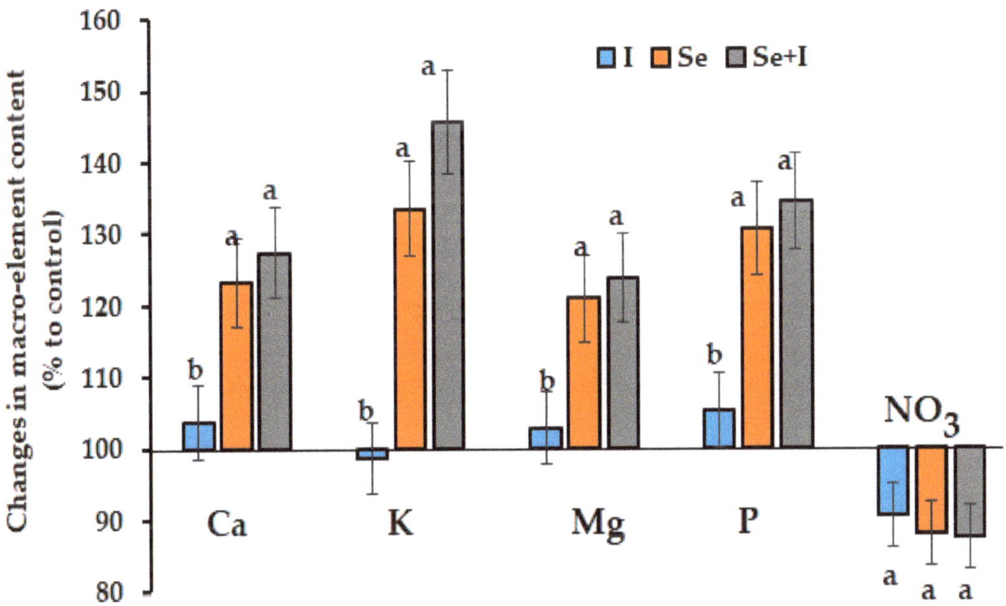

Figure 3. Changes in macroelement content in rocket leaves as affected by single and joint applications of selenium and iodine. Bars indicate standard deviations. For each element, values with the same letters do not differ statistically according to Duncan's test at $p < 0.05$ (n = 3).

Regarding the microelements in rocket leaves (Figure 4), Se concentration reached its peak in the third cycle, exceeding the values recorded in both the first and second cycles by approximately 16%. The same pattern was observed for iodine content, with the third cycle showing a 60% higher concentration than the first and second cycles.

The data presented in Figure 4A indicate a slight stimulation effect of I on Se accumulation both in the control and Se-treated plants. The similar beneficial effect of Se on iodine content was recorded in all three crop cycles. Controversial results of Se/I interaction in plants, reflecting the possibility of both synergistic and antagonistic relationships between the two microelements [1,11,49], suggest the need for further investigations aimed at revealing the factors affecting the related phenomenon.

In the present study, the selenium concentration in rocket leaves was the highest under the Se + I treatment, followed by the Se and I applications, with the lowest value corresponding to the untreated control. The biofortification levels of Se was equal to 50 in the case of joint Se/I supply, exceeded 44 upon single Se spraying, and reached 4.5 under the I treatment without endogenous Se. Iodine showed the peaked content in rocket leaves supplied with Se + I, followed by the I treatment, whereas it was not detected at all upon Se application and in the untreated control.

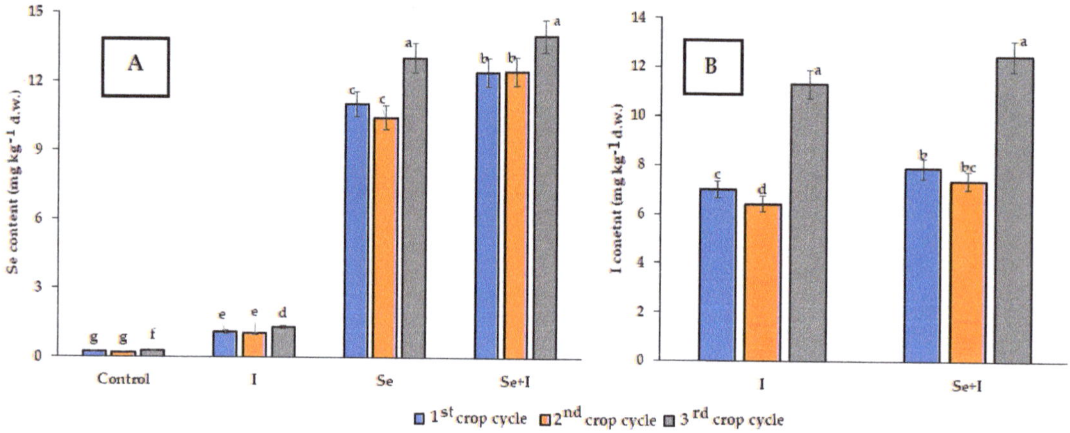

Figure 4. Content of Se (**A**) and I (**B**) in wall rocket leaves in the first, second, and third crop cycles under Se/I supply. Values with the same letters do not differ statistically according to Duncan's test at $p < 0.05$. Iodine concentrations in the control and Se-treated plants were below the detected level (n = 3).

In general, the increase in Se/I accumulation in the leaves grown in the third crop cycle, compared to the second and the first one, may be related to the improvement of the plant's antioxidant status under Se/I supply.

Changes in Zn, Fe, Cu, and Mn contents as a result of Se/I biofortification in different crop cycles are presented in Table 6 and Figure 5. In this respect, the zinc accumulation in the leaves harvested in the third cycle was about 1.2 times higher compared to those detected in the first and second cycles. Contrastingly, iron concentration significantly decreased in the third cycle, showing 2.5 times lower values than those measured in the first and second cycles. The copper content displayed a similar trend to iron, with the lowest value recorded in the leaves during the third cycle, 57% lower than those related to the first and second cycles. The manganese content was the highest in the leaves during the second cycle, followed by the first and the third cycles.

Surprisingly, the Se/I biofortification of perennial wall rocket resulted in the increased contents of both the macroelements and Zn, Fe, and Cu (Table 6, Figure 6).

Indeed, Zn and Fe concentrations were higher under the Se + I treatment, exceeding the untreated control by 44% and 370%, respectively. The copper content reached its maximum in the I treatment, 3.2 times higher than the control. A different pattern was observed for Mn, as the untreated leaves showed the highest concentration, exceeding the biofortification treatments by 30% to 43%.

The beneficial effect of single Se and I supply on mineral composition of wall rocket leaves was in agreement with previous findings in other plant species [72,73], but the joint Se/I effect on *D. tenuifolia* element accumulation was much higher than those recorded in different crops [8,13,15].

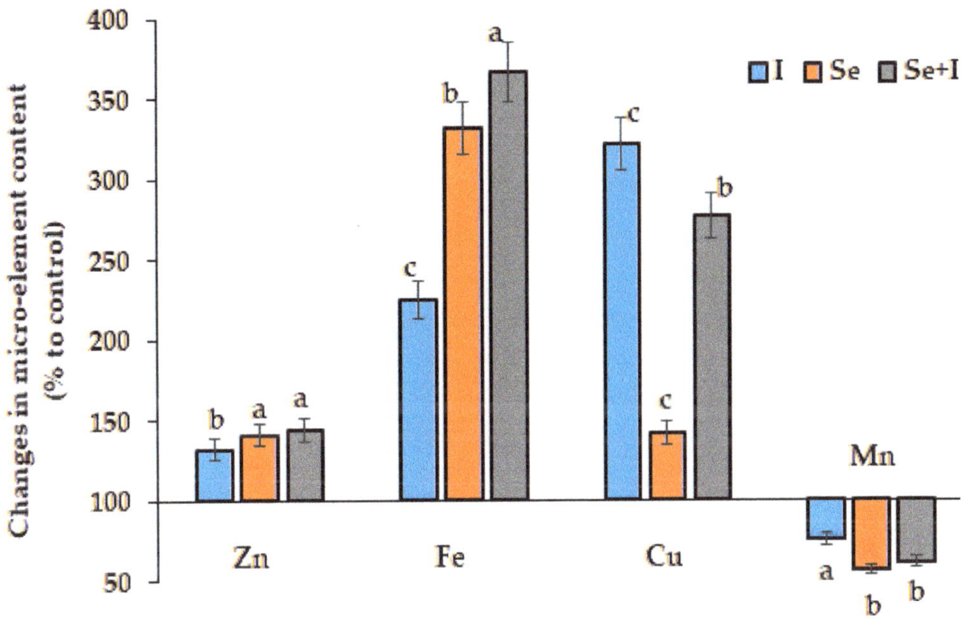

Figure 5. Changes in microelement content in rocket leaves as affected by single and joint applications of Se and I. Bars indicate standard deviations. For each element, values with the same letters do not differ statistically according to Duncan's test at $p < 0.05$ (n = 3).

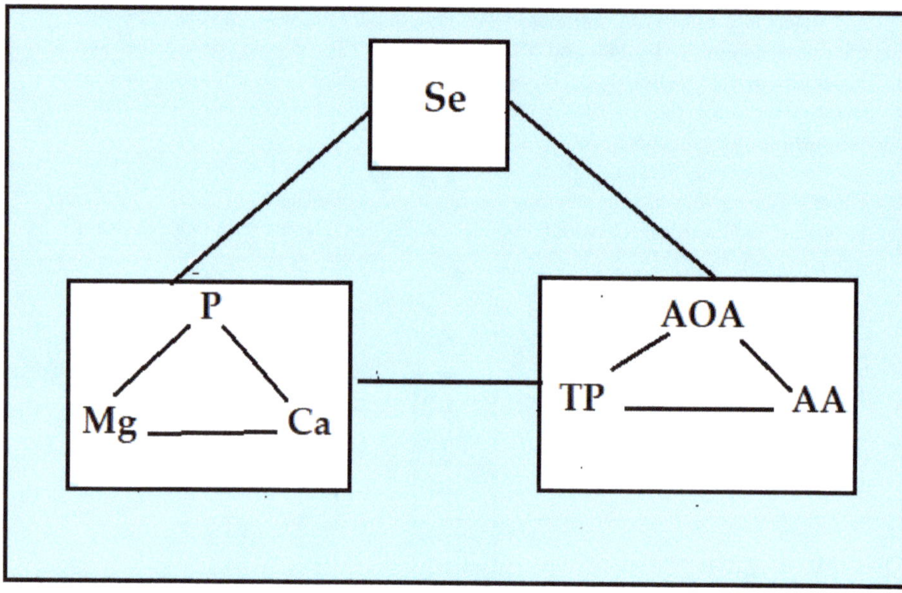

Figure 6. Relationship between selenium levels, antioxidant status, and macroelement content in perennial wall rocket.

3.4. Correlations Between the Variables

The relationship between wall rocket weight, mineral composition, and antioxidant status confirmed the importance of Se/I supply on crop yield and on the nutritional value of the product (Table 7).

The role of selenium in activating the antioxidant metabolism in plants was previously demonstrated in different plants [74,75] and is reflected by a strong correlation between Se, polyphenol, and ascorbic acid content, as well as the total antioxidant activity in wall rocket leaves ($r > 0.815$; $p < 0.001$). According to the obtained data, the biofortification of wall rocket with Se and I is characterized by significant correlations of Se both with the antioxidant parameters and with the contents of Ca, P, and Mg (Table 7, Figure 3).

A more complex interaction was revealed between Se and plant minerals. Indeed, the foliar supply of tea plants with glucosamin selenium was shown to increase macro- and microelement accumulation along with the enhancement of plant antioxidant status [76]. On the contrary, the Se treatment of maize resulted in Ca and P level increase, K concentration decrease, and no effects on Mg [77]. Se biofortification did not significantly affect the mineral content in oats [78], whereas Zinc, Ca, and Mg significantly increased under Se foliar supply in wheat plants [79]. In our research, the conditions of wall rocket Se/I biofortification greatly fostered the plant quality in terms of both antioxidant and mineral composition, which are reportedly affected by selenium dose, the method of supply, and genetic and environmental factors [22,79].

The data presented in Table 7 show that no significant correlations between iodine, Se, and other elements arose, suggesting the different mechanisms of Se and I uptake by plants. Indeed, KI is absorbed via chloride and Na-K/Cl transporters, while selenate via sulphate transporters [1]. In general, the beneficial effect of Se/I biofortification on wall rocket yield is connected with the positive correlation between plant yield, potassium, and proteins (Table 7); i.e., Se increased protein and K content, thus promoting perennial wall rocket growth.

3.5. *Mineral Consumption Levels*

The significant increase in antioxidant properties as well as macro- and microelement contents under Se/I biofortification represents an interesting outcome with the perspective to produce functional food with innovative characteristics (Table 8). The consumption levels of macro- and microelements associated with 50 g of fresh perennial wall rocket leaves suggest the importance of Se/I biofortified plants as a remarkable source of Se, I, Ca, and K. Iron ingestion, attaining 24% of RDA, should not be taken into consideration as only 7–9% of non-heme iron is absorbed by leafy vegetables [80]. Despite the great differences in Fe, Cu, Mn, Se, and I between the first and the third crop cycles, Se + I biofortified plants grown in the first crop cycle can provide, referring to RDA, up to 24% of Fe, 9.0% of K, 15% of Ca, 25.9% of I, and 88.8% of Se (Table 8). Furthermore, Cu, Zn, Mn, Mg, and P contents in wall rocket leaves may not be taken into account due to their low concentrations, providing less than 5% of the corresponding RDA values. On the contrary, significantly higher consumption levels of Se and I can be obtained from the third crop cycle, with values reaching 100% and 41.6% of RDA, respectively.

Table 7. Coefficients of correlation between the examined parameters of wall rocket leaves (n = 12).

	Ca	Mg	K	NO_3	Se	I	Zn	Fe	Cu	Mn	DM	SS	Protein	AOA	TP	AA	Yield
P	0.893 a	0.903 a	0.367	−0.553 f	0.925 a	0.232	0.703 c	0.420	−0.046	−0.509	0.826 a	0.727 c	0.603 f	0.797 a	0.766 a	0.827 a	0.117
	Ca	0.848 a	0.161	−0.582 f	0.838 a	0.273	0.751 b	0.255	−0.138	−0.512	0.706 c	0.626 e	0.453	0.800 a	0.762 a	0.782 a	−0.041
		Mg	0.221	−0.583 f	0.842 a	0.258	0.717 c	0.250	−0.124	−0.492	0.739 b	0.636 e	0.491	0.797 a	0.814 a	0.747 b	−0.009
			K	0.361	0.617 b	−0.053	−0.006	0.762 a	0.431	0.177	0.528	0.345	0.688 d	0.182	0.140	0.385	0.617 e
				NO_3	−0.446	−0.391	−0.733 b	0.161	0.220	0.733 b	−0.442	−0.510	−0.175	−0.600 f	−0.603 f	−0.500	0.267
					Se	0.168	0.656 e	0.509	0.024	−0.503	0.872 a	0.686 d	0.753 b	0.815 a	0.750 b	0.842 a	0.291
						I	0.515	0.159	0.443	−0.300	0.324	0.515	0.270	0.415	0.247	0.345	0.08
							Zn	0.178	0.035	−0.734 b	0.664 d	0.747 b	0.504	0.830 a	0.710 c	0.670 d	0.005
								Fe	0.615 f	−0.006	0.612 f	0.501	0.720 c	0.223	0.214	0.504	0.602 f
									Cu	0.161	0.284	0.343	0.470	−0.032	−0.122	0.195	0.473
										Mn	−0.431	−0.506	−0.462	−0.622 f	−0.484	−0.487	−0.214
											DM	0.837 a	0.783 a	0.760 a	0.620 f	0.836 a	0.271
												SS	0.721 c	0.727 c	0.581 f	0.747 b	0.284
													Protein	0.580 f	0.402	0.705 c	0.739 b
														AOA	0.751 b	0.788 a	0.100
															TP	0.619 f	−0.123
																AA	0.244

(a) $p < 0.001$; (b) 0.002; (c) $p < 0.005$; (d) $p < 0.01$; (e) $p < 0.02$; (f) $p < 0.05$; DM: dry matter; SS: soluble solids; AOA: total antioxidant activity; TP: polyphenols; AA: ascorbic acid.

Table 8. Consumption of mineral elements corresponding to 50 g of fresh wall rocket leaves biofortified with Se + I, compared to the dietary allowance level (RDA), in %.

Element	RDA (mg)	Treatment			
		Control	Iodine	Selenium	Iodine + Selenium
Fe	8	5.7 (3.9) *	14.2 (6.3)	22.3 (6.6)	24.0 (7.3)
Cu	2	1.8 (1.1)	6.1 (2.2)	2.7 (1.5)	5.3 (1.9)
Zn	10	0.5 (0.6)	0.6 (0.7)	0.7 (0.8)	0.7 (0.8)
Mn	2.3	3.9 (2.1)	2.1 (1.7)	1.6 (1.1)	1.7 (1.3)
Se	0.070	1.8 (2.1)	7.9 (9.4)	78.9 (93.4)	88.8 (100)
I	0.150	0	23.4 (37.7)	0	25.9 (41.6)
Ca	1000	14.4 (13.4)	12.7 (14.1)	14.7 (17.1)	15.0 (17.7)
K	3510	6.2 (3.3)	6.7 (4.0)	8.1 (5.1)	9.0 (5.5)
Mg	350	4.5 (4.7)	4.6 (5.3)	5.3 (5.9)	5.4 (6.2)
P	800	1.5 (1.6)	1.6 (1.7)	2.0 (2.2)	2.0 (2.2)

* Values in parenthesis refer to the third crop cycle.

In general, it may be inferred that the obtained Se/I biofortified wall rocket plants may be useful for decreasing the risks of cardiovascular diseases, osteoporosis, high blood pressure due to the high levels of Ca, K, and Se [73,81], as well as for enhancing both human organism immunity and Se/I status.

4. Conclusions

The foliar Se/I biofortification of *Diplotaxis tenuifolia* represents an interesting technology to boost yield, antioxidant status, Se, and I, as well as Ca, Mg, K, P, Fe, Cu, and Zn accumulation in the leaves of this species. The mentioned treatments allowed us to create a new functional food with high Se and I levels, useful to prevent and cure human Se/I deficiency, and protect against cardiovascular diseases and cancer due to the high levels of Se, I, K, and Ca. The present research has brought the perspective of valorising perennial wall rocket in terms of the industrial production of health beneficial products under sustainable management.

Author Contributions: A.V.T.: Investigation, Data curation, Formal analysis, Software, and Writing—original draft; N.G.: Methodology, Data curation, Formal analysis, Software, and Writing—original draft; S.D.P.: Writing—review and editing, Validation, and Funding acquisition; A.S.: Methodology and Writing, review and editing; R.P.: Methodology and Writing—review and editing; O.C.M.: Methodology, Formal analysis, and Writing—original draft; E.C.: Investigation, Formal analysis, and Software; V.C.: Investigation; G.C.: Conceptualization, Data curation, Writing—review and editing, Supervision, and Validation. All authors have read and agreed to the published version of the manuscript.

Funding: MIPAAF: National Horticultural Strategy 2018–2022. Sustainable Rocket Management.

Data Availability Statement: The data that support the findings of this study are available on request.

Conflicts of Interest: The authors declare that they have no conflict of interest.

References

1. Golubkina, N.; Moldovan, A.; Kekina, H.; Kharchenko, V.; Sekara, A.; Vasileva, V.; Skrypnik, L.; Tallarita, A.; Caruso, G. Joint Biofortification of Plants with Selenium and Iodine: New Field of Discoveries. *Plants* **2021**, *10*, 1352. [CrossRef]
2. The Iodine Global Network. *Global Scorecard of Iodine Nutrition in 2020 in the General Population Based on School-Age Children (SAC)*; IGN: Ottawa, ON, Canada, 2021.
3. Schomburg, L.; Köhrle, J. On the importance of selenium and iodine metabolism for thyroid hormone biosynthesis and human health. *Mol. Nutr. Food Res.* **2008**, *52*, 1235–1246. [CrossRef] [PubMed]
4. Lyons, G. Biofortification of Cereals with Foliar Selenium and Iodine Could Reduce Hypothyroidism. *Front. Plant Sci.* **2018**, *9*, 730. [CrossRef] [PubMed]

5. Avnee; Sood, S.; Chaudhary, D.R.; Jhorar, P.; Rana, R.S. Biofortification: An approach to eradicate micronutrient deficiency. *Front. Nutr.* **2023**, *10*, 1233070. [CrossRef] [PubMed]
6. Izydorczyk, G.; Ligas, B.; Mikula, K.; Witek-Krowiak, A.; Moustakas, K.; Chojnacka, K. Biofortification of edible plants with selenium and iodine—A systematic literature review. *Sci. Total Environ.* **2021**, *754*, 141983. [CrossRef] [PubMed]
7. Smolén, S.; Kowalska, I.; Sady, W. Assessment of biofortification with iodine and selenium of lettuce cultivated in the NFT hydroponic system. *Sci. Hortic. Amst.* **2014**, *166*, 9–16. [CrossRef]
8. Smolén, S.; Skoczylas, Ł.; Rakoczy, R.; Ledwożyw-Smolen, I.; Kopec, A.; Piatkowska, E. Mineral composition of field-grown lettuce (*Lactuca sativa* L.) depending on the diversified fertilization with iodine and selenium compounds. *Acta Sci. Pol. Hortorum Cultus* **2015**, *14*, 97–114.
9. Smolén, S.; Skoczylas, Ł.; Ledwożyw-Smolén, I.; Rakoczy, R.; Kopéc, A.; Piatkowska, E.; Kapusta-Duch, J. Biofortification of carrot (*Daucus carota* L.) with iodine and selenium in a field experiment. *Front. Plant Sci.* **2016**, *7*, 730. [CrossRef]
10. Smolén, S.; Baranski, R.; Ledwożyw-Smoleń, I.; Skoczylas, Ł.; Sady, W. Combined biofortification of carrot with iodine and selenium. *Food Chem.* **2019**, *300*, 125202. [CrossRef] [PubMed]
11. Germ, M.; Kacjan-Maršić, N.; Kroflič, A.; Jerše, A.; Stibilj, V.; Golob, A. Significant accumulation of iodine and selenium in chicory (*Cichorium intybus* L. var. foliosum Hegi) leaves after foliar spraying. *Plants* **2020**, *9*, 1766. [CrossRef] [PubMed]
12. Smolén, S.; Kowalska, I.; Skoczylas, Ł.; Liszka-Skoczylas, M.; Grzanka, M.; Halka, M.; Sady, W. The effect of salicylic acid on biofortification with iodine and selenium and the quality of potato cultivated in the NFT system. *Sci. Hort.* **2018**, *240*, 530–543. [CrossRef]
13. Golubkina, N.; Kekina, H.; Caruso, G. Yield, quality and antioxidant properties of Indian mustard (*Brassica juncea* L.) in response to foliar biofortification with selenium and iodine. *Plants* **2018**, *7*, 80. [CrossRef]
14. Golubkina, N.; Gomez, L.; Kekina, H.; Hallam, R.; Tallarita, A.; Cozzolino, E.; Torino, V.; Koshevarov, A.; Cuciniello, A.; Maiello, R. Joint selenium-iodine supply and arbuscular mycorrhizal fungi inoculation affect yield and quality of chickpea seeds and residual biomass. *Plants* **2020**, *9*, 804. [CrossRef] [PubMed]
15. Golubkina, N.; Moldovan, A.; Fedotov, M.; Kekina, H.; Kharchenko, V.; Folmanis, G.; Alpatov, A.; Caruso, G. Iodine and Selenium Biofortification of Chervil Plants Treated with Silicon Nanoparticles. *Plants* **2021**, *10*, 2528. [CrossRef]
16. Golob, A.; Kroflič, A.; Jerše, A.; Kacjan Maršić, N.; Šircelj, H.; Stibilj, V.; Germ, M. Response of Pumpkin to Different Concentrations and Forms of Selenium and Iodine, and their Combinations. *Plants* **2020**, *9*, 899. [CrossRef] [PubMed]
17. Golob, A.; Novak, T.; Kacjan Maršić, N.; Šircelj, H.; Stibilj, V.; Jerše, A.; Kroflič, A.; Germ, M. Biofortification with selenium and iodine changes morphological properties of *Brassica oleracea* L. var. gongylodes) and increases their contents in tubers. *Plant Physiol. Biochem.* **2020**, *150*, 234–243. [CrossRef] [PubMed]
18. Hu, L.; Fan, H.; Wu, D.; Wan, J.; Wang, X.; Huang, R.; Liu, W.; Shen, F. Assessing bioaccessibility of Se and I in dual biofortified radish seedlings using simulated in vitro digestion. *Food Res. Int.* **2019**, *119*, 701–708. [CrossRef] [PubMed]
19. Germ, M.; Stibilj, V.; Šircelj, H.; Jerše, A.; Kroflič, A.; Golob, A.; Kacjan Maršić, N. Biofortification of common buckwheat microgreens and seeds with different forms of selenium and iodine. *J. Sci. Food Agric.* **2019**, *99*, 4353–4362. [CrossRef] [PubMed]
20. Zhu, Y.G.; Huang, Y.; Hu, Y.; Liu, Y.; Christie, P. Interactions between selenium and iodine uptake by spinach (*Spinacia oleracea* L.) in solution culture. *Plant Soil* **2004**, *261*, 99–105. [CrossRef]
21. Jerše, A.; Kacjan Maršić, N.; Kroflič, A.; Germ, M.; Šircelj, H.; Stibilj, V. Is foliar enrichment of pea plants with iodine and selenium appropriate for production of functional food. *Food Chem.* **2018**, *267*, 368–375. [CrossRef] [PubMed]
22. Duborská, E.; Šebesta, M.; Matulová, M.; Zvěřina, O.; Urík, M. Current Strategies for Selenium and Iodine Biofortification in Crop Plants. *Nutrients* **2022**, *14*, 4717. [CrossRef]
23. Nascimento, V.L.; Souza, B.C.O.Q.; Lopes, G.; Guilherme, L.R.G. On the Role of Iodine in Plants: A Commentary on Benefits of This Element. *Front. Plant Sci.* **2022**, *13*, 836835. [CrossRef]
24. Schiavon, M.; Ertani, A.; Parrasia, S.; Vecchia, F.D. Selenium accumulation and metabolism in algae. *Aq. Toxicol.* **2017**, *189*, 1–8. [CrossRef]
25. Rakoczy-Lelek, R.; Smolen, S.; Grzanka, M.; Ambroziak, K.; Pitala, J.; Skoczylas, Ł.; Liszka-Skoczylas, M.; Kardasz, H. Effectiveness of Foliar Biofortification of Carrot With Iodine and Selenium in a Field Condition. *Front. Plant Sci.* **2021**, *12*, 656283. [CrossRef] [PubMed]
26. Cerretani, L.; Comandini, P.; Fumanelli, D.; Scazzina, F.; Chiavaro, E. Evaluation of iodine content and stability in recipes prepared with biofortified potatoes. *Int. J. Food Sci. Nutr.* **2014**, *65*, 797–802. [CrossRef] [PubMed]
27. Tessier, E.; Amouroux, D.; Abril, G.; Lemaire, E.; Donald, O.F.X. Formation and volatilisation of alkyl-iodides and -selenides in macrotidal estuaries. *Biogeochemistry* **2002**, *59*, 183–206. [CrossRef]
28. Zhang, H.; Zhao, Z.; Nie, B.; Lyu, C.; Liu, X. Selenium loss and changes in product quality during cooking of selenium enriched potato tubers. *J. Food Comp. Anal.* **2021**, *96*, 103728. [CrossRef]
29. Bertolino, F.A.; Stege, P.W.; Salinas, E.; Raba, M.G.A.J. Electrochemical Study of the Antioxidant Activity and the Synergic Effect of Selenium with Natural and Synthetic Antioxidants. *Anal. Lett.* **2010**, *43*, 2078–2090. [CrossRef]
30. Andrés, C.M.C.; Pérez de la Lastra, J.M.; Juan, C.A.; Plou, F.J.; Pérez-Lebeña, E. Antioxidant Metabolism Pathways in Vitamins, Polyphenols, and Selenium: Parallels and Divergences. *Int. J. Mol. Sci.* **2024**, *25*, 2600. [CrossRef]

31. Maia, L.B.; Maiti, B.K.; Moura, I.; Moura, J.J.G. Selenium—More than Just a Fortuitous Sulfur Substitute in Redox Biology. *Molecules* **2024**, *29*, 120. [CrossRef] [PubMed]
32. Finley, G.F. Selenium and glucosinolates in cruciferous vegetables: Metabolic interactions and implications for cancer chemoprevention in humans. *Acta Hortic.* **2007**, *744*, 171–180. [CrossRef]
33. Caruso, G.; Formisano, L.; Cozzolino, E.; Pannico, A.; El-Nakhel, C.; Rouphael, Y.; Tallarita, A.; Cenvinzo, V.; De Pascale, S. Shading Affects Yield, Elemental Composition and Antioxidants of Perennial Wall Rocket Crops Grown from Spring to Summer in Southern Italy. *Plants* **2020**, *9*, 933. [CrossRef]
34. Caruso, G.; El-Nakhel, C.; Rouphael, Y.; Comite, E.; Lombardi, N.; Cuciniello, A.; Woo, S.L. *Diplotaxis tenuifolia* (L.) DC. Yield and Quality as Influenced by Cropping Season, Protein Hydrolysates, and Trichoderma Applications. *Plants* **2020**, *9*, 697. [CrossRef] [PubMed]
35. Caruso, G.; Parrella, G.; Giorgini, M.; Nicoletti, R. Crop systems, quality and protection of *Diplotaxis tenuifolia*. *Agriculture* **2018**, *8*, 55. [CrossRef]
36. Guijarro-Real, C.; Prohens, J.; Rodríguez-Burruezo, A.; Fita, A. Consumers acceptance and volatile profile of wall rocket (*Diplotaxis erucoides*). *Food Res. Int.* **2020**, *132*, 109008. [CrossRef] [PubMed]
37. Hall, M.K.D.; Jobling, J.J.; Rogers, G.S. Some perspectives on rocket as a vegetable crop a review. *Veg. Crops Res. Bull.* **2012**, *76*, 21–41. [CrossRef]
38. Romano, R.; Pizzolongo, F.; De Luca, L.; Cozzolino, E.; Rippa, M.; Ottaiano, L.; Mormile, P.; Mori, M.; Di Mola, I. Bioactive Compounds and Antioxidant Properties of Wild Rocket (*Diplotaxis Tenuifolia* L.) Grown under Different Plastic Films and with Different UV-B Radiation Postharvest Treatments. *Foods* **2022**, *11*, 4093. [CrossRef] [PubMed]
39. Caruso, G.; De Pascale, S.; Nicoletti, R.; Cozzolino, E.; Rouphael, Y. Productivity, nutritional and functional qualities of perennial wall-rocket: Effects of pre-harvest factors. *Folia Hort.* **2019**, *31*, 71–80. [CrossRef]
40. Teliban, G.C.; Precupeanu, C.; Munteanu, N.; Burducea, M.; Cojocaru, A.; Vlăduț, N.V.; Popa, L.D.; Stoleru, V. The effect of the crop cycle and differential fertilization on a perennial wall-rocket crop established in polytunnel. *Acta Hortic.* **2024**, *1391*, 479–486. [CrossRef]
41. Bell, L.; Oruna-Concha, M.J.; Wagstaff, C. Identification and quantification of glucosinolate and flavonol compounds in rocket salad (*Eruca sativa, Eruca vesicaria* and *Diplotaxis tenuifolia*) by LC-MS: Highlighting the potential for improving nutritional value of rocket crops. *Food Chem.* **2015**, *172*, 852–861. [CrossRef] [PubMed]
42. Schiavon, M.; Nardi, S.; Pilon-Smits, E.A.H.; Dall'Acqua, S. Foliar selenium fertilization alters the content of dietary phytochemicals in two rocket species. *Front. Plant Sci.* **2022**, *13*, 987935. [CrossRef] [PubMed]
43. Dall'Acqua, S.; Ertani, A.; Pilon-Smits, E.A.H.; Fabrega-Prats, M.; Schiavon, M. Selenium Biofortification Differentially Affects Sulfur Metabolism and Accumulation of Phytochemicals in Two Rocket Species (*Eruca Sativa* Mill. and *Diplotaxis Tenuifolia*) Grown in Hydroponics. *Plants* **2019**, *8*, 68. [CrossRef]
44. Nascimento, C.S.; Nascimento, C.S.; Lopes, G.; Carrasco, G.; Gratão, P.L.; Cecílio Filho, A.B. Biofortified Rocket (*Eruca sativa*) with Selenium by Using the Nutrient Film Technique. *Horticulturae* **2022**, *8*, 1088. [CrossRef]
45. Aksu, G.; Temel, E.; Altay, H. Comparison of iodine application methods in Rocket Plant. *Agron. Res.* **2017**, *15*, 5–10.
46. Danso, O.P.; Asante-Badu, B.; Zhang, Z.; Song, J.; Wang, Z.; Yin, X.; Zhu, R. Selenium Biofortification: Strategies, Progress and Challenges. *Agriculture* **2023**, *13*, 416. [CrossRef]
47. Lawson, P.G.; Daum, D.; Czauderna, R.; Meuser, H.; Härtling, J.W. Soil versus foliar iodine fertilization as a biofortification strategy for field-grown vegetables. *Front. Plant Sci.* **2015**, *6*, 450. [CrossRef] [PubMed]
48. AOAC Association Official Analytical Chemists. *The Official Methods of Analysis of AOAC International; 22 'Vitamin C'*; AOAC: Rockville, MD, USA, 2012.
49. Golubkina, N.; Kekina, H.; Molchanova, A.; Antoshkina, M.; Nadezhkin, S.; Soldatenko, A. *Plants Antioxidants and Methods of Their Determination*; Infra-M: Moscow, Russia, 2020. (In Russia)
50. Walker, J.M. The Lowry Method for Protein Quantitation. In *The Protein Protocols Handbook*; Springer Protocols Handbooks; Humana Press: Totowa, NJ, USA, 2009; Chapter 2; pp. 7–10.
51. Alfthan, G.V. A micromethod for the determination of selenium in tissues and biological fluids by single-test-tube fluorimetry. *Anal. Chim. Acta* **1984**, *165*, 187–194. [CrossRef]
52. Jerše, A.; Kacjan Marši´c, N.; Šircelj, H.; Germ, M.; Kroflic, A.; Stibilj, V. Seed soaking in I and Se solutions increases concentrations of both elements and changes morphological and some physiological parameters of pea sprouts. *Plant. Physiol. Biochem.* **2017**, *118*, 285–294. [CrossRef] [PubMed]
53. Riyazuddin, R.; Singh, K.; Iqbal, N.; Nisha, N.; Rani, A.; Kumar, M.; Khatri, N.; Siddiqui, M.H.; Yasheshwar; Kim, S.T.; et al. Iodine: An emerging biostimulant of growth and stress responses in plants. *Plant Soil* **2022**, *486*, 119–133. [CrossRef]
54. Khan, Z.; Thounaojam, T.C.; Chowdhury, D. The role of selenium and nano selenium on physiological responses in plant: A review. *Plant Growth Regul.* **2023**, *100*, 409–433. [CrossRef]
55. Bandehagh, A.; Dehghanian, Z.; Gougerdchi, V.; Hossain, M.A. Selenium: A Game Changer in Plant Development, Growth, and Stress Tolerance, via the Modulation in Gene Expression and Secondary Metabolite Biosynthesis. *Phyton* **2023**, *92*, 2301–2324. [CrossRef]

56. Mishra, N.; Jiang, C.; Chen, L.; Paul, A.; Chatterjee, A.; Shen, G. Achieving abiotic stress tolerance in plants through antioxidative defense mechanisms. *Front. Plant Sci.* **2023**, *14*, 1110622. [CrossRef] [PubMed]
57. Ressurreição, S.; Salgueiro, L.; Figueirinha, A. Diplotaxis Genus: A Promising Source of Compounds with Nutritional and Biological Properties. *Molecules* **2024**, *29*, 2612. [CrossRef]
58. Pasini, F.; Verardo, V.; Caboni, M.F.; D'Antuono, L.F. Determination of glucosinolates and phenolic compounds in rocket salad by HPLC-DAD–MS: Evaluation of Eruca sativa Mill. and *Diplotaxis tenuifolia* L. genetic resources. *Food Chem.* **2012**, *133*, 1025–1033. [CrossRef]
59. Kim, J.K.; Park, S.U. Quercetin and its role in biological functions: An updated review. *EXCLI J.* **2018**, *17*, 856–863.
60. Ciriello, M.; Campana, E.; Colla, G.; Rouphael, Y. An Appraisal of Nonmicrobial Biostimulants' Impact on the Productivity and Mineral Content of Wild Rocket (*Diplotaxis tenuifolia* (L.) DC.) Cultivated under Organic Conditions. *Plants* **2024**, *13*, 1326. [CrossRef] [PubMed]
61. Golubkina, N.; Antoshkina, M.; Bondareva, L.; Sekara, A.; Campagna, E.; Caruso, G. Effect of Foliar Application of Sodium Selenate on Mineral Relationships in Brassicaceae Crops. *Horticulturae* **2023**, *9*, 535. [CrossRef]
62. Strzetelski, P.; Smolen, S.; Rożek, S.; Sady, W. The effect of diverse iodine fertilization on nitrate accumulation and content of selected compounds in radish plants (*Raphanus sativus* L.). *Acta Sci. Pol. Hortorum Cultus* **2010**, *9*, 65–73.
63. Santamaria, P.; Elia, A.; Serio, F. Effect of solution nitrogen concentration on yield, leaf element content, and water and nitrogen use efficiency of three hydroponically-grown rocket salad genotypes. *J. Plant Nutr.* **2002**, *25*, 245–258. [CrossRef]
64. Cavaiuolo, M.; Ferrante, A. Nitrates and Glucosinolates as Strong Determinants of the Nutritional Quality in Rocket Leafy Salads. *Nutrients* **2014**, *6*, 1519–1538. [CrossRef] [PubMed]
65. European Commission Commission. Regulation (EU) No 1258/2011 of 2 December 2011 amending Regulation (EC) No. 1881/2006 as regards maximum levels for nitrates in foodstuffs. *Off. J. Eur. Union* **2011**, *L320*, 15–17.
66. Karwowska, M.; Kononiuk, A. Nitrates/Nitrites in Food-Risk for Nitrosative Stress and Benefits. *Antioxidants* **2020**, *9*, 241. [CrossRef]
67. Schlagenhauf, U. On the Role of Dietary Nitrate in the Maintenance of Systemic and Oral Health. *Dent. J.* **2022**, *10*, 84. [CrossRef]
68. Pinaffi-Langley, A.C.D.C.; Dajani, R.M.; Prater, M.C.; Nguyen, H.V.M.; Vrancken, K.; Hays, F.A.; Hord, N.G. Dietary Nitrates from Plant Foods: A Conditionally Essential Nutrient for Cardiovascular Health. *Adv. Nutr.* **2024**, *15*, 100158. [CrossRef]
69. Pilon-Smits, E.A.H.; Quinn, C.F. Selenium metabolism in plants. In *Cell Biology of Metals and Nutrients*; Hell, R., Mendel, R.R., Eds.; Springer: Berlin/Heidelberg, Germany, 2010.
70. Medrano-Macias, J.; Leija-Martinez, P.; Gonzalez-Orales, S.; Juarez-Maldonado, A.; Benavides-Mendoza, A. Use of iodine to biofortify and promote growth and stress tolerance in crops. *Front. Plant Sci.* **2016**, *7*, 1146. [CrossRef] [PubMed]
71. Ding, Z.; Johanningsmeier, S.D.; Price, R.; Reynolds, R.; Truong, V.D.; Payton, S.C.; Breidt, F. Evolution of nitrate and nitrite content in pickled fruit and vegetable products. *Food Control* **2018**, *90*, 304–311. [CrossRef]
72. Al Alawi, A.M.; Majoni, S.W.; Falhammar, H. Magnesium and Human Health: Perspectives and Research Directions. *Int. J. Endocrinol.* **2018**, *2018*, 9041694. [CrossRef] [PubMed]
73. Pu, F.; Chen, N.; Xue, S. Calcium intake, calcium homeostasis and health. *Food Sci. Hum. Wellness* **2016**, *5*, 8–16. [CrossRef]
74. Lanza, M.G.D.B.; dos Reis, A.R. Roles of selenium in mineral plant nutrition: ROS scavenging responses against abiotic stresses. *Plant Physiol. Biochem.* **2021**, *164*, 27–43. [CrossRef]
75. Golubkina, N.; Kharchenko, V.; Moldovan, A.; Antoshkina, M.; Ushakova, O.; Sękara, A.; Stoleru, V.; Murariu, O.C.; Tallarita, A.V.; Sannino, M.; et al. Effect of Selenium and Garlic Extract Treatments of Seed-Addressed Lettuce Plants on Biofortification Level, Seed Productivity and Mature Plant Yield and Quality. *Plants* **2024**, *13*, 1190. [CrossRef] [PubMed]
76. Xu, J.; Zhang, Y.; Zhang, M.; Wei, X.; Zhou, Y. Effects of foliar selenium application on Se accumulation, elements uptake, nutrition quality, sensory quality and antioxidant response in summer-autumn tea. *Food Res. Int.* **2024**, *175*, 113618. [CrossRef] [PubMed]
77. Hawrylak-Nowak, B. Effect of selenium on selected macronutrients in maize plants. *J. Elementol.* **2008**, *13*, 513–519.
78. Ducsay, L.; Zapletalova, A.; Hozlar, P.; Cerny, I.; Varga, L.; Slepcan, M. Effects of selenium on macro- and micro nutrients and selected qualitative parameters of oat (*Avena sativa* L.). *Potravinarstvo* **2020**, *14*, 619–624. [CrossRef]
79. Xia, Q.; Yang, Z.; Shui, Y.; Liu, X.; Chen, J.; Khan, S.; Wang, J.; Gao, Z. Methods of Selenium Application Differentially Modulate Plant Growth, Selenium Accumulation and Speciation, Protein, Anthocyanins and Concentrations of Mineral Elements in Purple-Grained Wheat. *Front. Plant Sci.* **2020**, *11*, 1114. [CrossRef]
80. van Wonderen, D.; Melse-Boonstra, A.; Gerdessen, J.C. Iron Bioavailability Should be Considered when Modeling Omnivorous, Vegetarian, and Vegan Diets. *J. Nutr.* **2023**, *153*, 2125–2132. [CrossRef] [PubMed]
81. He, F.J.; MacGregor, G.A. Beneficial effects of potassium on human health. *Physiol. Plant.* **2008**, *133*, 725–735. [CrossRef]

horticulturae

Article

Effects of Exogenous Organic Matter on Soil Nutrient Dynamics and Its Role in Replacing Chemical Fertilizers for Vegetable Yield and Quality

Juan Wang [1,2], Xinyue Li [1], Anquan Chen [1], Yan Li [1,2], Mengyun Xue [3,*] and Shaoyuan Feng [1,*]

1 College of Hydraulic Science and Engineering, Yangzhou University, Yangzhou 225009, China; wangjuan@yzu.edu.cn (J.W.); mx120230658@stu.yzu.edu.cn (X.L.); mx120200610@stu.yzu.edu.cn (A.C.); liyan7986@126.com (Y.L.)
2 Modern Rural Water Resources Research Institute, Yangzhou University, Yangzhou 225009, China
3 Institute of Farmland Irrigation, Chinese Academy of Agricultural Sciences, Xinxiang 453002, China
* Correspondence: 15993777561@163.com (M.X.); syfeng@yzu.edu.cn (S.F.); Tel.: +86-159-9377-7561 (M.X.); +86-139-1214-4756 (S.F.)

Abstract: Searching for a low-cost soil amendment that can reduce the reliance on chemical fertilizers while maintaining crop yields is a vital issue to sustainable agricultural development. In this study, bio-organic matter derived from harmless disposal of livestock and poultry carcasses was used to discuss its potential for substitute chemical fertilizer. An incubation experiment was conducted by incorporating bio-organic matter into the soil at the rate of 0% (CK), 1%, 2%, 3%, 4%, 5%, 6% and 7% (T1, T2, T3, T4, T5, T6, T7) of soil mass to investigate the effects of bio-organic matter on soil physical properties and the nutrient release dynamics. A pot experiment was conducted with three treatments: 150 mg·kg^{-1} nitrogen from compound fertilizer (CK), 150 mg·kg^{-1} nitrogen from bio-organic matter (B1) and 300 mg·kg^{-1} nitrogen from bio-organic matter (B2), to evaluate the potential of bio-organic matter as a substitute for chemical fertilizers in influencing the yield and quality of Chinese cabbage (*Brassica chinensis* L.). Results showed that in the incubation experiment, bio-organic matter addition reduced soil bulk density of 1.5% to 8.9% and increased soil porosity by ranging from 1.5% to 10.9%. The soil physical properties were significantly improved when addition rates $\geq$ 4% (by soil mass). The interaction effects of addition rate and incubation time had a significant effect on soil nutrients. In the pot experiment, substitution of chemical fertilizer with bio-organic matter did not reduce the yield, and the increasing application rate of bio-organic matter led to significantly higher soluble protein, soluble sugar and total phenol content of vegetables. Additionally, nitrite content in the vegetables was slightly lower with bio-organic matter compared to that under CK. It is concluded that bio-organic matter derived from the harmless disposal of livestock and poultry carcasses is feasible; it has great potential to partially or entirely replace chemical fertilizers, thereby contributing to realizing chemical fertilizer reduction.

Keywords: harmless disposal of livestock and poultry carcasses; soil nutrient dynamics; soil properties; vegetable yield and quality

1. Introduction

With the increasing scarcity of agricultural resources and growing environmental pressures, the rising demand for food continues to place immense stress on agricultural production. Soil fertility plays a crucial role in agricultural productivity and ecosystem sustainability. By supplying essential nutrients, fertilizers effectively promote crop growth and development, helping to meet the ever-rising demand for food [1]. However, the excessive use of chemical fertilizers has led to soil degradation, reduced microbial activity, and long-term environmental damage. For instance, the excessive use of fertilizers, particularly nitrogen fertilizers, can result in soil acidification, disrupting the pH balance of the soil and degrading the soil microorganism environment, ultimately affecting crop growth [2]. Furthermore, nitrogen that is not absorbed by crops can leach into groundwater, contaminating water sources and contributing to eutrophication, while also posing serious health risks to humans [3]. Therefore, it is a main challenge for sustainable agricultural to reduce chemical fertilizer utilization at the same time as maintaining crop yield and quality.

In recent years, considerable progress has been made in the research field of reducing chemical fertilizer utilization. Duan et al. [4] found that replacing 75% of chemical fertilizers with organic fertilizers increased grain yield, number of ears and 1000-grain weight of winter wheat, with an average yield increase of 21.2% in a field study. Han et al. [5] demonstrated that substituting 20% of nitrogen fertilizer with bio-organic fertilizer significantly enhanced the economic benefits of vineyards, with a two-year average increase in income of 33.59% compared to that under CK. Li et al. [6] showed that using microbial organic fertilizers as a partial substitute for chemical fertilizers can increase the activity of nitrate reductase and nitrite reductase, improving nitrogen use efficiency and yield of winter wheat. Similarly, Zhu et al. [7] investigated the effects of replacing chemical fertilizers partially with straw-based microbial fertilizers on maize growth and soil properties, and found that the application of straw-based microbial fertilizers significantly increased soil organic carbon content, and a most notable effect on maize yield was measured with a 50% nitrogen fertilizer replacement. Additionally, combining soil amendments with chemical fertilizers has proven to reduce fertilizer utilization efficiently. Zhang et al. [8] found that compared to conventional nitrogen application without biochar, the total dry matter accumulation in rice was increased by 14.79% under 10% reduced nitrogen combined with 12.5 t/ha of biochar. In summary, the application of organic and microbial fertilizers along with soil amendments can effectively reduce the reliance on traditional chemical fertilizers without compromising crop growth and yield. However, the cost of commercial organic fertilizers is generally high, and the addition of soil amendments like biochar will boost the input cost of the agriculture industry; these all present a significant barrier to their widespread adoption. Therefore, identifying a low-cost external soil amendment that can partially or entirely replace chemical fertilizers is the key to advancing fertilizer reduction efforts.

In China, about 2% to 12% of pigs, cattle and sheep, and 12% to 20% of poultry died annually due to diseases [9]. In recent years, the state has been committed to speeding up the supervision system for the harmless treatment of dead livestock and poultry, and requires strengthening the harmless treatment. The utilization of harmless disposal of livestock and poultry carcasses is also encouraged in accordance with the relevant laws and regulations [10]. The by-products generated from the harmless disposal of livestock and poultry carcasses are abundant and rich in organic matter, making them a natural bio-organic fertilizer with great potential. The harmless disposal process involves crushing the carcasses of diseased animals and putting them into a harmless treatment tank, with the addition of sawdust, straw and other auxiliary materials for high-temperature, high-pressure sterilization. This process will eliminate pathogens carried by the animal carcasses, along with harmful bacteria that could pose a threat to the environment, ensuring that when these products are used as external bio-organic matter in soil, they do not cause secondary pollution [11], and they can significantly improve the physical and chemical properties of the soil. Zhai et al. [12] found that the application of the harmless treated products from animals in combination with chemical fertilizers significantly increased chlorophyll content by 22.85% in rapeseed, enhanced photosynthetic efficiency by 21.98% and increased the above-ground dry matter accumulation by 221.05%, ensuring the safety of rapeseed cultivation. Our previous research has also revealed that this type of bio-organic matter exhibits a soil improvement effect similar to biochar application [13]. Therefore, this study aims (1) to investigate the effects of bio-organic matter from harmless disposal of livestock and poultry carcasses on soil physical properties and the nutrient release process; (2) to examine the impact of partially substituting chemical fertilizers with bio-organic matter on soil fertility and the yield and quality of Chinese cabbage (*Brassica chinensis* L.); (3) to explore the potential of using bio-organic matter derived from the harmless disposal of livestock carcasses as a substitute for chemical fertilizers. The findings of this study would provide a theoretical basis for the resource utilization of bio-organic matter derived from the harmless disposal of livestock and poultry carcasses, reducing their potential environmental risks. Additionally, it offers new insights for reducing fertilizer usage.

2. Materials and Methods

2.1. Tested Soil and Materials

The tested soil was collected from the Agricultural Soil and Water Environment and Ecology experimental site at Yangzhou University. The soil was collected from the top 0–25 cm in a "Z" shape. Large gravel was removed, and the soil was air-dried and subsequently sieved through a 2 mm sieve. The soil particle size composition was measured using a Malvern laser particle size analyzer (MS–3000, Malvern Instruments Ltd., Malvern, UK). The proportions of clay (particle size < 0.002 mm), silt (particle size 0.002–0.02 mm), and sand (particle size 0.02–2 mm) were 4.38%, 32.76%, and 62.86%, respectively. The test soil texture can be classified as a sandy loam according to the International Soil Classification System [14]. The soil bulk density was 1.37 $g \cdot cm^{-3}$, the soil available nitrogen content was 32.08 $mg \cdot kg^{-1}$, the available phosphorus content was 10.54 $mg \cdot kg^{-1}$, the soil available potassium content was 45.02 $mg \cdot kg^{-1}$ and the soil organic matter content was 7.20 $g \cdot kg^{-1}$. The saturated water content was 38.05%, the electronic conductivity was 0.17 $mS \cdot cm^{-1}$, and the saturated hydraulic conductivity was 8.07×10^{-4} $cm \cdot s^{-1}$.

The chemical fertilizer used in the study was compound fertilizer produced from Stanley Fertilizer Corporation (Linyi, China). The total nutrient content of the fertilizer was ≥45%, and percentages of the N-P-K ratio were 15:15:15.

The bio-organic matter used in this study was the product of composting livestock carcasses after harmless treatment [15]. The process of harmless disposal of livestock carcasses was to first put the livestock carcasses into the pre-crushing machine and crush them into small pieces. When the temperature in the treatment tank was >140 °C and the pressure in the tank was ≥0.5 MPa, sawdust and straw were added as auxiliary materials and autoclaved for 4 h. When the temperature in the treatment tank dropped below 70 °C, composite microbial fungi were introduced to initiate degradation, which lasted for 6–8 h [16]. The elemental composition is presented in Table 1.

Table 1. The elemental composition of tested bio-organic matter (mass fraction).

Name	Appearance	pH	Size (mm)	Water Content (%)	Nitrogen (N)/%	Carbon (C)/%	Hydrogen (H)/%	Sulfur (S)/%
Bio-organic matter	Filamentous powder	6.5	2.0	10	7.28	50.91	7.24	0.63

2.2. The Incubation Experiment

The incubation experiment was conducted from November 2021 to January 2022 at the Agricultural Water, Soil, Environment and Ecology Laboratory, located on the southern campus of Yangzhou University. The experiment consisted of three incubation cycles, each lasting 15 days, for a total duration of 45 days.

The incubation buckets had an inner diameter of 12 cm and a height of 9 cm, with each bucket holding approximately 500 g of soil. The bio-organic matter was added in the following proportions of 1%, 2%, 3%, 4%, 5%, 6%, and 7% (by soil mass), labeled as T1, T2, T3, T4, T5, T6, and T7, respectively. After thoroughly mixing the soil with the bio-organic matter, the mixture was packed into the incubation buckets in layers and compacted. Pure soil without the addition of bio-organic matter served as the control (CK), and each treatment was replicated 9 times. An initial irrigation amount of 125 mL of water was slowly added to the buckets to maintain soil moisture content of 25% by mass. The daily water loss was recorded by weighing the buckets every day at 9:00 AM, and water was replenished to maintain soil moisture around 25%. Each incubation cycle lasted 15 days, with a total of three cycles. At the end of each cycle, 3 buckets were destructively sampled, and the soil was air-dried, ground and sieved. The soil physical properties (bulk density, porosity) and nutrient content (organic matter, available nitrogen, available phosphorus and available potassium) were measured.

2.3. The Pot Experiment

The pot experiment was conducted in a greenhouse of the Agricultural Water, Soil, Environment and Ecology Laboratory from May 2022 to June 2022, and the experiment period lasted for 45 days.

The pots had a capacity of an approximate 10 L with an inner diameter of 23 cm and a height of 25 cm. Small holes were drilled at the bottom of the pot for free drainage. To prevent the upper fine particles from plugging the apertures, 2 layers of geotextile were set at the bottom before packing the soil. Prior to the experiment, the collected soil was air-dried and sieved via a 2 mm stainless steel sieve. The amounts of fertilizer and soil were calculated, weighed and mixed thoroughly to minimize variability before the pots were filled. The soil was packed into pots by every 5 cm layer to achieve a bulk density of approximately 1.37 $g{\cdot}cm^{-3}$. The total soil depth of the pot was 20 cm, with a 5 cm gap for retaining water. Each layer was lightly raked before the next layer was packed to minimize discontinuities between layers.

The test crop used in the study was the economic crop *Brassica chinensis* L., commonly known as non-heading Chinese cabbage, which belongs to the *Brassicaceae* family. A total of 27 seeds were sown in each pot, 9 plants were kept after emergence, and plants were harvested 45 days later. Three treatments were set up as compound fertilizer with equal nitrogen application of 150 $mg{\cdot}kg^{-1}$ (CK), bio-organic matter with equal nitrogen application of 150 $mg{\cdot}kg^{-1}$ (B1), and bio-organic matter with equal nitrogen application of 300 $mg{\cdot}kg^{-1}$ (B2). After conversion, the application amounts of compound fertilizer were 1 $g{\cdot}kg^{-1}$ and the amounts of of bio-organic matter were 2.06 $g{\cdot}kg^{-1}$ and 4.12 $g{\cdot}kg^{-1}$, respectively. Three replicates were conducted for each treatment, and the experimental design is shown in Table 2. In order to better explore the effect of bio-organic matter on plant growth, the bio-organic matter application rate for B1 in the pot trial was close to that of T5 in the incubation trial, which showed a better result.

Table 2. The design of pot trial.

Treatments	Compound Fertilizer Application Rate ($g{\cdot}kg^{-1}$)	Bio-Organic Matter Application Rate ($g{\cdot}kg^{-1}$)	Equal Nitrogen Application Rate ($mg{\cdot}kg^{-1}$)
CK	1.00	/	150
B1	/	2.06	150
B2	/	4.12	300

Initially, 860 mL of water was irrigated for each, and subsequent irrigation was scheduled 2 to 3 times per week. The pots were weighed, and water was replenished based on the amount lost to maintain the original weight. After the harvest of Chinese cabbage, soil moisture and nutrient content were measured.

2.4. Measurement

Soil Physical Properties: The soil bulk density was determined using a 100 cm^3 core sampler (Cangzhou Litt Instrument Equipment Co., Ltd., Cangzhou, China) through the core method. Soil porosity was calculated based on the dried soil samples. The accumulative irrigation water amount was calculated based on daily irrigation water amount used to maintain a constant weight of the pot.

Soil Chemical Properties: Soil available nitrogen (AN) was tested by using the alkaline diffusion method. Soil available phosphorus (AP) was tested by using the $NaHCO_3$ extraction–molybdenum antimony anti-colorimetric method. Soil available potassium

(AK) was extracted with neutral ammonium acetate and tested with a flame photometer (ICPS-7500, Shimadzu, Japan). Soil organic matter (SOM) was tested by means of oxidation with potassium dichromate–sulfuric acid solution [17]. Chinese cabbage yield was obtained by measuring the fresh weight of the whole Chinese cabbage plant population in each pot and then converted by the pot area in t/ha.

2.5. Data Analysis

Microsoft Excel 2010 (Microsoft Corporation, Redmond, WA, USA, 2010) was used for data calculation and processing. The significant differences among treatments were determined by Duncan's test at levels of 95%, using the SPSS 20.0 statistical program (International Business Machines Corporation, Armonk, NY, USA, 2011). The analysis of variance (ANOVA) was performed with incubation time and addition rate as the main effects, including one- and two-way interactions. The figures were plotted and exported with Origin 8.0 (OriginLab Corporation, Northampton, MA, USA, 2007).

3. Results

3.1. The Results of Soil Incubation Experiment

3.1.1. Soil Bulk Density and Soil Porosity

The changes in soil bulk density and porosity under different treatments are illustrated in Figure 1. With the increasing addition rates of bio-organic matter, there was a decreasing trend in soil bulk density, with a reduction of 1.5% to 8.9%. Specifically, when the bio-organic matter addition rate was ≥4% (under T4, T5, T6 and T7 treatments), the soil bulk density was significantly lower than that under CK.

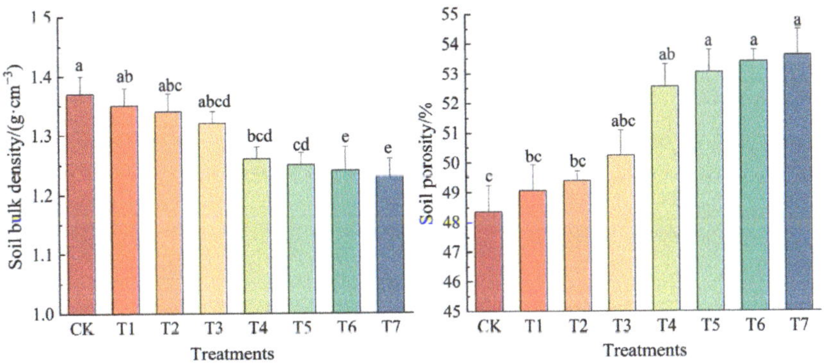

Figure 1. Soil bulk density and soil porosity under different treatments. Notes: The different lowercase letters indicate significant difference (at $p < 0.05$) according to Duncan's test, the same as below. Notes: CK, T1, T2, T3, T4, T5, T6 and T7 represent the treatments with bio-organic matter addition rate at 0%, 1%, 2%, 3%, 4%, 5%, 6%, and 7% (by soil mass), respectively, the same as below. The errors bars shown in the figures are standard error, the same as below.

Regarding soil porosity, there was an increasing trend as the bio-organic matter addition rate increased, with an improvement ranging from 1.5% to 10.9%. When the addition rate was ≥4% (under T4-T7 treatments), soil porosity was significantly higher than that under CK.

3.1.2. Soil Available Nitrogen

The effects of bio-organic matter addition and its rate on soil available nitrogen content dynamics are shown in Figure 2. Overall, there is a clear trend where increasing the addition rate of bio-organic matter leads to a corresponding rise in the soil available nitrogen content, and higher addition rates result in greater nitrogen levels. The enhancement effect of high bio-organic matter addition is particularly pronounced, with significantly higher nitrogen content levels under T4, T5, T6 and T7 compared to that under T1, T2 and T3 across all incubation periods. CK, which received no bio-organic matter, exhibited soil available nitrogen levels of 48.25 mg·kg^{-1}, 48.25 mg·kg^{-1} and 45.84 mg·kg^{-1} at the 15th, 30th and 45th days after incubation (DAI), respectively. In contrast, treatments with bio-organic matter demonstrated increases ranging from 177.5% to 850.0%, 240.0% to 860.0%, and 268.4% to 1063.1% at 15th, 30th and 45th DAI, respectively.

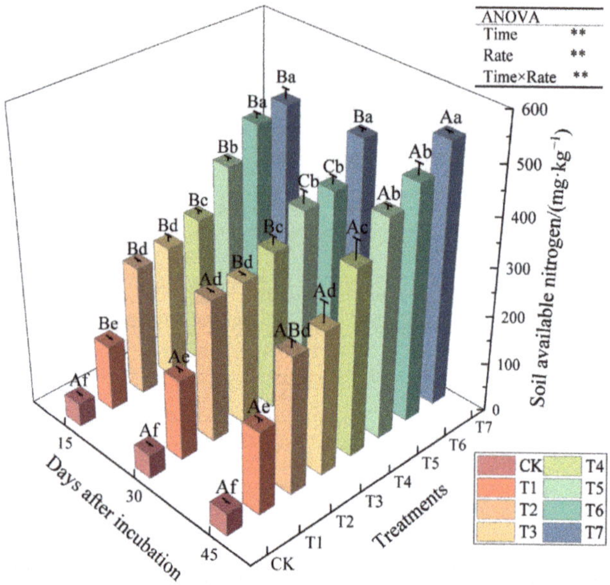

Figure 2. Soil available nitrogen content (mg·kg^{-1}) under different treatments in three periods. Notes: Different uppercase letters indicate significant differences ($p < 0.05$) in soil available nitrogen content at various sampling times under the same addition rate according to Duncan's test, while lowercase letters indicate significant differences ($p < 0.05$) in soil available nitrogen content across different addition rates at the same time according to Duncan's test. ** means extremely significant at $p < 0.01$.

From a temporal perspective, soil available nitrogen content exhibited a slight but consistent increase over time. Treatments with bio-organic matter addition had significantly higher soil available nitrogen levels at 45th DAI compared to 15th DAI ($p < 0.05$). For treatments T1 and T2, a significant increase in soil available nitrogen content was observed after 30th DAI ($p < 0.05$). However, for treatments with higher addition rates, while an increase was also noted at 30th DAI, statistical significance was only achieved after a longer incubation period of 45 days.

The interaction analysis showed that the soil available nitrogen was significantly affected by both incubation time and the addition rate of the bio-organic matter.

3.1.3. Soil Available Phosphorus

The dynamics of soil available phosphorus content under different addition rates of bio-organic matter are shown in Figure 3. In terms of addition rates, the overall trend indicated that the soil available phosphorus was gradually increased with higher addition rates of bio-organic matter. At the 15th day after incubation (DAI), the available phosphorus contents under T7 and T6 were significantly higher than others ($p < 0.05$). At the 30th DAI, the available phosphorus content under T7, T6 and T5 reached 42.43 mg·kg^{-1}, 38.93 mg·kg^{-1} and 37.35 mg·kg^{-1}, respectively, which were significantly higher than those under other treatments ($p < 0.05$). At the 45th DAI, the available phosphorus content under

T7, T6, T5 and T4 treatments was even more significantly higher than others ($p < 0.05$), representing increases of 561.77%, 318.50%, 306.94% and 265.25% compared to that under CK.

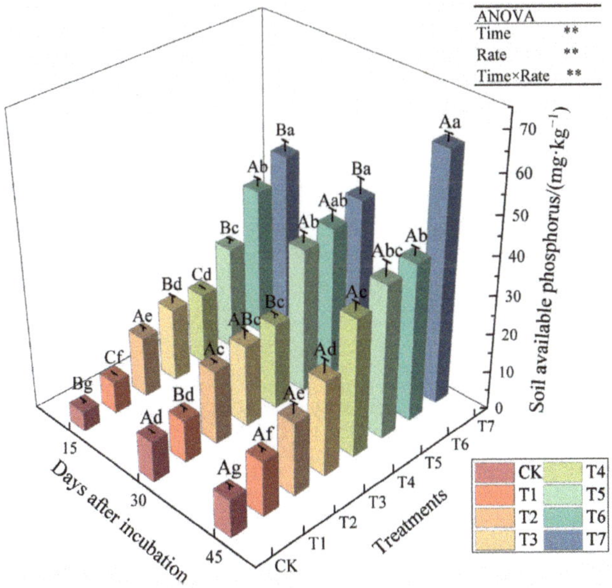

Figure 3. Soil available phosphorus content (mg·kg^{-1}) under different treatments in three periods. Notes: Different uppercase letters indicate significant differences ($p < 0.05$) in soil available phosphorus content at various sampling times under the same addition rate according to Duncan's test, while lowercase letters indicate significant differences ($p < 0.05$) in soil available phosphorus content across different addition rates at the same time according to Duncan's test. ** means extremely significant at $p < 0.01$.

As the incubation period was extended, soil available phosphorus content was steadily increased, and reached a significant level ($p < 0.05$) at the 45th DAI compared to those at 15th DAI, except for those under T2 and T6.

The main effect of incubation time, addition rate and the interaction effect of time × rate significantly affected soil available phosphorus content ($p < 0.01$).

3.1.4. Soil Available Potassium

The dynamics of soil available potassium content under different addition rates of bio-organic matter are shown in Figure 4. As the addition rate of bio-organic matter increased, the available potassium content was continually enhanced, with higher addition rates leading to greater potassium levels. Treatments with high bio-organic matter addition rates (T6 and T7) showed a significant increase in available potassium content at the early incubation stage (15 days) compared to others. By 30th DAI, the available potassium content in T5 reached 416.12 mg·kg^{-1}, comparable to that of T6 with 421.95 mg·kg^{-1}. At the end of the incubation period, the available potassium content in all treatments significantly increased with the bio-organic matter addition rate ($p < 0.05$). Compared to CK, the available potassium contents under T4, T5, T6, and T7 were increased by 149.54%, 159.95%, 168.29%, and 198.61%, respectively.

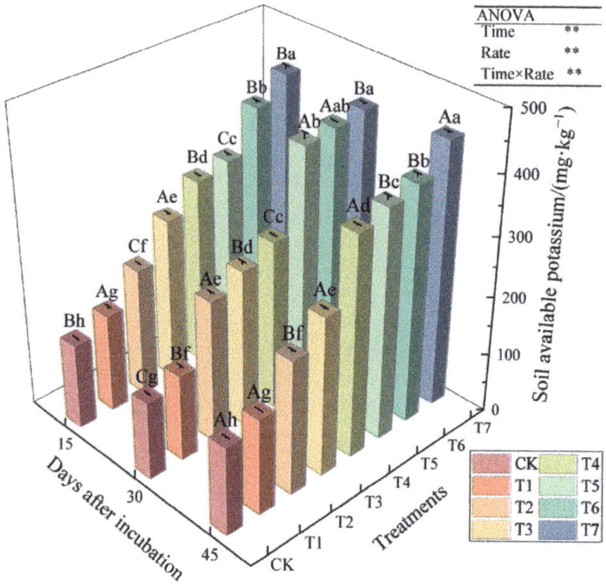

Figure 4. Soil available potassium content (mg·kg^{-1}) under different treatments in three periods. Notes: Different uppercase letters indicate significant differences ($p < 0.05$) in soil available potassium content at various sampling times under the same addition rate according to Duncan's test, while lowercase letters indicate significant differences ($p < 0.05$) in soil available potassium content across different addition rates at the same time according to Duncan's test. ** means extremely significant at $p < 0.01$.

The available potassium content in the soil fluctuated significantly over the incubation period, without showing a clear trend. It was initially decreased and then increased under T1, T3, and T4, while under T2, T5 and T6, it showed an initial increase followed by a decrease. For the T7 treatment, the available potassium content remained relatively stable during the early stages, and then increased significantly towards the end of the incubation period.

The interaction effect of incubation time × addition rate had a significant impact on soil available potassium content ($p < 0.01$).

3.1.5. Soil Organic Matter Content

The dynamics of soil organic matter content under different addition rates of bio-organic matter are shown in Figure 5. Similar to the trends observed in available nutrients, the soil organic matter content consistently increased with the addition rate of bio-organic matter. The soil organic matter contents under T6 and T7 were significantly higher than others. Compared to CK, soil organic matter treated with bio-organic matter addition increased by 221.1–942.5%, 73.1–334.5% and 102.4–514.9% at 15th, 30th, and 45th DAI, respectively. Unlike the available nutrients, the increase in soil organic matter did not reach significant levels over time. The results of two-way ANOVA indicated that the main effect of addition rate and the interaction effect of incubation time × addition rate significantly affected soil organic matter ($p < 0.01$). However, the main factor of incubation time did not significantly impact soil organic matter content, suggesting that the effects of the addition rate of bio-organic matter on soil organic matter content were similar across different incubation stages.

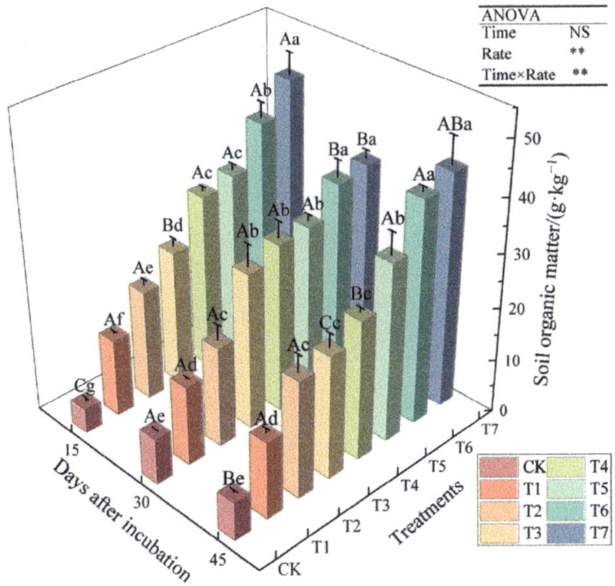

Figure 5. Soil organic matter content (g·kg^{-1}) under different treatments in three periods. Notes: Different uppercase letters indicate significant differences ($p < 0.05$) in soil organic matter content at various sampling times under the same addition rate according to Duncan's test, while lowercase letters indicate significant differences ($p < 0.05$) in soil organic matter content across different addition rates at the same time according to Duncan's test. NS means not significant ($p > 0.05$). ** means extremely significant at $p < 0.01$.

3.1.6. Analysis of Incubation Experiment Results

A linear regression model $y = ax + b$ was fitted to soil nutrient contents (available nitrogen, available phosphorus, available potassium, and organic matter) y and bio-organic matter addition rate x at the 45th day after incubation. The results are shown in Table 3.

Table 3. Linear fitting parameters of soil nutrients.

Soil Nutrients	*a*	*b*	R^2
Soil available nitrogen	66.202	97.007	0.9643
Soil available phosphorus	7.0123	6.3981	0.9252
Soil available potassium	45.586	140.22	0.9573
Soil organic matter	5.0866	8.4264	0.9749

The coefficient of determination R^2 values were all greater than 0.9, with $p < 0.05$, indicating good fitting results. This suggested that soil nutrient content increased with the addition rate of bio-organic matter and that a linear function can effectively estimate soil nutrient content y based on bio-organic matter addition x.

3.2. *The Results of Pot Experiment*

3.2.1. Soil Temperature

Soil temperature is an important indicator of soil structure and the water heat status. The changes in soil temperature under the treatments of compound fertilizer (CK), bio-organic matter with low application rate (B1) and bio-organic matter with high application rate (B2) are shown in Figure 6. Overall, throughout the entire growth period of Chinese cabbage, the soil temperature with bio-organic matter remained consistently higher than those treated with compound fertilizer. From the second day after sowing, soil temperatures with high bio-organic matter application were persistently higher than those treated with the low application rate. Compared to CK, soil temperature increased by approximately 0.9 °C to 3.4 °C under B1, and it increased by 1.6 °C to 3.7 °C under B2 throughout the entire growth period. In terms of soil temperature fluctuations, it showed smaller variations under bio-organic matter treatments compared to compound fertilizer treatment, especially in the mid-to-late growth stages, where the temperature remained more stable with bio-organic matter application.

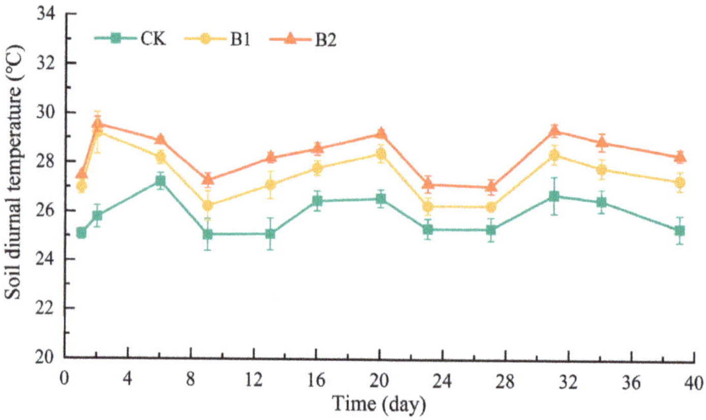

Figure 6. The soil temperature dynamics during the growth period. Notes: CK represents compound fertilizer with equal nitrogen application of 150 mg·kg^{-1}, and B1 and B2 represent bio-organic matter with equal nitrogen application of 150 mg·kg^{-1} and 300 mg·kg^{-1}, respectively, the same as below.

3.2.2. Accumulative Irrigation Water Amount

The accumulative irrigation water amount of Chinese cabbage during the growing period is shown in Figure 7. During the initial growth stage (1–16 days after sowing), the cumulative water consumption curves under all treatments were similar, indicating no significant difference in water consumption during the early stage. The accumulative irrigation water amounts under CK, B1 and B2 were 22.9 mm, 22.9 mm and 22.7 mm on the 16th day after sowing, respectively. Then, as the plants entered a more vigorous growth phase, obvious differences in accumulative irrigation water amount appeared. In the end of the pot experiment, the accumulative irrigation water amounts were 58.3 mm, 73.3 mm and 88.9 mm, respectively. Compared to CK, the total irrigation water amount was increased by 25.73% under B1 and 52.49% under B2, respectively.

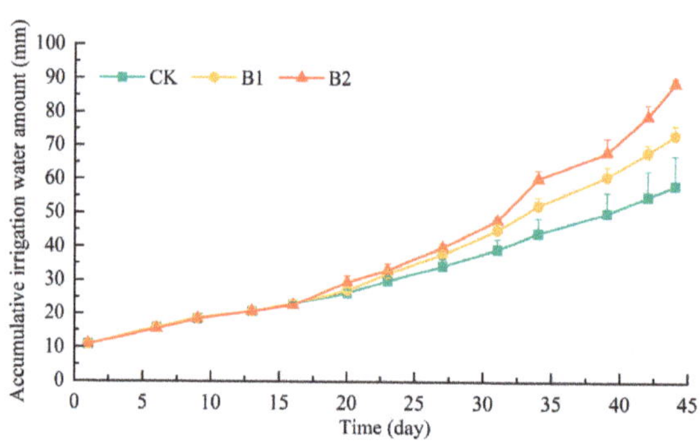

Figure 7. Accumulative irrigation water amount under different treatments.

3.2.3. Soil Nutrient Content After Harvest

Soil available nitrogen, available phosphorus, available potassium and organic matter are key indicators for assessing soil fertility. After harvest, pot soil samples were collected for measuring the nutrient contents to analyze the effects of different treatments on soil nutrients. The results are listed in Table 4.

Table 4. Soil nutrient indexes under different treatments. Notes: Different lowercase letters indicate significant differences ($p < 0.05$) in soil nutrient indexes under different treatments (CK, B1 and B2) according to Duncan's test.

Treatments	Soil Available Nitrogen ($mg \cdot kg^{-1}$)	Soil Available Phosphorus ($mg \cdot kg^{-1}$)	Soil Available Potassium ($mg \cdot kg^{-1}$)	Soil Organic Matter ($g \cdot kg^{-1}$)
CK	65.14 ± 7.24 a	25.18 ± 1.57 a	118.11 ± 2.14 a	8.15 ± 0.64 b
B1	44.63 ± 9.11 b	12.44 ± 1.60 b	46.69 ± 1.20 b	8.88 ± 0.91 ab
B2	49.46 ± 2.08 ab	13.68 ± 0.85 b	51.16 ± 3.15 b	9.31 ± 0.28 a

Compared to CK, the soil available nitrogen contents under B1 and B2 were decreased by 31.48% and 24.08%, respectively. With increased bio-organic matter application, there was a slight rise in soil available nitrogen, but it did not reach a significant level. The soil available phosphorus and available potassium content followed similar patterns to soil available nitrogen. The soil available phosphorus contents under B1 and B2 were 12.44 $mg \cdot kg^{-1}$ and 13.68 $mg \cdot kg^{-1}$, respectively, which represented a significant decrease of 50.59% and 45.66% compared to that under CK with 25.18 $mg \cdot kg^{-1}$. Compared to soil with compound fertilizer, the soil available potassium with the application of bio-organic matter was significantly reduced, with decreases of 60.47% and 56.68% under B1 and B2, respectively. There was no significant difference in available potassium between the soils treated with high and low application rates of bio-organic matter after harvest. After harvesting, the soil organic matter content under B2 was the highest and significantly greater than that under CK. Overall, bio-organic matter can effectively supplement soil organic matter levels, achieving a comparable effect to that treated with compound fertilizers even at a lower application rate. However, available nutrient content after harvest was significantly lower than that applied with compound fertilizer, which may impact the growth of subsequent crops.

3.2.4. The Yield and Quality of the Chinese Cabbage

The yield and quality are key indicators for assessing crop growth and are crucial factors influencing the economic efficiency of agricultural production. The yield of Chinese cabbage and the contents of soluble protein, soluble sugar, total phenols and nitrite are presented in Table 5.

Table 5. Crop quality indexes under different treatments. Notes: Different lowercase letters indicate significant differences ($p < 0.05$) in crop quality indexes under different treatments (CK, B1 and B2) according to Duncan's test.

Treatments	Yield (t/ha)	Soluble Protein ($mg \cdot g^{-1}$)	Soluble Sugar (%)	Total Phenols ($mg \cdot g^{-1}$)	Nitrite ($mg \cdot kg^{-1}$)
CK	23.862 ± 0.962 a	4.08 ± 0.01 c	3.17 ± 0.34 b	9.41 ± 1.17 b	1.82 ± 0.20 a
B1	24.810 ± 3.318 a	4.55 ± 0.02 b	4.52 ± 0.31 a	11.94 ± 0.34 a	1.71 ± 0.14 a
B2	28.882 ± 2.759 a	4.78 ± 0.10 a	5.08 ± 0.19 a	12.78 ± 1.25 a	1.51 ± 0.31 a

Compared to the yield under CK, the yields under B1 and B2 were increased by 3.97% and 21.04%, respectively, with the highest yield observed with high application rate of bio-organic matter. However, the differences among the treatments were not statistically significant. This indicates that using bio-organic matter as a substitute for chemical fertilizers does not result in yield reduction.

Compared to CK, the soluble protein content of Chinese cabbage significantly increased with bio-organic matter application by 11.52% and 17.16% under B1 and B2, respectively. Furthermore, the soluble protein content under B2 was significantly higher than that under B1. This indicates that substituting chemical fertilizers with bio-organic matter can effectively enhance the soluble protein content of Chinese cabbage, and that higher application rates lead to even greater increases in soluble protein content. The soluble sugar contents of Chinese cabbage under B1 and B2 were significantly higher than that under CK. However, there was no significant difference in soluble sugar content between the B1 and B2 treatments. Compared to CK, the total phenol content of Chinese cabbage was significantly increased by 26.88% and 35.81% under B1 and B2, respectively, while there was no significant difference in total phenol content between B1 and B2. Overall, the use of bio-organic matter as a substitute for chemical fertilizers significantly improved the quality of Chinese cabbage. The soluble protein content increased substantially with higher application rates, while the contents of soluble sugars and total phenols showed little sensitivity to the application levels of bio-organic matter.

Excessive nitrite in food can pose serious health risks to humans [18]. Following the use of bio-organic matter as a substitute for chemical fertilizers, the nitrite content in Chinese cabbage showed no significant difference compared to that with chemical fertilizers, and it even decreased slightly. Moreover, as the application rate of bio-organic matter increased, the nitrite content in the plants decreased. This indicates that using bio-organic matter as a substitute for chemical fertilizers is available for the cultivation of Chinese cabbage, effectively minimizing the risk of nitrite exceeding safe levels.

3.2.5. Correlation Analysis of Soil Nutrients with Yield and Quality of Chinese Cabbage

To further determine the relationship between soil nutrients and quality and yield of Chinese cabbage under different treatments, and to illustrate the degree of association between these indicators, a correlation analysis was conducted. The results are shown in Figure 8.

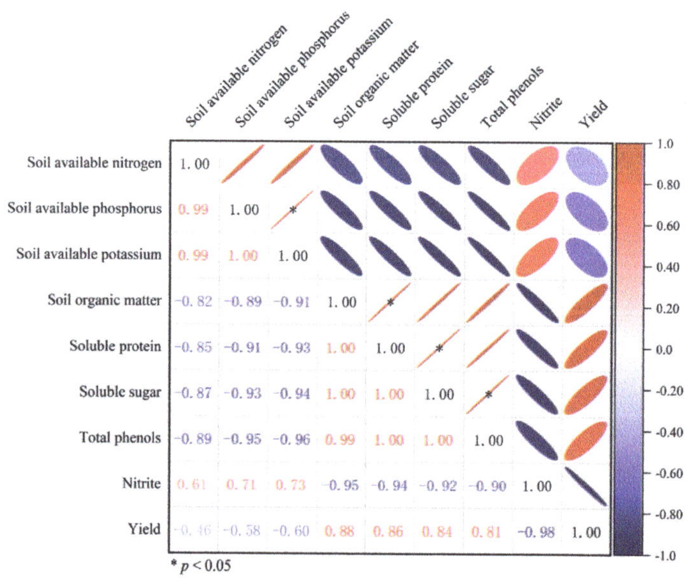

Figure 8. Correlation analysis of soil nutrients with yield and quality indices. Notes: * means significant at $p < 0.05$.

Soil organic matter was significantly positively correlated with soluble protein content and reached a significant level ($p < 0.05$). Soluble sugar was extremely significantly positively correlated with soluble protein and total phenols ($p < 0.05$). These results indicate that enhancing soil organic matter content contributed to improving crop quality and increasing the nutritional value of the produce, without affecting nitrite levels in Chinese cabbage, thereby ensuring that the crop was safe for consumption.

4. Discussion

Well-structured soil with appropriately sized pores results in a looser, more stable aggregate that can retain sufficient moisture, thereby supporting healthy root growth and facilitating better water and nutrient uptake, which in turn promotes the healthy growth of crops [19,20]. The application of organic fertilizers has been shown to improve soil physical structure effectively. Liu et al. [21] reported that in the black soils of Northeast China, organic fertilizer application significantly enhanced soil structure, increasing the proportion of high-quality medium and large aggregates. Similarly, Chen et al. [22] found that replacing chemical fertilizers with bio-organic fertilizers reduced soil bulk density and improved porosity. This study reached comparable conclusions that as the rate of bio-organic matter application increased, soil bulk density showed a decreasing trend. This may be attributed to the characteristics of bio-organic matter, which, as a product of harmless disposal of livestock carcasses, has low bulk density and a loose texture. So it effectively regulates soil bulk density and soil porosity, and then would improve soil water and air permeability. Additionally, bio-organic matter is rich in organic substances and beneficial soil microbes, contributing to optimized soil structure. However, in view of the relatively short duration of this incubation experiment, future research should focus on the long-term effects of bio-organic matter on soil aggregate formation and stability.

The contents of available nitrogen, available phosphorus, available potassium, and organic matter in soil are key indicators of its nutrient-supplying capacity. Applying organic fertilizers or external organic substances can improve soil fertility. Pu et al. [23] investigated the effect of chloropicrin fumigation on soil remediation and found that the application of bio-organic fertilizer increased nitrogen, phosphorus, and potassium levels in soil compared to the untreated control. Similarly, Wen et al. [24] observed that co-application of biochar and organic fertilizers enhanced the content of soil organic matter, available nitrogen, phosphorus, and potassium content.

In the incubation experiment of this study, the interaction effect of bio-organic matter application rate and incubation time had a significant impact on soil available nitrogen, phosphorus and potassium content ($p < 0.01$). This may be due to the enhancing effect of bio-organic matter on soil physical structure. As the application rate of bio-organic matter increases, the ability to absorb and retain nutrients also improves, leading to a more pronounced accumulation of readily available nutrients. Additionally, bio-organic matter can provide an abundant carbon and energy source for soil microorganisms, which stimulates microbial colonization and activity. The increase in microbial activity supports the decomposition of soil organic matter and the release of nutrients. With higher application rates and extended incubation time, nutrient mineralization and release are further promoted, leading to increased levels of available nitrogen, phosphorus, and potassium in the soil. For crops with short growth period, increasing the application rate of bio-organic matter (when pre-application is not feasible) can help rapidly elevate soil nutrient levels to meet crop needs. Conversely, for crops with longer growth period, a moderate application rate may be more appropriate, as available phosphorus and potassium gradually release over time to support sustained crop growth.

In this study, after crop harvest, soil organic matter content with bio-organic matter application was significantly higher compared to those treated with conventional chemical fertilizer. However, the levels of available nutrients under the bio-organic matter were notably lower than those under conventional chemical fertilizer. Liu et al. [25] investigated the effects of organic fertilizer and soil amendments on soil fertility and the growth of continuously cropped sweet potatoes; their study demonstrated that after harvest, the available nutrient levels with sole application of bio-organic fertilizer showed varying degrees of reduction in comparison to baseline soil.

This may be partly due to the fact that the mineralization of bio-organic matter into immediately available nutrients needed by crops is slow. Additionally, bio-organic matter application improves soil pore structure, indirectly enhancing the fluxes of water, nutrients, air, and heat within the soil. These improvements support the development of plant root systems, thus promoting overall plant growth. Vigorous crop growth leads to increased absorption and consumption of available nutrients, resulting in significantly lower levels of available nutrients in the soil after harvest compared to the control. It indicates that while the products of the harmless treatment of livestock and poultry carcasses are rich in organic matter and can effectively increase soil organic matter content, their release of readily available nutrients is relatively slow. Then, the slow release may induce stress in crops by limiting the immediate nutrient uptake needed for optimal growth. Therefore, in future research on the use of bio-organic matter as a substitute for chemical fertilizers, the nutrient release rate will be a critical factor to address. From the perspective of application rate, post-harvest levels of available nutrients were slightly higher with higher bio-organic matter application. This may be attributed to the increased availability of organic matter for microbial decomposition as application rates rise. Consequently, both the decomposition rate and nutrient release rate accelerate, leading to a higher accumulation of readily available nutrients in the soil. Based on the results of the incubation experiment, which reveal the nutrient release patterns of bio-organic matter, it was observed that, without the crop growing, soil available nutrients increased significantly with higher application rates of bio-organic matter. Additionally, soil available nutrient levels rose markedly over the incubation period. Excessive accumulation of available nutrients, particularly nitrogen, in the soil can increase the leaching risk. Therefore, it is inadvisable to increase the application rate of bio-organic matter unreasonably. Instead, application rates should be carefully designed, taking into account the combined effects on the soil environment and plant needs, as well as the potential for nutrient loss. An alternative approach would be to apply bio-organic matter to the soil some time before planting, allowing nutrient availability to peak in alignment with the crop's critical nutrient demand period, thereby ensuring an adequate nutrient supply for optimal growth and development. In this study, bio-organic products derived from harmless disposal of livestock and poultry carcasses were used as a complete substitute for compound fertilizers. In future research, a partial substitution approach could be explored to identify optimal replacement ratios. This would allow for a gradual reduction in chemical fertilizer usage at the same time as ensuring crop growth and yield.

Crop yield and quality are closely related to the field conditions of moisture, nutrients, atmosphere and temperature. The application of external soil amendments can effectively enhance soil physical and chemical properties. In this study, replacing chemical fertilizer with bio-organic matter did not result in yield reduction. On one hand, bio-organic matter likely reduces soil bulk density and increases soil porosity, enhancing soil structure and improving its water-holding and aeration capacity. This optimizes the water and soil ecological environment for the growth of Chinese cabbage. On the other hand, bio-organic matter is rich in mineral nutrients with a fertilizing effect, which can enhance nutrient availability in the rhizosphere soil. Additionally, bio-organic matter application may raise soil temperature, promoting water uptake during the later stages of crop growth and supporting overall crop development.

The soluble sugar, soluble protein and total phenol content are key crop nutritional indicators. Soluble sugar, as an energy source for respiratory metabolism in fruits and vegetables, indicates maturity level and storage quality through their concentration. Soluble protein content reflects the crop's high nutritional quality and its role in nutritional supplementation. Total phenol content can indicate the antioxidant capacity of a crop [26–28]. Xiong [29] found that bio-organic fertilizer can increase the sugar content, sugar-to-acid ratio and total phenolic content in wine grapes. Zhang et al. [30] found that the addition of organic fertilizers, such as fish bone powder, significantly improved the nicotine, reducing sugar and total sugar content in tobacco leaves. In this study, the soluble sugar, soluble protein and total phenol content of Chinese cabbage treated with bio-organic matter were significantly higher than those treated with compound fertilizer. Moreover, as the application rate of bio-organic matter increased, these quality indicators exhibited a positive upward trend. This may be attributed to the improvement of soil physical–chemical properties by bio-organic matter, which creates a favorable soil environment, thereby enhancing the accumulation of nutrients for crop growth. Excessive intake of nitrites poses potential health risks to humans. Therefore, it is crucial to fully consider the safety of agricultural products during crop cultivation. Li et al. [31] found that the application of bio-organic fertilizer significantly reduced the nitrite content in water celery. In this study, the nitrite content in Chinese cabbage plants under all treatments met national food safety standards (The content of nitrite in fresh vegetables should not exceed 4 $mg \cdot kg^{-1}$). Compared to compound fertilizer, the bio-organic matter application resulted in a slight reduction in nitrite content. This suggests that as a potential partial or complete substitute for chemical fertilizers, bio-organic matter not only eliminates the risk of exceeding nitrite limits in agricultural products but also exhibits a similar effect to existing organic fertilizers in reducing nitrite content in crops.

5. Conclusions

Through incubation and pot experiments, we investigated the nutrient release characteristics of the bio-organic matter, its effects on soil physical properties and the impacts of substituting chemical fertilizers with bio-organic matter on the yield and quality of Chinese cabbage. The main conclusions are as follows:

(1) The incubation experiment revealed that bio-organic matter addition reduced soil bulk density and increased soil porosity. The soil physical properties were significantly improved when addition rates $\geq 4\%$ (by soil mass), with the improvements becoming more pronounced as the addition rate increased. The interaction effect of bio-organic matter addition rates and incubation time had a significant effect on soil available nitrogen, available phosphorus, available potassium and organic matter content. At the 45th day after incubation, a linear relationship was observed between soil nutrient content and the addition rate of bio-organic matter.

(2) The pot experiment demonstrated that bio-organic matter effectively increased soil temperature and conserved soil moisture, with these effects being more pronounced at higher application rates. Compared to traditional chemical fertilizer, bio-organic matter increased the accumulative irrigation water amount during the crop growth period, showing a positive correlation with application rate. After harvest, the content of available nutrients in soils treated with bio-organic matter was lower than that under chemical fertilizer. However, the organic matter content was higher under bio-organic matter, with significantly higher levels observed at high application rates compared to the control.

(3) The substitution of chemical fertilizer with bio-organic matter not only maintained but also increased the yield, significantly improving its quality attributes, including higher levels of soluble protein, soluble sugar and total phenols. Correlation analysis revealed a significant positive relationship between soil organic matter content and soluble protein levels, indicating that the application of bio-organic matter can substantially improve the quality and nutritional value of Chinese cabbage. Moreover, the decreased nitrite content in the plants confirms that based on its fertilizing efficacy and its impact on crop yield and quality, the partial or complete replacement of chemical fertilizers with bio-organic matter derived from the harmless disposal of livestock and poultry carcasses is feasible.

Future research should focus on elucidating the nutrient release dynamics of bio-organic matter and aligning them with the critical nutrient uptake stages of crops to enhance its nutrient-supplying capacity. Additionally, investigations should prioritize determining the optimal proportion of bio-organic matter as a substitute for traditional fertilizers to achieve fertilizer reduction while maintaining crop growth and yield. Subsequent studies can further explore possible long-term accumulation of undesirable elements in the soil and the variability of results between different soil types.

Author Contributions: Conceptualization, J.W. and S.F.; methodology, M.X. and S.F.; validation, J.W. and A.C.; formal analysis, J.W. and X.L.; investigation, X.L. and A.C.; resources, M.X. and S.F.; data curation, X.L. and A.C.; writing-original draft preparation, J.W. and X.L.; writing—review and editing, J.W. and Y.L.; visualization, X.L.; supervision, S.F.; project administration, M.X. and S.F.; funding acquisition, J.W. and Y.L. All authors have read and agreed to the published version of the manuscript.

Funding: This research was funded by the National Natural Science Foundation of China (51609209, 52409073); the Key Laboratory Program of the Shangqiu Station of National Field Agro-ecosystem (Grants No. SQZ-2023-01); the Postgraduate Research and Technology Innovation Program of Jiangsu Province (No. KYCX24_3764).

Data Availability Statement: The data that support this study cannot be publicly shared due to ethical or privacy reasons but may be shared upon reasonable request to the corresponding author if appropriate.

Acknowledgments: Special thanks to Yuhua Shan for his support and help with this manuscript.

Conflicts of Interest: The authors declare no conflicts of interest.

References

1. Ishfaq, M.; Wang, Y.; Xu, J.; Hassan, M.U.; Yuan, H.; Liu, L.; He, B.; Ejaz, I.; White, P.J.; Cakmak, I.; et al. Improvement of nutritional quality of food crops with fertilizer: A global meta-analysis. *Agron. Sustain. Dev.* **2023**, *43*, 74. [CrossRef]
2. Guo, J.; Liu, X.; Zhang, Y.; Shen, J.; Han, W.; Zhang, W.; Christie, P.; Goulding, K.W.T.; Vitousek, P.M.; Zhang, F. Significant Acidification in Major Chinese Croplands. *Science* **2010**, *327*, 1008–1010. [CrossRef] [PubMed]
3. Cen, S.; Zeng, Y.; Li, X.; Zhuang, X.; Ji, F. Nitrogen Flow in Rural Production and Living System and Its Environmental Impact Characteristics. *Jiangsu J. Agric. Sci.* **2021**, *37*, 1224–1233. (In Chinese with English Abstract)
4. Duan, C.; Li, J.; Zhang, B.; Wu, S.; Fan, J.; Feng, H.; He, J.; Siddique, K.H.M. Effect of Bio-Organic Fertilizer Derived from Agricultural Waste Resources on Soil Properties and Winter Wheat (*Triticum aestivum* L.) Yield in Semi-Humid Drought-Prone Regions. *Agric. Water Manag.* **2023**, *289*, 108539. [CrossRef]
5. Han, X.; Li, J.; Zhang, J.; Guo, Y.; Zhang, L.; Ji, Y. Effects of Bio-Organic Fertilizer Replacing Part of Chemical Fertilizer on Grape Growth and Soil Fertility. *J. Agric. Sci. Technol.* **2024**, *26*, 195–205. (In Chinese with English Abstract)
6. Li, Y.; Wei, J.; Ma, L.; Wu, X.; Zheng, F.; Cui, R.; Tan, D. Enhancing Wheat Yield Through Microbial Organic Fertilizer Substitution for Partial Chemical Fertilization: Regulation of Nitrogen Conversion and Utilization. *J. Soil Sci. Plant Nutr.* **2024**, *24*, 935–943. [CrossRef]
7. Zhu, L.; Chen, J.; Chen, R.; Xu, S.; Liu, T. Effects of Straw Bacterial Manure Substitution for Chemical Fertilizer on Maize Growth and Soil Characteristics. *J. N. Agric. Sci.* **2021**, *46*, 26–29. (In Chinese with English Abstract)
8. Zhang, Z.; Zhou, L.; Li, H.; Kong, F.; Lv, X. Effect of Water-Saving Irrigation and Nitrogen Reduction Combined with Biochar Application on Photosynthetic Characteristics and Water and Nitrogen Utilization of Rice. *Trans. Chin. Soc. Agric. Mach.* **2024**, *55*, 386–395+438. (In Chinese with English Abstract)
9. Liu, Y.; Hou, J.; Li, B.; Li, M.; Cao, J.; Zhang, R.; Wu, G. The Pilot Study of Harmless Disposing Equipment of Diseased Dead Animals. *J. Environ. Eng. Technol.* **2018**, *8*, 662–670. (In Chinese with English Abstract)
10. Ministry of Agriculture and Rural Affairs, People's Republic of China. Notice from the General Office of the Ministry of Agriculture and Rural Affairs on Further Strengthening the Disposal of Dead Livestock and Poultry in a Harmless Manner. Available online: https://www.gov.cn/zhengce/zhengceku/202411/content_6984686.htm (accessed on 10 November 2024).
11. Gu, X. Thoughts on Harmless Treatment and Comprehensive Utilization of Livestock and Poultry Breeding Waste. *China Anim. Ind.* **2023**, *3*, 73–75. (In Chinese with English Abstract)
12. Zhai, Z.; Xiong, B.; Zhang, L.; Jiang, B.; Li, C.; Pang, H.; Li, Y. Effects of Organic Manure from Animal Carcasses on *Brassiacampestris* L. Growth and Soil Heavy Metal Accumulation. *J. Agro-Environ. Sci.* **2016**, *35*, 985–991. (In Chinese with English Abstract)
13. Shi, D. Research and Application of Reduced Nitrogen Fertilization Combined with Bio-Organic Fertilizer and Biochar for Soil Fertility Enhancement and Crop Growth Promotion. Master's Thesis, Yangzhou University, Yangzhou, China, 2024. (In Chinese with English Abstract)
14. Schad, P.; Huyssteen, C.; Michéli, E. World Reference Base for Soil Resources 2014. In *International Soil Classification System for Naming Soils and Creating Legends for Soil Maps*; Organización de las Naciones Unidas para la Alimentación y la Agricultura (FAO): Rome, Italy, 2014.
15. Shinjo, H.; Fujita, H.; Gintzburger, G.; Kosaki, T. Soil Aggregate Stability under Different Landscapes and Vegetation Types in a Semiarid Area in Northeastern Syria. *Soil Sci. Plant Nutr.* **2000**, *46*, 229–240. [CrossRef]

16. Wang, J.; Zhai, B.; Shi, D.; Chen, A.; Liu, C. How Does Bio-Organic Fertilizer Combined with Biochar Affect Chinese Small Cabbage's Growth and Quality on Newly Reclaimed Land? *Plants* **2024**, *13*, 598. [CrossRef] [PubMed]
17. Mei, L.; Zhang, N.; Wei, Q.; Cao, Y.; Li, D.; Cui, G. Alfalfa Modified the Effects of Degraded Black Soil Cultivated Land on the Soil Microbial Community. *Front. Plant Sci.* **2022**, *13*, 938187. [CrossRef]
18. Moazeni, M.; Heidari, Z.; Golipour, S.; Ghaisari, L.; Sillanpaa, M.; Ebrahimi, A. Dietary Intake and Health Risk Assessment of Nitrate, Nitrite, and Nitrosamines: A Bayesian Analysis and Monte Carlo Simulation. *Environ. Sci. Pollut. Res.* **2020**, *27*, 45568–45580. [CrossRef]
19. Huang, J.; Zhu, C.; Kong, Y.; Cao, X.; Zhu, L.; Zhang, Y.; Ning, Y.; Tian, W.; Zhang, H.; Yu, Y.; et al. Biochar Application Alleviated Rice Salt Stress via Modifying Soil Properties and Regulating Soil Bacterial Abundance and Community Structure. *Agronomy* **2022**, *12*, 409. [CrossRef]
20. Chen, W.; Bastida, F.; Liu, Y.; Zhou, Y.; He, J.; Song, P.; Kuang, N.; Li, Y. Nanobubble Oxygenated Increases Crop Production via Soil Structure Improvement: The Perspective of Microbially Mediated Effects. *Agric. Water Manag.* **2023**, *282*, 108263. [CrossRef]
21. Liu, K. Research on Using Bio-Organic Fertilizer to Improve the Fertility and Biological Activity of Black Soil in Northeast China. Master's Thesis, Nanjing Agricultural University, Nanjing, China, 2020. (In Chinese with English Abstract)
22. Chen, C.; Lv, Q.; Tang, Q. Impact of Bio-Organic Fertilizer and Reduced Chemical Fertilizer Application on Physical and Hydraulic Properties of Cucumber Continuous Cropping Soil. *Biomass Convers. Biorefinery* **2024**, *14*, 921–930. [CrossRef]
23. Pu, R.; Wang, P.; Guo, L.; Li, M.; Cui, X.; Wang, C.; Liu, Y.; Yang, Y. The Remediation Effects of Microbial Organic Fertilizer on Soil Microorganisms after Chloropicrin Fumigation. *Ecotox. Environ. Saf.* **2022**, *231*, 113188. [CrossRef] [PubMed]
24. Wen, H.; Du, S.; Cheng, G.; Wang, Z.; Hu, Y.; Zhou, H.; Wu, X. Influences of Combined Application of Biochar and Organic Fertilizer on Growth Quality, Soil Nutrients and Enzyme Activities of Chinese Cabbage. *Jiangsu Agric. Sci.* **2023**, *51*, 224–230. (In Chinese with English Abstract)
25. Liu, Y.; Lu, Y.; Wang, W.; Hu, Q.; Chu, F.; Li, Z. Effects of Organic Fertilizer and Soil Conditioner on the Growth and Development of Continuous Cropping Sweet Potato and Soil Fertility. *Crops* **2024**, *3*, 168–174. (In Chinese with English Abstract)
26. Zhao, A.; An, R.; Wang, X.; Han, Y.; Xie, H.; Li, P.; Li, G.; Hu, H. Effect of Atomized Slightly Acidic Electrolyzed Water Treatment during Vacuum Precooling on the Quality of Postharvest Chinese Little Greens during Low Temperature Circulation and Shelf Life. *Food Sci.* **2023**, *44*, 218–227. (In Chinese with English Abstract)
27. Cai, H.; You, M.; Ryall, K.; Li, S.; Wang, H. Physiological Response of Chinese Cabbage to Intercropping Systems. *Agron. J.* **2011**, *103*, 331–336. [CrossRef]
28. Wang, L.; Zhang, S.; Li, J.; Zhang, Y.; Zhou, D.; Li, C.; He, L.; Li, H.; Wang, F.; Gao, J. Identification of Key Genes Controlling Soluble Sugar and Glucosinolate Biosynthesis in Chinese Cabbage by Integrating Metabolome and Genome-Wide Transcriptome Analysis. *Front. Plant Sci.* **2022**, *13*, 1043489. [CrossRef]
29. Xiong, B. Effects of Different Application of Fertilization on Growth and Fruiting of Xiahei Grape. *S. China Fruits* **2018**, *47*, 124–126+129. (In Chinese with English Abstract)
30. Zhang, H.; Yang, C.; Rao, X.; Li, X.; Lu, J.; Liao, S. Effects of Different Organic Fertilizers on Nutrient Uptake, Yield and Quality of Tobacco Leaves in Different Parts of Cigar Tobacco. *Soil Fert. Sci. China* **2023**, *12*, 227–235. (In Chinese with English Abstract)
31. Li, W.; Huang, Z.; Wei, Y.; Wu, Y.; Yin, H.; Jiang, Z.; Wu, X.; Wang, K.; Guo, R.; Wang, D. Effects of Bioorganic Fertilizer on Growth and Quality of Oenanthe Javanica. *China Cucurbits Veg.* **2021**, *34*, 65–68. (In Chinese with English Abstract)

horticulturae

MDPI

Article

Lignin Sulfonate-Chelated Calcium Improves Tomato Plant Development and Fruit Quality by Promoting Ca^{2+} Uptake and Transport

Jiucheng Zhang [1,†], Minghui Du [1,2,†], Genzhong Liu [1], Fangfang Ma [1,*] and Zhilong Bao [1,*]

1 College of Horticulture Science and Engineering, Shandong Agricultural University, Tai'an 271018, China; zhangjcedu@163.com (J.Z.); duminghui2023@126.com (M.D.); 13657221786@163.com (G.L.)

2 College of Life Sciences, Zaozhuang University, Zaozhuang 277000, China

* Correspondence: fma@sdau.edu.cn (F.M.); zbao@sdau.edu.cn (Z.B.)

† These authors contributed equally to this work.

Abstract: Calcium (Ca) plays a key role in cell wall stabilization and various physiological processes. Ca fertilizers are widely used in agriculture to meet crop demands and improve yield and quality. However, traditional Ca fertilizers often suffer from low solubility, poor absorption, and mobility issues. Chelated Ca fertilizers offer enhanced efficiency and uptake. In this study, we compared the effects of lignosulfonate-chelated Ca (LS–Ca), EDTA-chelated Ca (EDTA–Ca), $Ca(NO_3)_2$, and alcohol sugar-chelated Ca (AS–Ca) with a Ca concentration of 15 mg/L on tomato growth. The results showed that LS–Ca increased the contents of chlorophyll a and b contents in leaves by 26% and 46%, respectively. The application of Ca fertilizers significantly enhanced Ca^{2+} uptake and transport, with the LS–Ca treatment achieving the highest utilization efficiency. Without altering fruit weight, the LS–Ca treatment increased the firmness of mature tomato fruits by 29%. Furthermore, the LS–Ca treatment improved fruit sweetness by 33%, with the total sugar content increasing by 45%, sucrose by 80%, reducing sugars by 64%, and titratable acidity by 18%. This study aims to compare the effects of different chelated Ca fertilizers on tomato cultivation and to explore optimal Ca supplementation strategies, thereby contributing to improvements in tomato cultivation practices and fruit quality.

Keywords: calcium; calcium content; fertilizers; calcium utilization efficiency; fruit firmness; fruit sweetness

1. Introduction

Tomatoes, with their increasing yield and quality demands, are particularly sensitive to Ca nutrition. Ca deficiency in tomatoes causes disorders, such as blossom-end rot, the necrosis of growth points, and abnormal floral structures, severely impacting yield and quality [1,2]. Calcium (Ca) plays a crucial role in enhancing the rigidity and stability of the cell wall, as well as in cell division and elongation. It is also a key component of both the cell wall and the cell membrane [3,4]. Acting as a secondary messenger, Ca is involved in signal transduction for various physiological processes, such as hormone regulation and stress responses, and regulates enzyme activities that affect plant metabolism [5].

Ca deficiency can lead to symptoms such as apical necrosis, leaf curling, poor root development, and delayed fruit growth [6–8]. In agricultural production, Ca fertilizers can neutralize soil acidity, improve soil pH and structure, promote beneficial microbial activity, and enhance fruit firmness, shelf life, and resistance to physiological disorders like fruit cracking and blossom-end rot [8–12]. Additionally, Ca application improves crop yield and quality and enhances resistance to diseases, drought, and salt stress [13,14]. Inorganic Ca fertilizers, such as calcium nitrate ($Ca(NO_3)_2$), calcium carbonate ($CaCO_3$), calcium sulfate ($CaSO_4$), and calcium phosphate ($Ca_3(PO_4)_2$), are widely used due to their high purity, fast action, multifunctionality, and affordability. However, challenges related to their use include soil acidification, rapid dissolution, uneven crop absorption, and environmental sensitivity [8,10,15].

Chelated Ca fertilizers, which stabilize Ca^{2+} using chelating agents, offer higher bioavailability and stability [14]. EDTA-chelated Ca (EDTA–Ca), for example, binds Ca^{2+} to ethylenediaminetetraacetic acid (EDTA) and promotes flowering in tomatoes, although excessive application can cause chlorosis and necrosis in older leaves [16,17]. Polyols, such as xylitol and mannitol, also serve as carriers for Ca^{2+}, improving its mobility and uptake within plants. The foliar application of sorbitol–Ca fertilizer, for instance, enhances Ca^{2+} absorption and boosts peanut (*Arachis hypogaea* L.) yields [18], while alcohol sugar-chelated Ca (AS–Ca) treatments significantly improve grape berry size and shape index [19].

Lignin, a natural polymer in plant cell walls, exhibits excellent chelating ability [14,20]. Lignosulfonates, derived from sulfite pulping processes, are low-cost materials containing hydrophilic sulfonate and electroactive methoxy phenol groups, offering potential for improving nutrient bioavailability and reducing production costs [21,22]. Chelated Ca fertilizers, including lignosulfonate–Ca (LS–Ca), have demonstrated superior Ca^{2+} transport and enhanced crop growth [23,24].

Ca absorption depends on soil Ca^{2+} availability and xylem transport driven by transpiration. The accumulation of Ca in tomato fruits is dependent on the rate of xylem sap flow, which is influenced by transpiration and growth rates [25,26]. The capacity of other Ca pools within the plant can affect calcium accumulation in tomatoes and may lead to Ca-related physiological disorders such as blossom-end rot [27]. Early in fruit development, both the xylem and phloem contribute to nutrient transport; however, as the fruit matures and transpiration decreases, xylem's role is gradually supplanted by that of phloem. The low mobility of Ca^{2+} in the phloem leads to localized Ca deficiency in the fruit tissues [28–30]. Therefore, enhancing Ca^{2+} transport in soluble chelated forms may effectively address late-stage deficiencies.

Due to the high stability of complexes, nutrient elements can be absorbed and transported in a chelated form [31]. After the foliar application of amino acid-chelated iron fertilizers, the average iron concentration in all tested rice varieties increased by 14% compared to that of the control [32]. Recent studies have indicated that the use of sugar alcohol-chelation technology to chelate Ca^{2+} can convert Ca from an inorganic to an organic state, thereby influencing calcium absorption and transformation [31]. However, the role of LS–Ca fertilizers in the absorption and transport of Ca^{2+} in tomatoes remains unknown.

We systematically analyzed plant growth and fruit-quality differences and monitored Ca^{2+} distribution across tissues during different developmental stages. This study aims to identify effective chelated Ca supplementation strategies for tomato cultivation, providing technical guidance for improving tomato quality and yield.

2. Materials and Methods

2.1. Plant Materials and Growth Conditions

Tomato cultivar 'Micro-TOM' seeds were germinated and sown in the soil at a greenhouse. The greenhouse conditions were maintained at a 16 h/8 h light/dark cycle with temperatures of 25 °C during the day and 20 °C at night, a relative humidity of 60%, and a light intensity of 300 $\mu mol\ m^{-2}\ s^{-1}$. The experiment was conducted in the autumn of 2023 at the State Key Laboratory of Wheat Breeding, Shandong Agricultural University, Tai'an, China.

Ca fertilizer treatments for tomato plants were applied three times, at 7, 14, and 30 days after transplanting, through both foliar spraying and root irrigation. The four types of chelated calcium fertilizers used were, as follows: T1, a lignosulfonate-chelated Ca (LS–Ca, containing lignosulfonate ammonium and 15% $CaCl_2$); T2, an EDTA–Ca (containing EDTA chelating agent, 10% total Ca and 1% Ca^{2+}); T3, $Ca(NO_3)_2$; and T4, an alcohol sugar-chelated Ca (AS–Ca, containing alcohol sugar chelating agent and 14% total Ca). Water was used as the control treatment. The Ca concentration for each application was maintained at 15 mg/L.

During the experiment, regular self-pollination was performed and marked, with pollination occurring between 10:00 a.m. and 12:00 p.m. Relevant growth parameters, physiological measurements, and other required data were recorded at predetermined intervals. For determining the red ripe (RR) stage of fruit, marking was conducted at the breaker (Br) stage of fruit development, with Br + 8 uniformly used as the marker for RR [33].

2.2. Measurement of Plant Fresh and Dry Weight

Fresh tomato plants were cut at the junction between the root and shoot, and surface moisture was removed. The fresh weight of the samples was then measured using an electronic balance. For dry weight determination, the tissues were placed in a forced-air drying oven at 70 °C for three days until completely dried, after which the weight was recorded. Each treatment contained 5 biological replicates.

2.3. Measurement of Photosynthetic Rate, Chlorophyll, and Carotenoid Content

The net photosynthetic rate of tomato leaves (the five true leaves from the bottom) was measured using the CIRAS-3 portable photosynthesis system between 9:00 a.m. and 12:00 p.m. on the 17th day after planting (after second fertilization), with the light intensity set at 300 μmol m^{-2} s^{-1}. For the chlorophyll and carotenoid content analysis, 0.1 g of fresh tomato leaves (with veins removed) were ground into powder using liquid nitrogen and extracted with 80% acetone on a shaker at 400 rpm/min in darkness for 4 h. The extract was centrifuged at 8000 rpm/min and the supernatant was collected. Each treatment contained 4 biological replicates. The chlorophyll and carotenoid contents were quantified spectrophotometrically at wavelengths of 663 nm, 647 nm, and 470 nm, following the method described by Niu et al. (2022) [34]. The chlorophyll and carotenoid contents were derived using the following formula:

$$\text{chlorophyll a content} = \frac{\mathrm{V}(12.21 \times A_{663} - 2.81 \times A_{647})}{\mathrm{m}}$$

$$\text{chlorophyll b content} = \frac{\mathrm{V}(20.13 \times A_{647} - 5.03 \times A_{663})}{\mathrm{m}}$$

$$\text{carotenoid content} = \frac{\mathrm{V}(1000 \times A_{470} - 3.27 \times C_a - 104 \times C_b)}{\mathrm{m}}$$

A: absorbance, C_a: chlorophyll a concentration, C_b: chlorophyll b concentration, m: weight of sample (g), V: volume (L).

2.4. Measurement of Fruit Yield and Individual Fruit Weight

After the third fertilization, all fruits that had successfully set on each plant were harvested and weighed using a precision balance to calculate the total fruit yield per plant. The total fruit yield of the whole plant was calculated, and the biological replication between various treatments was 5 plants. Mature fruits were collected and weighed individually to determine the average fruit weight. Each treatment contained 30–40 biological replicates.

2.5. Measurement of Fruit Firmness and Brix

At the RR stage, the firmness of the fruit was measured using a penetrometer (YT-GY-4, Yuntang, Weifang, China) at a distance of 3 cm. The fruit juice was squeezed out and the soluble solids content of the fruit was determined using a refractometer (ATAGO PAL-1, Tokyo, Japan); the results were expressed in degrees °Brix. Each treatment contained 12 biological replicates.

2.6. Measurement of Total Sugar, Sucrose, Reducing Sugar, and Titratable Acid Contents

The contents of the total sugar, sucrose, and reducing sugar in the fruit were measured using the anthrone colorimetric method [35–37]. An 0.2 g dry sample was dissolved in 8 mL of 80% ethanol and heated at 85 °C for 30 min. The reaction mixture was then centrifuged at 3000 rpm for 5 min and the supernatant was collected. This extraction process was repeated three times and the supernatants were combined and diluted to a final volume of 50 mL. Each treatment contained 4 biological replicates.

The total sugar content was determined using the anthrone colorimetric method. To 0.1 mL of the supernatant, 3.5 mL of an 0.15% (w/v) anthrone reagent was added and heated at 90 °C for 15 min. The absorbance was measured at 620 nm using a glucose solution as a standard.

For the reducing sugar determination, 0.4 mL of the extract was diluted with 1.6 mL of ddH_2O and 1 mL of 1% (w/v) 3,5-dinitrosalicylic acid (containing 20% (v/v) 1N NaOH and 30% (w/v) sodium tartrate) was then added. The reaction was carried out at 95 °C for 5 min; absorbance was measured at 520 nm using a glucose solution as a standard.

For sucrose determination, 0.4 mL of the extract was mixed with 0.1 mL of 30% KOH (w/v) and heated at 95 °C for 10 min, followed by the addition of 3.5 mL of the anthrone reagent. The reaction was continued at 95 °C for 15 min; absorbance was measured at 620 nm using a sucrose solution as a standard.

Titratable acidity was determined by performing an acid-base titration [38]. A 5 g sample of the fruit juice or extract was diluted to a known volume with distilled water. A few drops of a phenolphthalein indicator were added to the solution. The sample was then titrated with a standardized 0.1 M NaOH solution until a pale-pink endpoint was reached. The volume of NaOH used was recorded, and the titratable acidity was calculated using the following formula:

$$\text{Titratable Acidity}(\%) = \frac{\mathrm{V} \times \mathrm{C} \times E_m}{\mathrm{m}}$$

V: volume of NaOH (mL), C: concentration of NaOH (mol/L), E_m: equivalent weight of acid, m: weight of sample (g).

2.7. Determination of Flavonoid Content in Fruit

An 0.1 g sample of fruit pulp was ground using liquid nitrogen and 1 mL of pre-chilled 65% ethanol was added and mixed thoroughly. The mixture was then extracted at 4 °C in the dark for 4 h, followed by centrifugation at 12,000 rpm for 20 min. A 50 μL aliquot of the supernatant was collected and sequentially mixed with 0.1 mL of 5% $NaNO_2$, 0.1 mL of 10% $Al(NO_3)_3$, and 0.4 mL of 2 mol/L NaOH. After standing for 15 min, the absorbance was measured at 510 nm. Rutin (Sigma Chemical, St. Louis, MO, USA) was used as the standard; the flavonoid content was expressed as mg of rutin per gram of fresh weight (mg rutin·g^{-1} FW) [39]. Each treatment contained 4 biological replicates.

2.8. Determination of Plant Ca and Mg Content

Tissue samples were collected on the 5th day after three Ca fertilizer treatments. The various organs (roots, stems, leaves, flowers, and fruits) of the tomato plants were separated, dried in a forced-air oven at 80 °C for 2 days and then ground into a fine powder. A total of 0.1 g of the dry sample was added to 5 mL of H_2SO_4 and treated overnight to ensure complete carbonization. The sample was then subjected to digestion in a digestion furnace at 300 °C, with the addition of 30% H_2O_2 dropwise during the heating process until the solution became clear. After filtration, the total Ca and Mg concentrations were measured using an atomic absorption spectrophotometer (flame atomic absorption method) [40,41]. Each treatment contained 3 biological replicates.

2.9. Determination of Soil Total Ca Content and Replaceable Ca Content

Soil was collected in autumn 2023 at the Horticultural Experiment Station of Shandong Agricultural University, Tai'an, China. Field soil was treated with/without 1% $CaCl_2$ and lignin sulfonate (LS, Figure 1A) for 14 days. For the determination of the total soil Ca content, 5 g of the dry soil samples was weighed; the determination method was the same as method 2.8. For the determination of soil replaceable Ca content, 5 g of the air-dried soil samples was added to 50 mL of a 1 mol/L ammonium acetate solution. The mixture was shaken on a shaker at 70 rpm/min for 1 h to ensure sufficient interaction between the Ca^{2+} and the exchange reagent. The extract was then filtered through filter paper and the filtrate was collected. The Ca concentration in the filtrate was determined using an atomic absorption spectrophotometer (flame atomic absorption method) [42]. Each treatment contained 3 biological replicates.

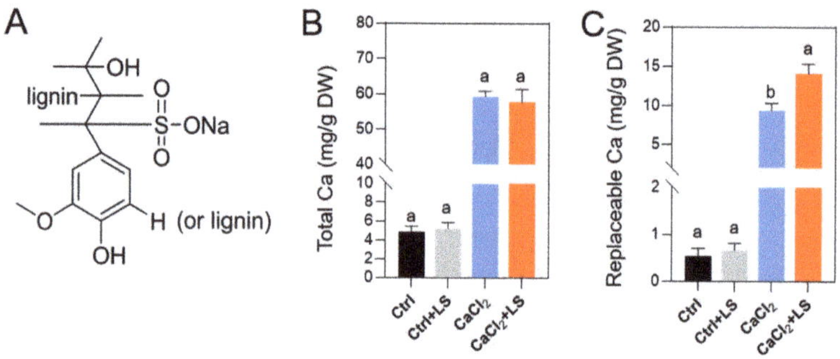

Figure 1. Effects of lignin sulfonates on Ca^{2+} chelation. Total Ca and replaceable Ca contents were determined from the soil: (**A**) chemical formula of sodium lignin sulfonate (LS); (**B**) differences in total soil Ca content; and (**C**) differences in soil-replaceable Ca content. Ctrl: without the application of $CaCl_2$; Ctrl + LS: without the application of $CaCl_2$ but the lignin sulfonates were added; $CaCl_2$: soil treated with 1% $CaCl_2$; $CaCl_2$ + LS: soil treated with 1% $CaCl_2$ and lignin sulfonates. Error bars represent standard deviations (n = 3). Different letters represent significantly different values at $p \leq 0.05$.

2.10. Determination of Ca Content in Xylem Sap

Tissue samples were collected on the 5th day after three Ca fertilizer treatments. The Ca content in the xylem sap was determined using the gravimetric method [43]. Tomato plants were cut 0.5 cm above the root, and a plastic film was shaped into a tube and sealed at both ends based on the required thickness. A small amount of defatted cotton was placed inside the plastic tube and tied with fine thread, then weighed. The defatted cotton was placed in contact with the cut surface and secured with string. After 6 h, the plastic tube was removed and weighed again. The difference between the two measurements was calculated. The absorbed cotton was placed in a centrifuge tube, eluted with 5% HNO_3, centrifuged, and the supernatant was collected. Finally, the volume was adjusted to 1 mL with 5% HNO_3. The Ca concentration was determined using an atomic absorption spectrophotometer (flame atomic absorption method). Each treatment contained 4 biological replicates.

2.11. Data Analysis

The statistical analysis was performed using the one-way ANOVA test plus Duncan's multiple range test with Data Processing System (DPS) software (version 7.05). Different letters indicate significant differences with $p \leq 0.05$.

3. Results

3.1. Lignosulfonate Maintains Soil Available Ca

To investigate the chelating ability of lignin sulfonates on soil Ca^{2+}, we measured the Ca content in the soil. The results showed that, without the application of $CaCl_2$, the lignin sulfonates (LS) treatment did not alter the total Ca content (Figure 1B) or the exchangeable Ca content (Figure 1C) in the soil. However, under the $CaCl_2$ treatment, the LS treatment did not significantly affect the total Ca content (Figure 1B) but significantly increased the replaceable Ca content (Figure 1C). This indicates that the LS treatment maintained a higher level of absorbable Ca^{2+}.

3.2. LS–Ca Fertilizer Promotes the Growth of Aboveground Parts of Tomato Plants

To investigate the effects of different Ca fertilizers on tomato growth, we conducted pot experiments using 'Micro-TOM' tomato seedlings as the model. Four Ca fertilizers—LS–Ca, EDTA–Ca, $Ca(NO_3)_2$, and SA–Ca—were applied via root irrigation and foliar spraying. The results showed that, compared to the control, the LS–Ca treatment significantly enhanced the aboveground growth of tomato plants (Figure 2A), with the fresh and dry weights of the aboveground parts increasing by 56% and 67%, respectively (Figure 2B,C). In contrast, no significant differences were observed in the fresh and dry weights of the underground parts between the LS–Ca treatment, the control, and other Ca fertilizers (Figure 2D,E).

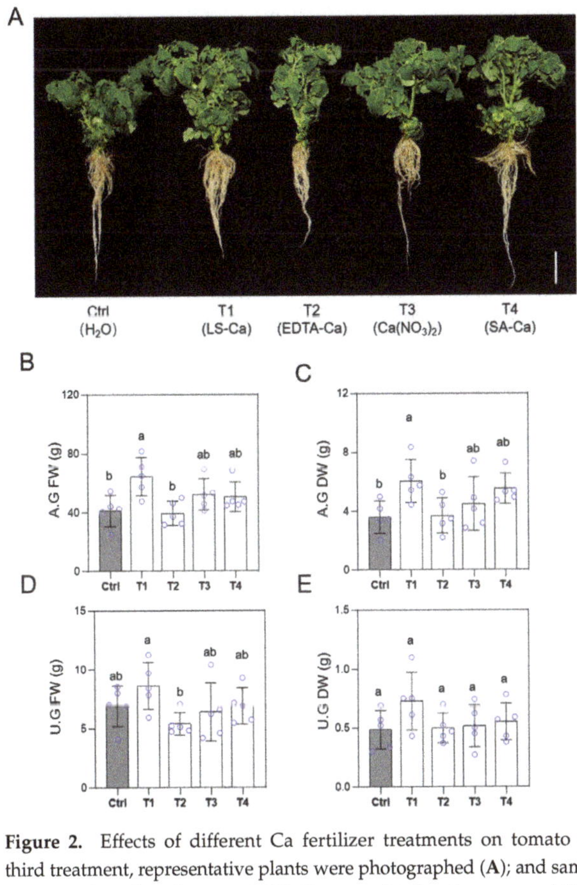

Figure 2. Effects of different Ca fertilizer treatments on tomato growth. At the end of the third treatment, representative plants were photographed (**A**); and sampled for fresh weights aboveground (**B**); and underground (**D**). Dry weights for aboveground (**C**) and underground (**E**) samples were measured after the samples were completely dried. Ctrl (H_2O), no fertilizer treatment; T1 (LS–Ca), lignosulfonate-chelated Ca fertilizer treatment; T2 (EDTA–Ca), EDTA-chelated Ca fertilizer treatment; T3 ($Ca(NO_3)_2$), $Ca(NO_3)_2$ fertilizer treatment; T4 (AS–Ca), alcohol sugar-chelated Ca fertilizer treatment. The Ca concentration of each fertilizer treatment was maintained at 15 mg/L. Error bars represent standard deviations (n = 5). Different letters represent significantly different values at $p \leq 0.05$.

3.3. LS–Ca Fertilizer Improves Light Use Efficiency in Tomatoes

To investigate the impact of Ca fertilizers on tomato photosynthesis, we measured the net photosynthetic rate and chlorophyll content in tomato leaves. The results showed no significant differences in the net photosynthetic rate among the Ca fertilizer treatments compared to the control (Figure S1). However, the LS–Ca treatment significantly increased the contents of chlorophyll a and chlorophyll b by 27% and 46%, respectively (Figure 3A,B). Additionally, the LS–Ca treatment reduced the carotenoid content in leaves by 29% (Figure 3C).

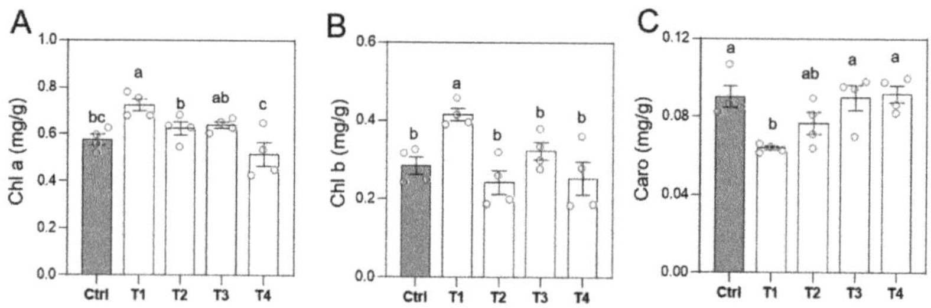

Figure 3. Effects of different Ca fertilizer treatments on tomato phenotypes after the second treatment: chlorophyll a (**A**); chlorophyll b (**B**); and carotenoid (**C**) contents in tomato leaves fertilized with different calcium sources. Ctrl, H_2O; T1, LS–Ca; T2, EDTA–Ca; T3, $Ca(NO_3)_2$; T4, AS–Ca treatment. Error bars represent standard deviations (n = 4). Different letters represent significantly different values at $p \leq 0.05$.

3.4. LS–Ca Fertilizer Enhances Ca^{2+} Uptake and Transport

To investigate the effects of various chelated Ca fertilizers on Ca uptake and transport in tomatoes, we measured the changes in Ca^{2+} concentrations in roots, stems, and leaves after each of the three fertilizer applications. The results revealed that, except for LS–Ca, Ca^{2+} concentrations in roots did not significantly change after the first and third applications for other treatments but increased significantly after the second application (Figure 4A–C). In contrast, the LS–Ca treatment significantly increased root Ca^{2+} concentrations across all three applications (Figure 4A), indicating faster root uptake and sustained Ca availability during late tomato-growth stages. Foliar application is a rapid and efficient method for Ca supplementation. Following the Ca fertilizer treatments, leaf Ca^{2+} concentrations increased significantly until the third application, after which the concentrations plateaued (Figure 4A–C). Additionally, the LS–Ca treatment notably enhanced Mg concentrations in roots and stems (Figure S2), suggesting that LS may also facilitate Mg^{2+} uptake and transport.

After the first two fertilizer applications, stem Ca^{2+} concentrations increased significantly (Figure 4A,B), indicating enhanced Ca^{2+} transport. Interestingly, following the third application, only the LS–Ca treatment resulted in a significant increase in stem Ca^{2+} concentrations compared to those of the control (Figure 4C), demonstrating its sustained efficacy in promoting Ca^{2+} transport within the plant. Moreover, the xylem sap analysis revealed that the LS–Ca treatment significantly elevated Ca^{2+} concentrations compared to those of the control and other treatments (Figure 4D,E). After the third application, xylem sap Ca^{2+} concentrations in the EDTA–Ca, $Ca(NO_3)_2$, and SA–Ca treatments showed no significant difference from those of the control, whereas LS–Ca concentrations were markedly higher (Figure 4F). These findings further confirm the role of lignosulfonate chelation in enhancing Ca^{2+} transport within the plant.

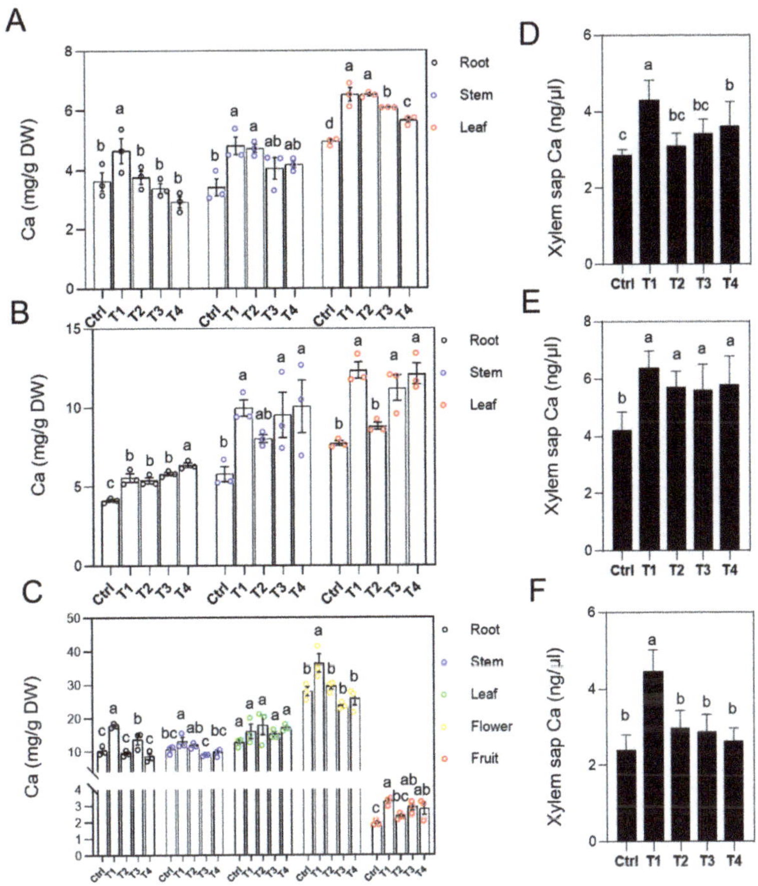

Figure 4. Effects of different Ca fertilizer treatments on Ca content in tomatoes: (**A**) after the first treatment, Ca contents in tomato roots, stems and leaves; (**B**) after the second treatment, Ca contents in tomato roots, stems and leaves; (**C**) after the third treatment, Ca contents in tomato roots, stems, leaves, flowers and fruits; and (**D–F**) Ca^{2+} concentrations in xylem sap during three times treatment (n = 3). Ctrl, H_2O; T1, LS–Ca; T2, EDTA–Ca; T3, $Ca(NO_3)_2$; T4, AS–Ca treatment. Error bars represent standard deviations. Different letters represent significantly different values at $p \leq 0.05$.

3.5. LS–Ca Fertilizer Enhances Tomato Fruit Quality

To investigate the effects of various Ca fertilizers on tomato fruit development and quality, red, ripe fruits were collected, and their yield and quality traits were analyzed. Across the treatments, while the Ca fertilizer applications advanced flowering times and maintained higher flower numbers after the third application (Figure S3), there were no significant differences in the total fruit weight per plant (Figure 5B). Compared to the control, the EDTA–Ca and $Ca(NO_3)_2$ treatments slightly increased individual fruit weights (Figure 5A,C). Notably, fruit firmness in the LS–Ca treatment was significantly higher than that of the control and other treatments (Figure 5D). In terms of fruit quality, the LS–Ca, $Ca(NO_3)_2$, and SA–Ca treatments significantly improved fruit sweetness, with the LS–Ca treatment leading to a 33% increase, while the $Ca(NO_3)_2$ and SA–Ca treatments resulted in 13% and 14% increases, respectively (Figure 5E). These findings highlight the superior efficacy of LS–Ca in enhancing both the structural integrity and taste quality of tomato fruits.

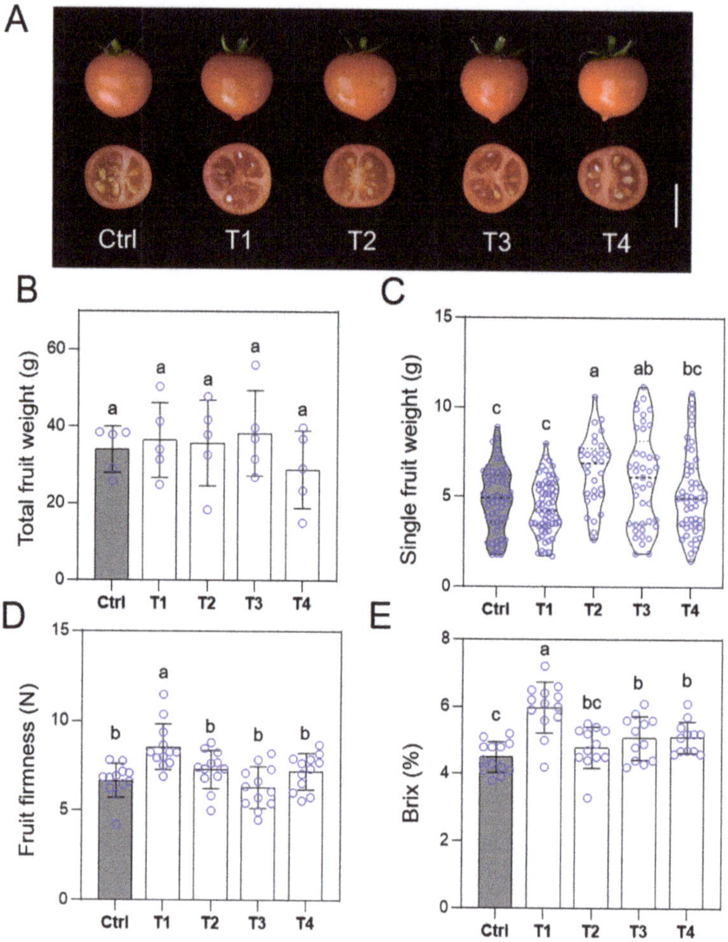

Figure 5. Effects of different Ca fertilizer treatments on fruit size, firmness, and brix of tomatoes after the third treatment: (**A**) representative red ripe fruit were photographed. Scale bar = 1 cm; (**B**) fruit yield (n = 5); (**C**) single fruit weight (n = 30–40); (**D**) fruit firmness (n = 12); and (**E**) soluble solid contents (n = 12). Ctrl, H_2O; T1, LS–Ca; T2, EDTA–Ca; T3, $Ca(NO_3)_2$; T4, AS–Ca treatment. Error bars represent standard deviations. Different letters represent significantly different values at $p \leq 0.05$.

To explore the variations in fruit sweetness, the sugar and acid contents in the fruits were analyzed. The results showed that, compared to the control, the LS–Ca treatment increased the total sugar, sucrose, and reducing sugar contents by 45%, 79%, and 64%, respectively (Figure 6A–C). The reduced sugar content was also elevated by 46%, 33%, and 30% under the EDTA–Ca, $Ca(NO_3)_2$, and SA–Ca treatments, respectively (Figure 6C). For the titratable acid content, the LS–Ca treatment resulted in an 18% increase compared to that of the control, whereas no significant differences were observed in the other treatments (Figure 6D). Additionally, the LS–Ca and EDTA–Ca treatments significantly enhanced the flavonoid contents of the fruits, with increases of 77% and 89%, respectively (Figure 6E). The LS–Ca treatment also increased the carotenoid contents in the fruits by 14% (Figure 6F). In summary, our findings indicate that the LS–Ca treatment enhanced fruit firmness and sweetness without affecting fruit weight and yield.

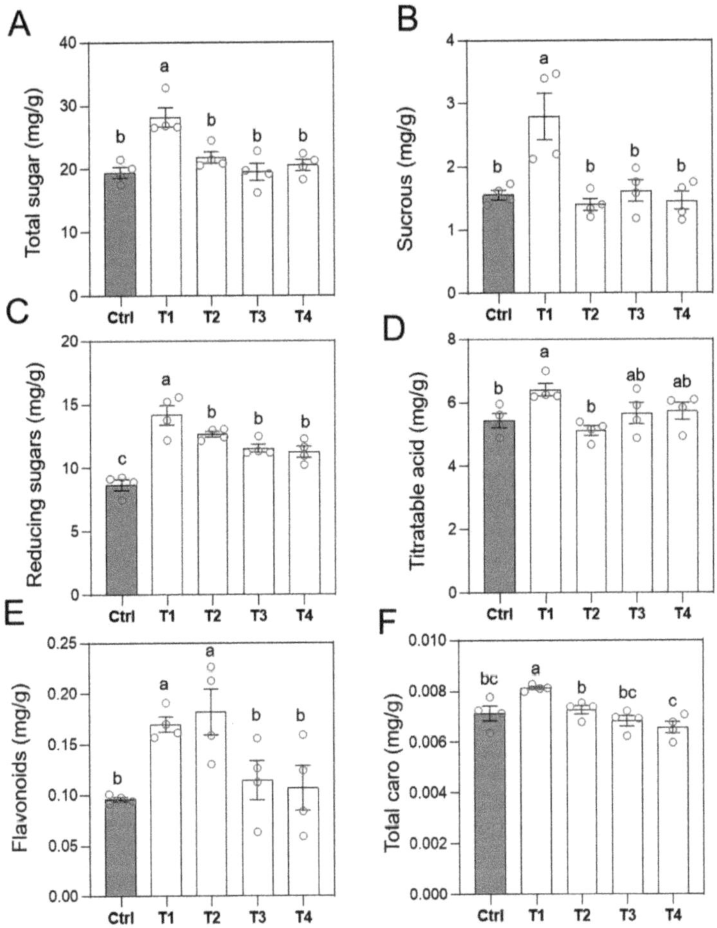

Figure 6. Effects of different Ca fertilizer treatments on tomato fruit quality after the third treatment: measurement of total sugar (**A**); measurement of reducing sucrose (**B**); measurement of reducing sugars (**C**); measurement of titratable acid (**D**); measurement of flavonoids (**E**); and measurement of carotenoids (**F**). Ctrl, H_2O; T1, LS–Ca; T2, EDTA–Ca; T3, $Ca(NO_3)_2$; T4, AS–Ca treatment. Error bars represent standard deviations (n = 4). Different letters represent significantly different values at $p \leq 0.05$.

4. Discussion

Lignin, a natural and renewable polyphenolic polymer, is deposited in plant cell walls and tightly cross-linked with cellulose and hemicellulose, providing rigidity to the cell wall [44]. A significant portion of Ca in plants is located in the cell wall, where it binds to carboxyl groups in pectin, particularly pectic acid, forming Ca–pectate complexes essential for cell wall integrity [45–47]. This synergy between lignin and Ca within plant cell walls ensures the adaptability and durability of the cell wall under various environmental conditions [48].

Lignin-derived materials effectively chelate elements such as Fe and Ca, promoting root development in rice (*Oryza sativa*) [21]. The sulfonic and hydroxyl groups on lignosulfonates (LS) bind to soil particles, enhancing their resistance to wind and water erosion [49]. Additionally, lignin-based slow-release fertilizers for Ca and Fe significantly improve the growth of plants such as holly (*Ilex* spp.), poplar (*Populus* spp.), and ginseng

(*Panax* spp.) [50]. In this study, an LS–Ca fertilizer was applied in tomato cultivation and significantly increased the content of chlorophyll a and b in leaves during the growth stages, particularly that of chlorophyll b (Figure 3). Chloroplasts, known as organelles with a high concentration of Ca^{2+}, are closely associated with photosynthesis [20]. Light-dependent Ca^{2+} uptake occurs in isolated chloroplasts, where Ca^{2+} is absorbed from the cytoplasm and stored either in the thylakoid lumen or bound by Ca^{2+}-binding proteins under light conditions [51–54]. With the combined effects of root and foliar applications, Ca^{2+} was highly enriched in leaves (Figure 4). However, despite the significant increase in chlorophyll a and b, there were no observable changes in net photosynthesis (Figure S1). Studies have demonstrated that Ca^{2+} optimizes photosynthetic performance and minimizes photooxidative damage. Under low light conditions, the photosynthetic electron transport system captures excitation energy to drive photochemical processes, while under high excitation pressure, Ca^{2+} contributes to the quenching of excess photons to prevent photodamage [55]. Ca has also been shown to regulate enzymatic activity, either catalyzing or inhibiting specific processes [56]. Low resting free [Ca^{2+}] in the stroma is essential for activating Calvin–Benson–Bassham (CBB) cycle enzymes, whereas high concentrations of Ca^{2+} can lead to the inactivation of fructose-1,6-bisphosphatase (FBPase) through the release of stored Ca^{2+} following light-to-dark transitions [56–58]. Thus, the increased Ca content may enhance leaf structural stability, improve stress resistance, or optimize resource allocation, without necessarily resulting in a measurable increase in net photosynthesis.

The availability of Ca in soil is significantly influenced by pH [59]. In alkaline soils, Ca readily precipitates as insoluble compounds, such as $CaCO_3$ and $Ca_3(PO_4)_2$, making it difficult for plants to absorb Ca even when it is abundant [59]. Acidic soils promote the release of Ca^{2+} but also lead to substantial Ca leaching during irrigation [60]. Moreover, microbial activity in acidic soils may be reduced, limiting the release of organically bound Ca [60]. Additionally, other cations present in acidic soils, such as hydrogen ions (H^+), ferrous ions (Fe^{2+}), and manganese ions (Mn^{2+}), can adversely affect Ca solubility [61]. Lignin, a natural chelating agent extracted from plants, contains abundant hydroxyl and carboxyl functional groups that form stable complexes with some ions such as iron, zinc, copper, and Ca [21]. These chelates prevent metal ions from precipitating or being fixed in the soil. Compared with other chelated Ca fertilizers, the LS–Ca treatment resulted in a significant increase in Ca content in tomato roots after the first fertilization (Figure 4), indicating that LS–Ca enhances the bioavailability of Ca to plants. This trend persisted throughout the later stages of fertilization. Even when total Ca content in the soil remained constant, LS–Ca treatments mobilized many free Ca^{2+}, suggesting that lignosulfonate-chelated Ca provides a slow and sustained release of Ca^{2+}. This slow-release characteristic enhances fertilizer efficiency by ensuring a steady nutrient supply and minimizing nutrient loss through leaching or precipitation. Consequently, LS–Ca offers an efficient approach to optimizing Ca availability in various soil conditions, addressing challenges associated with pH-induced Ca fixation or leaching [62]. Although previous studies have shown that EDTA chelating agents can enhance the absorption of trace elements [63], sugar alcohol-chelated Ca significantly improved photosynthesis and tuber quality in potatoes under drought conditions [64]. However, our findings suggest that LS–Ca provides a more sustained release of calcium (Figure 4), thereby enhancing absorption and increasing chlorophyll content (Figure 3). This is consistent with the work of Morales et al. (2023), who demonstrated the benefits of slow-release Ca fertilizers in promoting Ca^{2+} absorption and distribution [65].

Ca also plays a critical role in fruit development. During the early stages, it supports cell division and metabolism, while in the later stages, it contributes to cell-to-cell adhesion [29,66]. Within the cell wall, pectic acids bind to Ca to form Ca pectate, which serves as a structural backbone, preventing gel layer disintegration and enhancing cell wall strength [67]. Typically, Ca is absorbed by roots and transported through the xylem to various plant organs; however, only a small fraction reaches the fruit [68]. Efficient Ca transport within the plant is, therefore, critical for ensuring adequate Ca supplementation in fruits. The LS–Ca treatment not only enhanced Ca uptake but also significantly increased the Ca^{2+} concentration in xylem sap (Figure 4). This indicates that LS–Ca facilitates more efficient transport of Ca to various plant organs, including fruits. Consistent with expectations, the Ca content in the fruit significantly increased, contributing to enhanced fruit firmness. Interestingly, the sweetness of tomatoes treated with LS–Ca was higher than that of the control (Figure 5). Additionally, the contents of sugars, acids, and carotenoids in the fruit also showed varying degrees of increase (Figure 6). In the late stages of fruit ripening, the transport of water, sugars, and essential nutrients primarily occurs through the phloem [68–70]. However, because Ca^{2+} has limited mobility in the phloem, its supplementation in fruits largely depends on xylem transport [68]. The increased xylem transport of Ca^{2+} likely contributed to timely Ca accumulation in the fruit. This may have optimized the phloem's capacity for transporting sugars and acids to the fruit. Ca in the cell wall also delayed fruit softening, leading to a higher accumulation of sugars, acids, and other metabolites, ultimately improving fruit quality. Similar to our results, Liu et al. (2023) observed enhanced nutrient absorption and growth in rice treated with lignin-based chelating materials [21]. However, our study is the first to demonstrate the significant impact of LS–Ca on xylem transport (Figure 4) and fruit-quality enhancement in tomato cultivation (Figures 5 and 6).

Plants naturally produce approximately 150 billion tons of lignin annually. Additionally, the pulp and paper industry generates about 50 million tons of lignin as a byproduct each year [71]. This abundant availability makes lignin a low-cost resource. Transforming lignin into high-value controlled-release fertilizers not only enhances nutrient absorption by plants but also conserves resources and mitigates the environmental pollution caused by lignin incineration. As a chemically modified material, lignin serves as an excellent carrier for nutrients. Its ability to form controlled-release fertilizers stems from chemical interactions between lignin and nutrients, allowing for gradual nutrient release [72]. With advancements in the preparation techniques for controlled-release fertilizers, lignin-based fertilizers hold significant potential for agricultural sustainability. They can improve resource use efficiency, protect agricultural ecosystems, and contribute to sustainable agricultural growth. By utilizing lignin, a renewable and eco-friendly resource, the agricultural sector can address both productivity and environmental challenges, highlighting lignin's promising role in modern sustainable practices.

5. Conclusions

The application of LS–Ca increased Ca absorption and transport in tomato plants and improved fruit quality. Treatment with LS–Ca increases fruit hardness and sweetness, surpassing the effects of other Ca fertilizers. These results provide a theoretical basis for optimizing tomato cultivation practices and improving fruit quality through the application of LS–Ca fertilizers. These findings provide a theoretical foundation for optimizing tomato cultivation practices and improving fruit quality through the application of LS–Ca fertilizers. The demand for high quality and disease resistance in tomatoes is a huge challenge. A future trend for producers in the tomato industry may be able to integrate the advantages of various compound fertilizers and apply them to tomato cultivation.

Supplementary Materials: The following supporting information can be downloaded at: https://www.mdpi.com/article/10.3390/horticulturae10121328/s1, Figure S1: Effects of different Ca fertilizer treatments on tomato leaf Net Photosynthetic Rate (Pn) after the second treatment; Figure S2: Effects of different Mg fertilizer treatments on Mg content in tomato; Figure S3: Effects of different Ca fertilizer treatments on the number of flowers after the second treatment ($n \geq 7$).

Author Contributions: F.M. and Z.B. conceived and designed the experiments. J.Z. and M.D. performed most of experiments, and G.L. provided technical assistance. J.Z., F.M. and Z.B. analyzed the data. J.Z., F.M. and Z.B. wrote the manuscript. All authors have read and agreed to the published version of the manuscript.

Funding: This research was supported by the Taishan Scholar Foundation of Shandong Province (tsqn201909073 to F.M.; tsqn201812034 to Z.B).

Data Availability Statement: Data are contained within the article and Supplementary Materials.

Acknowledgments: We are grateful to Chunxiang You's lab for help with the metal element determination experiments.

Conflicts of Interest: The authors declare no conflicts of interest.

References

1. Topcu, Y.; Nambeesan, S.U.; van der Knaap, E. Blossom-end rot: A century-old problem in tomato (*Solanum lycopersicum* L.) and other vegetables. *Mol. Hortic.* **2022**, *2*, 1. [CrossRef] [PubMed]
2. Watanabe, T.; Tomizaki, R.; Watanabe, R.; Maruyama, H.; Shinano, T.; Urayama, M.; Kanayama, Y. Ionomic differences between tomato introgression line IL8–3 and its parent cultivar M82 with different trends to the incidence of blossom-end rot. *Sci. Hortic.* **2021**, *287*, 110266. [CrossRef]
3. Sajid, M.; Ullah, I.; Rab, A.; Shah, S.T.; Ahmad, N.; Ahmad, I.; Ali, A.; Basit, A.; Bibi, F.; Ahmad, M. Foliar application of calcium improves growth, yield and quality of tomato cultivars. *Pure Appl. Biol.* **2020**, *9*, 10–19. [CrossRef]
4. Kadir, S.A. Fruit quality at harvest of "Jonathan" apple treated with foliarly-applied calcium chloride. *J. Plant Nutr.* **2005**, *27*, 1991–2006. [CrossRef]
5. McMahon, M.J.; Kofranek, A.; Rubatzky, V.E.; Hartmann, H.T. *Plant Science: Growth, Development, and Utilization of Cultivated Plants*; Pearson: London, UK, 1983.
6. Ren, H.; Zhao, X.; Li, W.; Hussain, J.; Qi, G.; Liu, S. Calcium signaling in plant programmed cell death. *Cells* **2021**, *10*, 1089. [CrossRef]
7. Imenšek, N.; Sem, V.; Kolar, M.; Ivančič, A.; Kristl, J. The distribution of minerals in crucial plant parts of various elderberry (*Sambucus* spp.) interspecific hybrids. *Plants* **2021**, *10*, 653. [CrossRef]
8. Jing, T.; Li, J.; He, Y.; Shankar, A.; Saxena, A.; Tiwari, A.; Maturi, K.C.; Solanki, M.K.; Singh, V.; Eissa, M.A.; et al. Role of calcium nutrition in plant physiology: Advances in research and insights into acidic soil conditions—A comprehensive review. *Plant Physiol. Biochem.* **2024**, *210*, 108602. [CrossRef]
9. Vista, S.P.; Gaihre, Y.K.; Dahal, K.R. Plant Nutrient Availability in Acid Soil and Management Strategies. In *Climate Change and Soil-Water-Plant Nexus: Agriculture and Environment*; Rahman, M.M., Biswas, J.C., Meena, R.S., Eds.; National Singapore: Singapore, 2024; pp. 331–353.
10. Goulding, K.W.T. Soil acidification and the importance of liming agricultural soils with particular reference to the United Kingdom. *Soil Use Manag.* **2016**, *32*, 390–399. [CrossRef]
11. Moraes, M.T.d.; Debiasi, H.; Franchini, J.C.; Mastroberti, A.A.; Levien, R.; Leitner, D.; Schnepf, A. Soil compaction impacts soybean root growth in an oxisol from subtropical Brazil. *Soil Tillage Res.* **2020**, *200*, 104611. [CrossRef]
12. Tanveer, K.; Gilani, S.; Hussain, Z.; Ishaq, R.; Adeel, M.; Ilyas, N. Effect of salt stress on tomato plant and the role of calcium. *J. Plant Nutr.* **2020**, *43*, 28–35. [CrossRef]
13. Bangerth, F. Calcium-related physiological disorders of plants. *Annu. Rev. Phytopathol.* **1979**, *17*, 97–122. [CrossRef]
14. Souto, A.G.; Pessoa, A.M.; Sá, S.A.; Sousa, N.R.; Barros, E.S.; Morais, F.M.; Ferreira, F.N.; Silva, W.A.; Batista, R.O.; Silva, D.V.; et al. Potential of Ca-complexed in amino acid in attenuating salt stress in sour passion fruit seedlings. *Plants* **2024**, *13*, 2912. [CrossRef] [PubMed]
15. Rajan, M.; Shahena, S.; Chandran, V.; Mathew, L. Chapter 3—Controlled release of fertilizers—Concept, reality, and mechanism. In *Controlled Release Fertilizers for Sustainable Agriculture*; Lewu, F.B., Volova, T., Thomas, S., Rakhimol, K.R., Eds.; Academic Press: Cambridge, MA, USA, 2021; pp. 41–56.
16. de Souza Alonso, T.A.; Ferreira Barreto, R.; de Mello Prado, R.; Pereira de Souza, J.; Falleiros Carvalho, R. Silicon spraying alleviates calcium deficiency in tomato plants, but Ca-EDTA is toxic. *J. Plant Nutr. Soil Sci.* **2020**, *183*, 659–664. [CrossRef]
17. Norvell, W.A. Reactions of Metal Chelates in Soils and Nutrient Solutions. In *Micronutrients in Agriculture*; Wiley: Hoboken, NJ, USA, 1991; pp. 187–227.
18. Li, T.; Wei, Q.; Sun, W.; Tan, H.; Cui, Y.; Han, C.; Zhang, H.; Zeng, F.; Huang, M.; Yan, D. Spraying sorbitol-chelated calcium affected foliar calcium absorption and promoted the yield of peanut (*Arachis hypogaea* L.). *Front. Plant Sci.* **2022**, *13*, 1075488. [CrossRef] [PubMed]
19. Ma, T.; Hui, Y.; Zhang, L.; Su, B.; Wang, R. Foliar application of chelated sugar alcohol calcium fertilizer for regulating the growth and quality of wine grapes. *Int. J. Agric. Biol. Eng.* **2022**, *15*, 153–158. [CrossRef]
20. Guedes, L.R.; Cavalcante, L.F.; Souto, A.G.d.L.; Carvalho, L.H.M.; Cavalcante, Í.H.L.; Diniz Neto, M.A.; Lima, G.S.d.; Melo, T.d.S.; Henrique, J.C.G.d.S. Liquid fertilizers on photochemical efficiency and gas exchange in yellow passion fruit under saline stress. *Rev. Bras. Eng. Agric. Ambiental* **2023**, *27*, 153–158. [CrossRef]
21. Liu, Q.; Kawai, T.; Inukai, Y.; Aoki, D.; Feng, Z.; Xiao, Y.; Fukushima, K.; Lin, X.; Shi, W.; Busch, W.; et al. A lignin-derived material improves plant nutrient bioavailability and growth through its metal chelating capacity. *Nat. Commun.* **2023**, *14*, 4866. [CrossRef]

22. Kok, A.D.-X.; Wan Abdullah, W.M.A.N.; Tang, C.-N.; Low, L.-Y.; Yuswan, M.H.; Ong-Abdullah, J.; Tan, N.-P.; Lai, K.-S. Sodium lignosulfonate improves shoot growth of *Oryza sativa* via enhancement of photosynthetic activity and reduced accumulation of reactive oxygen species. *Sci. Rep.* **2021**, *11*, 13226. [CrossRef]
23. Abbas, A.; Wang, Z.; Zhang, Y.; Peng, P.; She, D. Lignin-based controlled release fertilizers: A review. *Int. J. Biol. Macromol.* **2022**, *222*, 1801–1817. [CrossRef]
24. Lu, J.; Cheng, M.; Zhao, C.; Li, B.; Peng, H.; Zhang, Y.; Shao, Q.; Hassan, M. Application of lignin in preparation of slow-release fertilizer: Current status and future perspectives. *Ind. Crops Prod.* **2022**, *176*, 114267. [CrossRef]
25. Ho, L.C.; Belda, R.; Brown, M.; Andrews, J.; Adams, P. Uptake and transport of calcium and the possible causes of blossom-end rot in tomato. *J. Exp. Bot.* **1993**, *44*, 509–518. [CrossRef]
26. Tonetto de Freitas, S.; McElrone, A.J.; Shackel, K.A.; Mitcham, E.J. Calcium partitioning and allocation and blossom-end rot development in tomato plants in response to whole-plant and fruit-specific abscisic acid treatments. *J. Exp. Bot.* **2014**, *65*, 235–247. [CrossRef]
27. Ho, L.C.; White, P.J. A cellular hypothesis for the induction of blossom-end rot in tomato fruit. *Ann. Bot.* **2005**, *95*, 571–581. [CrossRef] [PubMed]
28. Fischer, A.M. Nutrient remobilization during leaf senescence. In *Annual Plant Reviews Volume 26: Senescence Processes in Plants*; Wiley-Blackwell: Hoboken, NJ, USA, 2007; pp. 87–107.
29. Hocking, B.; Tyerman, S.D.; Burton, R.A.; Gilliham, M. Fruit calcium: Transport and physiology. *Front. Plant Sci.* **2016**, *7*, 569. [CrossRef]
30. Rogiers, S.; Keller, M.; Holzapfel, B.P.; Virgona, J.M. Accumulation of potassium and calcium by ripening berries on field vines of *Vitis vinifera* (L) cv. Shiraz. *Aust. J. Grape Wine Res.* **2000**, *6*, 240–243. [CrossRef]
31. Niu, J.; Liu, C.; Huang, M.; Liu, K.; Yan, D. Effects of foliar fertilization: A review of current status and future perspectives. *J. Soil Sci. Plant Nutr.* **2021**, *21*, 104–118. [CrossRef]
32. Yuan, L.; Wu, L.; Yang, C.; Lv, Q. Effects of iron and zinc foliar applications on rice plants and their grain accumulation and grain nutritional quality. *J. Sci. Food Agric.* **2013**, *93*, 254–261. [CrossRef]
33. Shinozaki, Y.; Nicolas, P.; Fernandez-Pozo, N.; Ma, Q.; Evanich, D.J.; Shi, Y.; Xu, Y.; Zheng, Y.; Snyder, S.I.; Martin, L.B.B.; et al. High-resolution spatiotemporal transcriptome mapping of tomato fruit development and ripening. *Nat. Commun.* **2018**, *9*, 364. [CrossRef]
34. Niu, C.; Wang, G.; Sui, J.; Liu, G.; Ma, F.; Bao, Z. Biostimulants alleviate temperature stress in tomato seedlings. *Sci. Hortic.* **2022**, *293*, 110712. [CrossRef]
35. Morris, D.L. Quantitative determination of carbohydrates with dreywood's anthrone reagent. *Science* **1948**, *107*, 254–255. [CrossRef]
36. Somogyi, M. Notes on sugar determination. *J. Biol. Chem.* **1952**, *195*, 19–23. [CrossRef]
37. Somogyi, M. Determination of blood sugar. *J. Biol. Chem.* **1945**, *160*, 69–73. [CrossRef]
38. Sadler, G.D.; Murphy, P.A. pH and titratable acidity. In *Food Analysis*; Nielsen, S.S., Ed.; Springer US: Boston, MA, USA, 2010; pp. 219–238.
39. Zhishen, J.; Mengcheng, T.; Jianming, W. The determination of flavonoid contents in mulberry and their scavenging effects on superoxide radicals. *Food Chem.* **1999**, *64*, 555–559. [CrossRef]
40. Zhang, J.-C.; Wang, X.-N.; Sun, W.; Wang, X.-F.; Tong, X.-S.; Ji, X.-L.; An, J.-P.; Zhao, Q.; You, C.-X.; Hao, Y.-J. Phosphate regulates malate/citrate-mediated iron uptake and transport in apple. *Plant Sci.* **2020**, *297*, 110526. [CrossRef] [PubMed]
41. Zhang, J.-C.; Wang, X.-F.; Wang, X.-N.; Wang, F.-P.; Ji, X.-L.; An, J.-P.; Yang, K.; Zhao, Q.; You, C.-X.; Hao, Y.-J. Abscisic acid alleviates iron deficiency by regulating iron distribution in roots and shoots of apple. *Sci. Hortic.* **2020**, *262*, 109018. [CrossRef]
42. Carter, M.R.; Gregorich, E.G. *Soil Sampling and Methods of Analysis*, 2nd ed.; CRC Press: Boca Raton, FL, USA, 2007.
43. Alexou, M.; Peuke, A.D. Methods for Xylem Sap Collection. In *Plant Mineral Nutrients: Methods and Protocols*; Maathuis, F.J.M., Ed.; Humana Press: Totowa, NJ, USA, 2013; pp. 195–207.
44. Rubin, E.M. Genomics of cellulosic biofuels. *Nature* **2008**, *454*, 841–845. [CrossRef]
45. Cosgrove, D.J. Growth of the plant cell wall. *Nat. Rev. Mol. Cell Biol.* **2005**, *6*, 850–861. [CrossRef]
46. Hepler, P.K.; Winship, L.J. Calcium at the cell wall-cytoplast interface. *J. Integr. Plant Biol.* **2010**, *52*, 147–160. [CrossRef]
47. Gilliham, M.; Dayod, M.; Hocking, B.J.; Xu, B.; Conn, S.J.; Kaiser, B.N.; Leigh, R.A.; Tyerman, S.D. Calcium delivery and storage in plant leaves: Exploring the link with water flow. *J. Exp. Bot.* **2011**, *62*, 2233–2250. [CrossRef]
48. Kang, X.; Kirui, A.; Dickwella Widanage, M.C.; Mentink-Vigier, F.; Cosgrove, D.J.; Wang, T. Lignin-polysaccharide interactions in plant secondary cell walls revealed by solid-state NMR. *Nat. Commun.* **2019**, *10*, 347. [CrossRef]
49. Ahmad, U.M.; Ji, N.; Li, H.; Wu, Q.; Song, C.; Liu, Q.; Ma, D.; Lu, X. Can lignin be transformed into agrochemicals? Recent advances in the agricultural applications of lignin. *Ind. Crop. Prod.* **2021**, *170*, 113646. [CrossRef]
50. Kevers, C.; Soteras, G.; Baccou, J.C.; Gaspar, T. Lignosulfonates: Novel promoting additives for plant tissue cultures. *Vitr. Cell. Dev. Biol.—Plant* **1999**, *35*, 413–416. [CrossRef]
51. Muto, S.; Izawa, S.; Miyachi, S. Light-induced Ca^{2+} uptake by intact chloroplasts. *FEBS Lett.* **1982**, *139*, 250–254. [CrossRef]
52. Kreimer, G.; Melkonian, M.; Holtum, J.A.M.; Latzko, E. Characterization of calcium fluxes across the envelope of intact spinach chloroplasts. *Planta* **1985**, *166*, 515–523. [CrossRef] [PubMed]

53. Kreimer, G.; Melkonian, M.; Latzko, E. An electrogenic uniport mediates light-dependent Ca^{2+} influx into intact spinach chloroplasts. *FEBS Lett.* **1985**, *180*, 253–258. [CrossRef]
54. Hochmal, A.K.; Schulze, S.; Trompelt, K.; Hippler, M. Calcium-dependent regulation of photosynthesis. *Biochim. Biophys. Acta (BBA)—Bioenerg.* **2015**, *1847*, 993–1003. [CrossRef]
55. Takahashi, S.; Badger, M.R. Photoprotection in plants: A new light on photosystem II damage. *Trends Plant Sci.* **2011**, *16*, 53–60. [CrossRef]
56. Hertig, C.; Wolosiuk, R.A. A dual effect of Ca^{2+} on chloroplast fructose-1,6-bisphosphatase. *Biochem. Biophys. Res. Commun.* **1980**, *97*, 325–333. [CrossRef]
57. Charles, S.A.; Halliwell, B. Action of calcium ions on spinach (*Spinacia oleracea*) chloroplast fructose bisphosphatase and other enzymes of the Calvin cycle. *Biochem. J.* **1980**, *188*, 775–779. [CrossRef]
58. Portis, A.R.; Heldt, H.W. Light-dependent changes of the Mg^{2+} concentration in the stroma in relation to the Mg^{2+} dependency of CO_2 fixation in intact chloroplasts. *BBA—Bioenerg.* **1976**, *449*, 434–446. [CrossRef]
59. Lawrence, G.; Siemion, J.; Antidormi, M.; Bonville, D.; McHale, M. Have sustained acidic deposition decreases led to increased calcium availability in recovering watersheds of the adirondack region of New York, USA? *Soil Syst.* **2021**, *5*, 6. [CrossRef]
60. Feng, C.; Cui, B.; Ge, H.; Huang, Y.; Zhang, W.; Zhu, J. Reinforcement of recycled aggregate by microbial-induced mineralization and deposition of calcium carbonate—Influencing factors, mechanism and effect of reinforcement. *Crystals* **2021**, *11*, 887. [CrossRef]
61. Ng, J.F.; Ahmed, O.H.; Jalloh, M.B.; Omar, L.; Kwan, Y.M.; Musah, A.A.; Poong, K.H. Soil nutrient retention and pH buffering capacity are enhanced by calciprill and sodium silicate. *Agronomy* **2022**, *12*, 219. [CrossRef]
62. Priva, E.; Sarkar, S.; Maji, P.K. A review on slow-release fertilizer: Nutrient release mechanism and agricultural sustainability. *J. Environ. Chem. Eng.* **2024**, *12*, 113211. [CrossRef]
63. Bloem, E.; Haneklaus, S.; Haensch, R.; Schnug, E. EDTA application on agricultural soils affects microelement uptake of plants. *Sci. Total Environ.* **2017**, *577*, 166–173. [CrossRef]
64. Wang, Y.; Qin, T.; Pu, Z.; Dekomah, S.D.; Yao, P.; Sun, C.; Liu, Y.; Bi, Z.; Bai, J. Foliar Application of chelated sugar alcohol calcium improves photosynthesis and tuber quality under drought stress in potatoes (*Solanum tuberosum* L.). *Int. J. Mol. Sci.* **2023**, *24*, 2216. [CrossRef]
65. Morales, J.; Martínez-Alcántara, B.; Bermejo, A.; Millos, J.; Legaz, F.; Quiñones, A. Effect of calcium fertilization on calcium uptake and its partitioning in citrus trees. *Agronomy* **2023**, *13*, 2971. [CrossRef]
66. Bemadac, A.; Jean-Baptiste, I.; Bertoni, G.; Morard, P. Changes in calcium contents during melon (*Cucumis melo* L.) fruit development. *Sci. Hortic.* **1996**, *66*, 181–189. [CrossRef]
67. Zhang, L.; Wang, J.-W.; Chen, J.-Y.; Song, T.; Jiang, Y.-G.; Zhang, Y.-F.; Wang, L.-J.; Li, F.-L. Preharvest spraying calcium ameliorated aroma weakening and kept higher aroma-related genes expression level in postharvest 'Nanguo' pears after long-term refrigerated storage. *Sci. Hortic.* **2019**, *247*, 287–295. [CrossRef]
68. Drazeta, L.; Lang, A.; Hall, A.J.; Volz, R.K.; Jameson, P.E. Causes and Eeffects of changes in xylem functionality in apple fruit. *Ann. Bot.* **2004**, *93*, 275–282. [CrossRef]
69. Rogiers, S.Y.; Greer, D.H.; Hatfield, J.M.; Orchard, B.A.; Keller, M. Solute transport into shiraz berries during development and late-ripening shrinkage. *Am. J. Enol. Vitic.* **2006**, *57*, 73. [CrossRef]
70. Choat, B.; Gambetta, G.A.; Shackel, K.A.; Matthews, M.A. Vascular function in grape berries across development and its relevance to apparent hydraulic isolation. *Plant Physiol.* **2009**, *151*, 1677–1687. [CrossRef] [PubMed]
71. Chen, J.; Fan, X.; Zhang, L.; Chen, X.; Sun, S.; Sun, R.-C. Research progress in lignin-based slow/controlled release fertilizer. *ChemSusChem* **2020**, *13*, 4356–4366. [CrossRef] [PubMed]
72. Kumar, A.; Singh, U.M.; Manohar, M.; Gaur, V.S. Calcium transport from source to sink: Understanding the mechanism(s) of acquisition, translocation, and accumulation for crop biofortification. *Acta Physiol. Plant.* **2014**, *37*, 1722. [CrossRef]

horticulturae

Article

Enhancing Fruit Retention and Juice Quality in 'Kinnow' (*Citrus reticulata*) Through the Combined Foliar Application of Potassium, Zinc, and Plant Growth Regulators

Iqra Arshad [1,2], Muhammad Saleem [1,2], Muhammad Akhtar [1,2], Muhammad Yousaf Shani [2,3], Ghulam Farid [1], Wacław Jarecki [4,*] and Muhammad Yasin Ashraf [1,*]

1 Soil and Environmental Sciences Division, Nuclear Institute for Agriculture and Biology, P.O. Box 128, Jhang Road, Faisalabad 38000, Pakistan; iqraniab14@gmail.com (I.A.); saleemneab@gmail.com (M.S.)
2 Nuclear Institute for Agriculture and Biology, College (NIAB-C), Pakistan Institutes of Engineering and Applied Science (PIEAS), Islamabad 45650, Pakistan
3 Plant Breeding and Genetics Division, Nuclear Institute for Agriculture and Biology (NIAB), Jhang Road, Faisalabad 38000, Pakistan
4 Department of Crop Production, University of Rzeszow, Zelwerowicza 4, 35-601 Rzeszow, Poland
* Correspondence: wjarecki@ur.edu.pl (W.J.); niabmyashraf@gmail.com (M.Y.A.)

Abstract: Improving fruit quality and reducing pre-harvest fruit drop are critical goals for *Citrus reticulata* production in Pakistan, where climatic and nutritional challenges affect yield and juice quality. This study evaluated the combined effects of plant growth regulators (salicylic acid and indole acetic acid) and nutrients (potassium and zinc) on fruit drop and juice volume in *Citrus reticulata* L. Field trials were conducted at three locations in Punjab, Pakistan (Layyah, Faisalabad, and Sargodha) using a randomized complete block design (RCBD) with five replications per treatment. Nutrients (K and Zn at 100 mg/L each) and growth regulators (SA at 100 mg/L and IAA at 5 mg/L) were applied individually or in combination at three growth stages. Statistical analyses, including PCA, ANOVA, and GGE biplot, were used to identify the most effective treatments for improving fruit juice quality and reducing fruit drop. The combined foliar application of SA + K + Zn was the most effective across all parameters, except fruit drop, juice citric acid contents, and juice pH, which were negatively affected. The highest juice potassium content was observed with K application. The PCA and GGE biplot analysis indicated that the Sargodha orchard performed best, with the SA + K treatment being the most effective there, while SA + K + Zn showed the best results at Layyah and Faisalabad for reducing fruit drop, enhancing juice volume, and improving fruit quality. However, individual fruit, juice, and juice nutrient contents traits analyses revealed that the most significant improvements in fruit and juice quality were observed at the Sargodha site instead of Layyah and Faisalabad. The treatment SA + K + Zn proved to be the most stable and consistent in enhancing citrus fruit and juice quality across all three selected locations. The findings suggest that adopting the SA + K + Zn treatment could be a practical approach for citrus farmers aiming to enhance crop yield and fruit quality, thereby supporting agricultural productivity and export potential in Pakistan.

Keywords: *Citrus reticulata*; fruit quality; pre-harvest fruit drop; plant growth regulators; foliar application

1. Introduction

Kinnow (*Citrus reticulata* L.), a perennial evergreen cultivar of the Rutaceae family, plays a vital role in strengthening the economy of developing countries like Pakistan, where it represents 75% of the nation's total citrus production [1]. Citrus fruits are highly sought after worldwide due to their delicious taste and flavor. In Pakistan, citrus production covers 177.22 thousand hectares, yielding approximately 2.29 million tons annually [2]. The country's citrus production has seen a steady rise from 0.445 million tons in 1970 to 2.30 million tons in 2020, growing at an annual rate of 4.14% [3]. While 'Kinnow' is cultivated across all provinces, Punjab contributes over 95% of the total crop yield, with the Sargodha region renowned for producing high-quality citrus fruits [4].

Adequate nutrient supply is critical for sustaining the long growth period required for optimal fruit development [5]. In Pakistan, however, the lack of proper nutrient management has led to issues such as poor fruit quality, smaller size, color defects, and premature fruit abscission [6]. Deficiencies in essential nutrients can hinder the production of plant growth hormones, resulting in decreased fruit size, quality, and color development [7].

The foliar application of plant growth regulators (PGRs) such as salicylic acid (SA) and indole acetic acid (IAA), along with nutrients like potassium (K) and zinc (Zn), plays a vital role in optimizing growth and improving the physio-biochemical processes in citrus crops across diverse agro-climatic regions of Pakistan [7]. Among these, potassium is essential for enhancing fruit growth, productivity, and overall development [8]. Zinc is equally crucial for improving fruit yield and quality, as its deficiency can cause cytological and morpho-anatomical changes that reduce flowering and fruit set, ultimately affecting citrus quality [9].

Auxins are commonly used in citrus to enhance juice quality and prevent excessive fruit drop [10]. Additionally, salicylic acid (SA) plays a key role in regulating physiological processes by stimulating glycine betaine production, nitrogen metabolism, photosynthesis, and antioxidant defenses, which help plants combat stress [11]. These properties protect citrus crops from various abiotic stresses, ensuring better plant health and productivity [12].

Fruit drop is a significant factor contributing to reduced citrus yields, with maximum drop occurring in the following three phases: post-blossom, June drop, and pre-harvest [13]. Harsh environmental conditions, such as drought, high temperatures, pest infestations, and strong winds, can exacerbate fruit abscission [14]. To mitigate excessive fruit drop, foliar treatments with growth regulators like auxins and gibberellins can be used, along with proper water and nutrient management for susceptible cultivars [15,16].

The novelty of this multi-location study lies in its comprehensive evaluation of the combined effects of two plant growth regulators, salicylic acid (SA) and indole acetic acid (IAA), with potassium (K) and zinc (Zn) nutrients on fruit drop control, fruit retention, yield, and juice volume in *Citrus reticulata* (Kinnow). Unlike previous studies, this research uniquely integrates plant growth regulators and nutrients across different growth stages and geographic locations, offering new insights into optimizing citrus fruit production and quality through tailored foliar treatments. The findings can potentially enhance sustainable citrus farming practices and improve yield and fruit quality in diverse agro-climatic conditions.

2. Materials and Methods

2.1. Multi-Location Experimental Sites and Soil Physio-Chemical Attributes

The current multi-locations study was conducted during the period 15 October 2019 to 25 March 2020 in three citrus orchards located in Layyah (site 1, 30.9693_ N, 70.9428_ E), Faisalabad (site 2, NIAB Orchard, 31.4504_ N, 73.1350_ E), and Sargodha (site 3, 32.0740_ N, 72.6861_ E). The experiment was laid out using a randomized complete block design (RCBD) with factorial arrangements and five replications per treatment. At each experimental site, 63 citrus plants (7 in each row) were selected. For the estimation of the physio-biochemical parameters, soil samples were randomly cored from each selected orchard up to 30 cm in depth following a "W" scheme with the help of an instrument called an auger. Soil attributes such as soil texture, pH, and electrical conductivity were estimated using pH (HI-8520, Hanna, Germany) and EC meters (WTW, Germany), respectively. While soil P contents were analyzed using the method of Olsen [17], K^+ contents were estimated by following the protocol established by Simard [18], Na contents by a flame photometer (Jenway PFP 7, UK), and Zn concentration was assessed by following the protocol of Soltanpour and Schwab [19]. Other soil parameters, such as soil saturation percentage, carbonates, bicarbonates, and the sodium adsorption ratio were measured using the method of Estefen et al. [20], as shown in (Table 1).

Table 1. Physio-chemical characteristics of soils (0–60 cm) and irrigation water of three citrus orchards.

Parameters	Layyah	Faisalabad	Sargodha
Soil texture	Sandy-loam	Clay-loam	Clay-loam
Soil saturation (%)	26.25	37.50	37.00
ECe (dS m^{-1})	1.74	0.93	1.85
pH	8.42	7.84	7.94
Sodium adsorption ratio (mmol ½ L-1/2)	1.98	0.96	2.02
Organic matter (%)	0.15	0.18	0.25
NO_3 -N (mg kg^{-1})	19.00	18.90	17.50
Available phosphorus (mg kg^{-1})	6.15	7.25	10.48
Sodium (mg kg^{-1})	94.50	72.00	141.00
Calcium + Magnesium (meq kg^{-1})	4.82	5.07	5.25
Zinc (mg kg^{-1})	0.33	0.68	0.50
Potassium (mg kg^{-1})	8.60	16.85	20.15
Carbonates (meq kg^{-1})	Nil	Nil	Nil
Bicarbonates (meq kg^{-1})	5.40	4.70	6.00
Irrigation water analysis			
EC (dS m^{-1})	1.58	1.02	1.40
pH	8.20	8.15	8.20
Residual sodium carbonate	1.20	1.50	1.70
Sodium (mg L^{-1})	78.0	54.0	75.0
Chloride (meq L^{-1})	2.20	2.00	1.90
Potassium (mg L^{-1})	9.00	7.00	10.0
Calcium + Magnesium (meq kg^{-1})	4.20	3.00	3.50
Carbonates (meq kg^{-1})	Nil	Nil	Nil
Bicarbonates (meq kg^{-1})	5.40	4.50	5.20
SAR (meq L^{-1})	2.34	1.92	2.47

Values represent the mean of the soil and irrigation water data taken during 2019.

Each selected citrus orchard was irrigated with tube-well water, and its physio-biochemical characteristics were estimated. Electrical conductivity and pH were determined with EC and pH meters. Meanwhile, K and Na contents were measured by flame photometer. Additionally, carbonate, bicarbonate, chloride, calcium + magnesium, residual sodium carbonate, and the sodium absorption ratio were estimated according to Estefen et al. [20], as reported in Table 1.

2.2. Treatments

In the current work, foliar treatments of Zn, K, SA, and IAA were applied alone or in combination. In each selected orchard, NPK fertilizers were applied after the harvest of the previous year's fruits (March 2019) with an appropriate ratio $N:P_2O_5:K_2O$ (100:50:75 kg ha^{-1}) as urea (N 46%), diammonium phosphate (DAP; P_2O_5 46%), and sulfate of potash (SOP; 50% K_2O). The selected citrus plants were foliar-sprayed with 100 mg L^{-1} SA, 5 mg L^{-1} IAA, 100 mg L^{-1} K, and 100 mg L^{-1} Zn, and their combinations were K + SA, Zn + SA, IAA + SA, and K + Zn + SA, whereas control plants were treated with normal distilled water. These foliar treatments were applied thrice at the flower initiation (first spray), fruit formation (second spray), and fruit-color initiation stages (third spray). All three selected orchards contained citrus trees that were 10 to 15 years old. Each orchard has 11 rows, with 11 trees per row (totaling 121 trees). Out of the 11 plants in each row, the two peripheral (edge) plants were excluded and considered non-experimental. The remaining nine plants (in each row) were randomly sprayed with a specific treatment and five replications per treatment were maintained. The spray was applied only to the foliar parts of the tree, ensuring that all leaves, flowers, and fruits were wetted with the treatment solution at each specific stage.

2.3. Growth and Yield Data Collection

Twenty fruit samples per plant were randomly collected from both treated and non-treated plants, and the fruit diameter, height, and width were assessed with the help of measuring tape and peel thickness by digital vernier caliper. The fruit shape index was estimated by using the protocol of Brewer et al. [21], fruit weight via weighing balance (JC-1202A, A&D, Company, USA), and fruit juice volume was evaluated by a graduated cylinder. Fruit dropping was computed by the formula given by Shani et al. [22] in the first week of September.

2.4. Juice Quality and Chemical Analysis

Juice volume was determined after mechanical extraction from fruits of equal size selected from all sites. Juice pH was determined by pH meter, juice EC was measured using an EC-meter (LF538, WTW, Germany), and total solids in the juice were determined by refractometer. Citric acid was estimated by titrating the juice against 0.1 N sodium hydroxide, and ascorbic acid by reducing 2,6-dichlorophenol indophenol with the juice. N, P, K, Ca, and Na contents in juice were also analyzed as described elsewhere [23].

2.5. Statistical Analysis

The data collected from the experiments were analyzed using the Statistix-10 software. A two-way analysis of variance (ANOVA) was conducted to assess the effects of different treatments on various traits. The least significant difference (LSD) test was used to compare treatment means at a 5% probability level. ANOVA helped (Table 2) determine the significance of each treatment across the selected sites. Additionally, principal component analysis (PCA) and GGE (genotype and genotype by environment interaction) biplot analysis were performed using "R software" (version 4.4.2). PCA helped reveal variations in the observed attributes, while the GGE biplot identified the most promising treatment for each location. Furthermore, the multi-trait genotype-ideotype distance index (MGIDI) was employed to select the most consistent and effective treatment across multilocation trials.

Table 2. Means of the sum of squares for all the investigated fruit and juice quality parameters in citrus orchards treated with Zn, K, and PGRs alone or in combination.

CO	DF	FSI	PT	FD	FW	JV	pH	EC	TSS	CA	AA	TSS/Acid	N	P	K
S	2	0.27 ***	71.67 ***	2 ***	29,578 ***	897 ***	0.026 ns	1.6 ***	29.9 ***	0.1 ***	1943.6 ***	329.6 ***	0.04 ***	0.07 ***	1.2 ***
T	8	0.01 ***	2.4 ***	451 **	9644 ***	801 ***	0.028 ns	0.2 ***	5.8 ***	0.1 ***	37.1 ***	38.7 ***	0.21 ***	0.77 ***	2.7 ***
S × T	16	0.002 ***	0.41 ***	19 ***	1145 ***	42.5 ***	0.024 ns	0.1 ***	0.9 ***	0.003 ***	6.9 ***	1.8 ***	0.01 ***	0.01 ***	0.08 ***
Error	108	0.001	0.02	0.2	119.3	3.7	0.03	0.02	0.01	0.0002	0.2	0.06	0.001	0.001	0.005
Total	134														

* Showing significant differences among all the applied drought stress treatments $p > 0.001$ ***, $p < 0.01$ **, ($0.01 < p > 0.05$) *, $p > 0.05$, ^{ns} (non-significant); CO, citrus orchard; DF, degree of freedom; FSI, fruit shape index; PT, peel thickness; FD, fruit dropping; FW, fruit weight; JV, juice volume; pH, juice pH; EC, juice EC; TSS, total soluble sugars in juice; CA, juice citric acid content; AA, juice ascorbic acid concentration; TSS/Acid, TSS/Acid ratio of fruit juice; N, fruit juice nitrogen concentration; P, fruit juice phosphorus contents; and K, fruit juice potassium content.

3. Results

3.1. Kinnow Fruit Quantity and Quality Analyses

The current experiment systematically demonstrated that the foliar application of nutrients (potassium "K" and zinc "Zn") combined with plant growth regulators (PGRs) significantly enhanced various fruit quality traits in *Citrus reticulata*. These improvements were consistent across all experimental sites (Sargodha, Faisalabad, and Layyah), although notable differences were observed between treatments and locations.

According to the fruit shape index [the ratio of the longitudinal diameter to the transverse diameter], foliar applications of PGRs and nutrients significantly ($p \leq 0.05$) improved the fruit shape index. The highest fruit shape index was recorded in fruits from plants treated with a combination of salicylic acid (SA) and potassium (K), followed by the SA + K + Zn, SA + IAA, SA, and K treatments. In contrast, the control group exhibited the lowest fruit shape index, followed by plants treated with SA + Zn, IAA, and Zn alone (Figure 1A). Among the locations, fruits from Sargodha displayed the best shape index, followed by those from Faisalabad and Layyah.

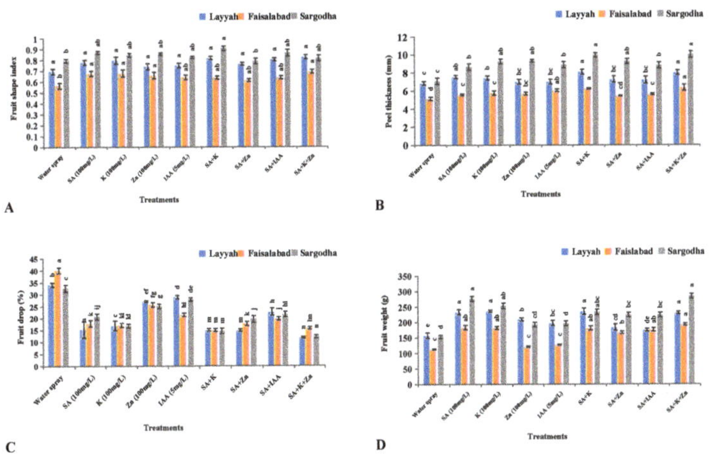

Figure 1. Effect of foliar application of plant growth regulators (PGRs) and nutrients on fruit shape index (**A**); peel thickness (**B**); fruit drop percent (**C**); and fruit weight (**D**); in citrus orchards at different selected locations. [Bars sharing similar alphabets did not differ significantly].

Peel Thickness: Peel thickness was significantly ($p \leq 0.05$) increased by the foliar application of PGRs and nutrients, with the control group having the thinnest peels (Figure 1B). Among the treated plants, the thickest peel was observed in fruits from trees sprayed with SA + K + Zn. Fruits from the Sargodha orchard had the thickest peels, followed by Layyah, while fruits from Faisalabad showed the thinnest peels.

Fruit Drop Rate: The highest fruit drop rates were observed in control plants across all sites, while the lowest rates were recorded in plants treated with SA + K + Zn. Conversely, the control group had the highest fruit drop rate (Figure 1C). Among the orchards, the Sargodha citrus orchard exhibited the lowest fruit drop, followed by Layyah and Faisalabad.

Fruit Weight: Fruit weight was significantly ($p \leq 0.05$) affected by foliar applications of PGRs and nutrients (Table 2). The heaviest fruits were produced by plants treated with SA + K + Zn, which were statistically similar to those from plants treated with K, SA, and SA + K. Among the locations, the Sargodha orchard produced the heaviest fruits, followed by Layyah and Faisalabad (Figure 1D). This experiment highlighted the effectiveness of foliar nutrient supplementation and plant growth regulators in improving key fruit quality parameters, with variations across sites influencing the outcomes.

3.2. Juice Quality Analyses

The foliar application of PGRs or nutrients significantly influenced fruit juice quality parameters except for juice pH (Table 2). However, a slight increase in fruit juice pH was observed in plants treated with SA + K + Zn and SA + K compared to other treatments. Variations in juice pH across the different sites were also non-significant, though juice from the fruit of the Layyah orchard exhibited marginally higher pH levels, followed by Faisalabad and Sargodha (Figure 2A). In contrast, the foliar application of PGRs and nutrients significantly ($p < 0.05$) impacted juice electrical conductivity (EC) (Table 2). The highest EC levels were recorded in fruits from trees treated with SA + K, which were statistically comparable to those treated with SA + K + Zn. The lowest EC was found in the juice from untreated control trees (Figure 2B). Variations in EC were also site-dependent, with the highest values recorded at the Layyah orchard, followed by Sargodha, and the lowest EC in juice recorded at the Faisalabad orchard (Figure 2B).

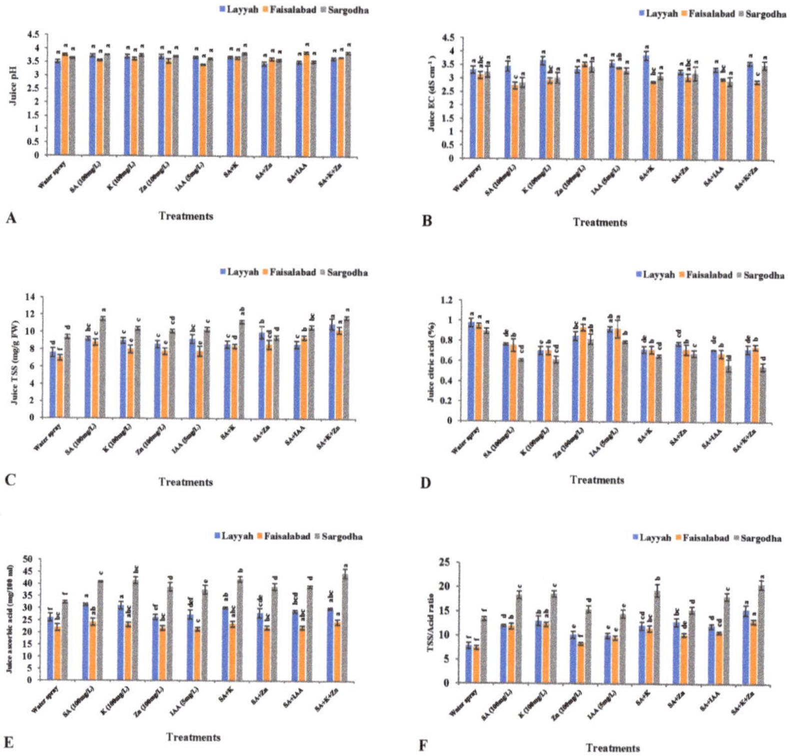

Figure 2. Effect of foliar application of plant growth regulators (PGRs) and nutrients on fruit juice pH (**A**); electrical conductivity (EC) (**B**); total soluble solids (TSS) (**C**); citric acid percentage (**D**); ascorbic acid concentration (**E**); and juice TSS/acid ratio (**F**) in fruit juice of citrus orchards in selected locations. [Bars sharing similar alphabets did not differ significantly].

Total Soluble Solids (TSS): TSS content in the fruit juice was significantly influenced by the foliar application of PGRs and nutrients like K and Zn across all orchards (Table 2). The highest TSS values were observed in juice from trees treated with SA + K + Zn, while untreated trees had the lowest TSS values. Significant site-wise differences in TSS were also noted, with the Sargodha orchard producing juice with the highest TSS content, and the Faisalabad orchard producing juice with the lowest TSS content (Figure 2C).

Citric Acid Content: Foliar applications also affected citric acid levels in the juice. The highest citric acid content was found in juice from untreated trees, whereas the lowest levels were recorded in juice from trees treated with SA + IAA, which was statistically similar to SA + K + Zn and K-treated trees (Figure 2D). Among the orchards, the Sargodha orchard had the lowest citric acid content, followed by Layyah and Faisalabad.

Ascorbic Acid Content: The ascorbic acid concentration in fruit juice was significantly enhanced by the foliar application of PGRs and nutrients (Table 2, Figure 3A). The highest ascorbic acid content was found in juice from trees treated with SA + K + Zn, which was statistically comparable to juice from trees treated with SA, SA + K, and K alone. The control group (untreated trees) had the lowest ascorbic acid levels. Site-wise, ascorbic acid content varied, with the highest levels found in juice from the Sargodha orchard, followed by Layyah and Faisalabad (Figure 3A).

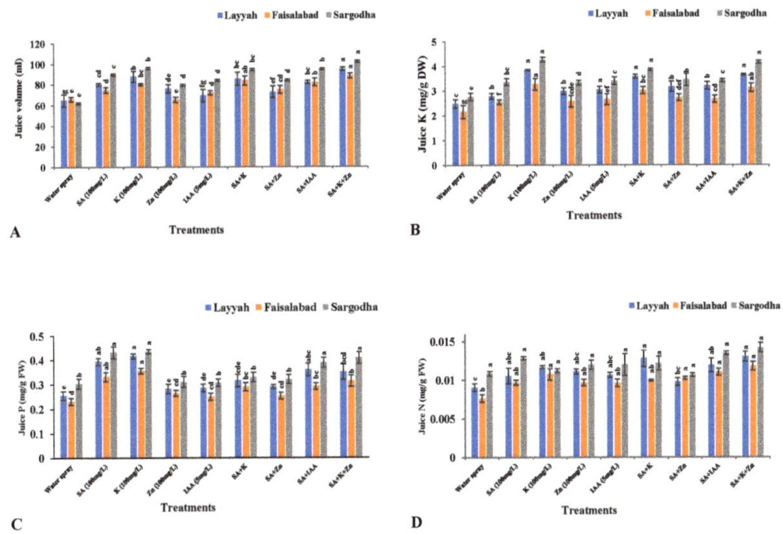

Figure 3. Effect of foliar application of plant growth regulators (PGRs) and nutrients on fruit juice volume concentration (**A**); fruit juice potassium contents (**B**); fruit juice phosphorus content (**C**); and fruit juice nitrogen contents (**D**) in citrus orchards of selected locations. [Bars sharing similar alphabets did not differ significantly].

TSS/Acid Ratio: The TSS/acid ratio, a crucial quality indicator, was significantly influenced by foliar applications of PGRs and nutrients (Table 2). The highest TSS/acid ratio was observed in juice from plants treated with SA + K + Zn, while the lowest ratio was found in juice from untreated plants (Figure 3D). Among the locations, the Sargodha orchard yielded the highest TSS/acid ratio, while the Faisalabad orchard had the lowest (Figure 3D).

Fruit juice volume: A significant increase in fruit juice volume was observed after the foliar application of PGRs and nutrients (Table 2). The highest juice volume was recorded in fruits from plants treated with SA + K + Zn, while the lowest volume was found in fruits from untreated control plants. There were also significant site-to-site variations ($p < 0.05$) in fruit juice volume, fruits with the Sargodha orchard producing the highest juice volume, followed by Layyah and Faisalabad (Figure 3A).

3.3. Fruit Juice Nutrient Contents

The foliar application of PGRs or nutrients significantly ($p \leq 0.05$) affected nitrogen (N), phosphorus (P), and potassium (K) contents, specifically. The highest fruit juice potassium (K) contents were observed in fruits from trees sprayed with K alone, closely followed by those treated with SA + K + Zn and SA + K (Figure 3B). In contrast, the lowest K content was found in juice from untreated trees. Site-to-site variations were also significant, with fruits from the Sargodha orchard having the highest juice K content, while those from the Faisalabad orchard had the lowest juice K content (Figure 3B).

Fruit juice phosphorus (P) contents in fruit juice were similarly enhanced by foliar applications of PGRs and nutrients (Table 2, Figure 3C). The highest fruit juice P contents were found in juice from fruits of trees sprayed with K alone, with statistically similar values observed for trees treated with SA. The lowest P contents were recorded in juice from untreated trees (Figure 3C). Site-specific differences were also significant, with the highest P contents found in fruit juice from the Sargodha orchard, followed by Layyah and Faisalabad orchards (Figure 4B).

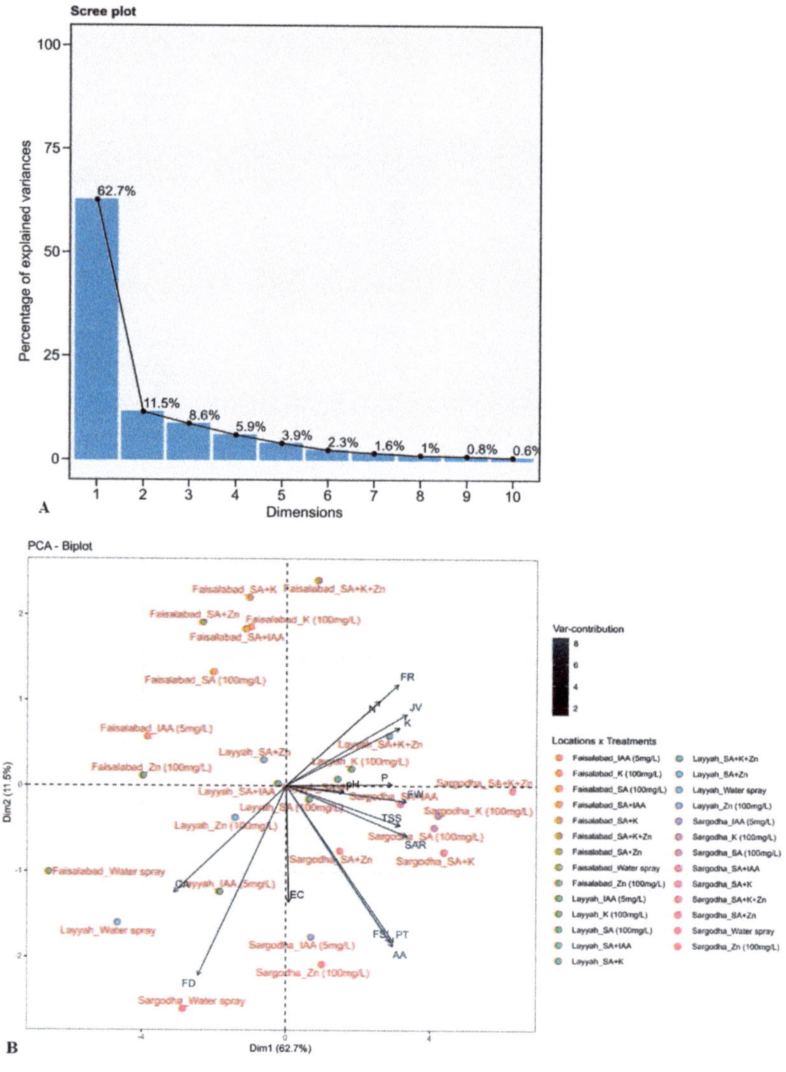

Figure 4. (**A**) Scree plot representing the contribution of principal components towards cumulative variability and (**B**) depicting the principal component analysis of studied traits under varying treatments alone or in combination.

Nitrogen (N) contents in fruit juice also increased significantly following the foliar application of PGRs and nutrients. The maximum N contents were recorded in fruit juice from trees treated with SA + K + Zn, followed by SA + IAA and SA + K. The lowest N contents were found in juice from untreated trees, closely followed by those treated with SA + Zn and SA alone (Figure 3D). Similarly to other parameters, juice N content varied across locations, with the highest levels found in juice from fruits of the Sargodha orchard, followed by Layyah and Faisalabad.

3.4. Principal Component Analysis

To better understand the variation in fruit quality traits resulting from the different individual and combined treatments, a Principal Component Analysis (PCA) was conducted. The analysis focused on the first two principal components (PCs) due to their higher contribution to the total cumulative variability. Specifically, PC-1 accounted for 62.7% of the total variability, while PC-2 contributed an additional 11.5%. These two components together captured a substantial portion of the data's variation, making them key in interpreting the results (Figure 4A).

PC-1 was primarily associated with several important fruit quality traits, including nitrogen content (N), juice volume (JV), total soluble solids (TSS), fruit weight (FW), peel thickness (PT), and fruit shape index (FSI). A strong, synergistic positive interaction was observed among these traits, indicating that improvements in one were generally accompanied by improvements in others. For instance, treatments that increased fruit weight often also enhanced juice volume, nitrogen content, and fruit shape index, suggesting a coherent pattern of positive relationships among these key parameters (Figure 4B). On the other hand, PC-2 was mainly characterized by traits such as citric acid (CA) and fruit drop (FD). Interestingly, these traits exhibited antagonistic relationships with the fruit quality traits in PC-1. This suggests that while certain treatments enhanced juice volume and other positive attributes, they might simultaneously reduce citric acid content and fruit drop (Figure 4B).

Regarding the site-specific effects, the combined foliar supplementation of salicylic acid (SA), potassium (K), and zinc (Zn) at the Layyah site demonstrated a highly favorable interaction in boosting pivotal traits such as nitrogen content (N), juice volume (JV), and potassium content (K) in the juice. This highlights Layyah as a site where the combination of SA + K + Zn was particularly effective in improving the quality of the fruits (Figure 4B). At the Sargodha site, a different set of trends emerged. The foliar application of SA + Indole acetic acid (IAA) significantly enhanced fruit weight (FW), underscoring the effectiveness of this treatment in promoting heavier fruits. Additionally, the application of SA alone resulted in a marked improvement in the total soluble solids (TSS) content of the fruit juice, emphasizing the role of SA in enhancing the sweetness and overall quality of the fruit produced at this site (Figure 4B).

These findings underscore the importance of both the specific combination of treatments and orchard location in influencing fruit quality traits. The PCA provided a clear separation of these relationships, helping clarify how certain traits were positively or negatively influenced by the treatments in different contexts (Figure 4).

3.5. GGE Biplot Analysis

The GGE biplot analysis, introduced by Yan and Wu [24], was employed to evaluate the effects of different treatments on citrus fruit cultivation across multiple sites, including Sargodha, Layyah, and Faisalabad. This analysis helps to clearly understand how effectively each treatment improves physiological traits and fruit quality. Moreover, GGE biplot analysis is useful for identifying the most suitable site for citrus cultivation. The average environmental coordinate (AEC) system was also used to assess the overall performance of each treatment applied. In the GGE biplot, the first principal component (PC-1), which accounts for 72.27% of the total variance, represents fruit quality, while the second principal component (PC-2), contributing 19.3% of the variance, reflects the stability of each treatment across the different citrus orchards. The results indicated that the foliar application of SA + K was most effective in enhancing fruit quality at the Sargodha site. In contrast, at the Faisalabad and Layyah sites, the combination of SA + K + Zn significantly improved fruit quality and juice concentration (Figure 5A).

The mean vs. stability analysis of the GGE biplot revealed that treatments positioned closest to the mean line contributed the most to improving citrus fruit quality. The exogenous application of SA + K at Sargodha played a crucial role in enhancing fruit attributes, as it was located near the arrow on the line. Conversely, the lowest impact on fruit quality was observed in plants sprayed with water across all sites (Figure 5B). The dotted line on the biplot represents the mean line, illustrating the stability of each treatment across the different citrus orchards. The analysis further revealed that the SA + K + Zn treatment was the most effective in improving fruit quality traits at the Faisalabad and Layyah orchards. However, at the Sargodha orchard, the SA + K treatment yielded the most desirable fruit quality outcomes. In contrast, water spray had a minimal impact on improving fruit quality parameters at any of the selected sites (Figure 5B). Additionally, foliar supplementation with SA + K proved to be the most effective in enhancing citrus fruit quality, followed closely by IAA application. However, treatments with SA + Zn, SA alone, Zn, and IAA individually did not produce the desired improvements in fruit quality and juice volume.

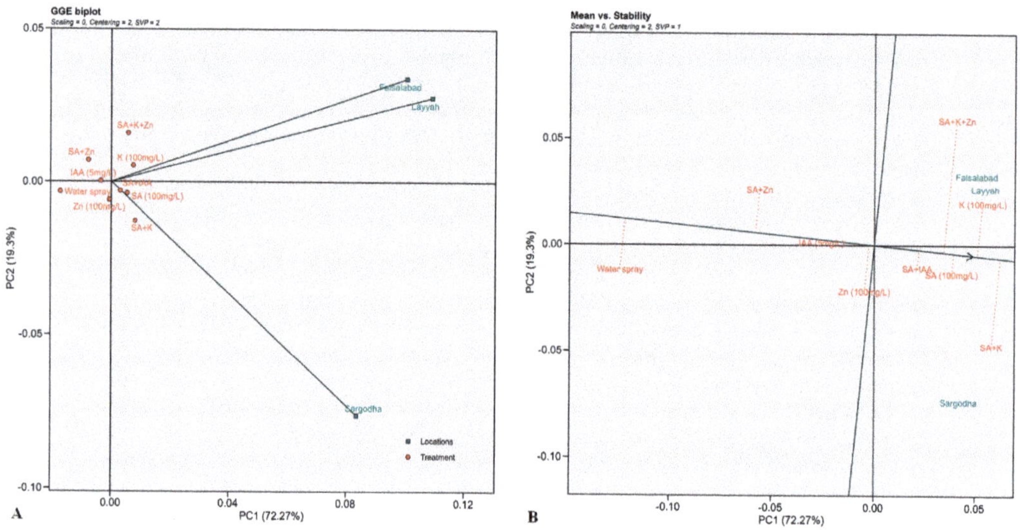

Figure 5. (**A**) GGE plot among two principal components (PC-1 and PC-2) demonstrating the stability of each treatment in selected sites. (**B**) "Mean vs. Stability" representing GGE biplot unveiling the effectiveness of each treatment at multiple locations.

3.6. GGE Biplot (Which-Won-Where

The GGE biplot (which-won-where, constructed using R software, identifies the best-performing treatments at specific sites. The polygon representation helps differentiate crossover from non-crossover treatments [25]. This analysis is useful for selecting suitable environmental locations for 'Kinnow' cultivation, as well as for identifying PGRs or nutrients that can improve citrus fruit and juice quality and yield in Pakistan. Treatments within the same sector indicate the best choices for each orchard. Each sector of the biplot contains one "winning" treatment at the vertex of the polygon, signifying the treatment most favorable for enhancing citrus fruit quality at the respective location [26]. The biplot shows that the SA + K + Zn and K treatments are in a common sector and are suitable for improving citrus fruit quality in Faisalabad and Layyah. However, the SA + K treatment emerged as the winning treatment for the Sargodha site, proving essential in regulating fruit quality parameters (Figure 6). In contrast, the treatments IAA and Zn did not associate with any sector across the locations. Their effects were minimal and fell below the grand mean of other treatments. However, the proximity of the IAA treatment to the origin of the biplot suggests it has general stability across sites. This scenario illustrates the interaction between treatments and locations, as each site has a specific winning treatment. Positive interactions occurred between treatments and locations within the same sector, while opposite sectors revealed antagonistic interactions [27].

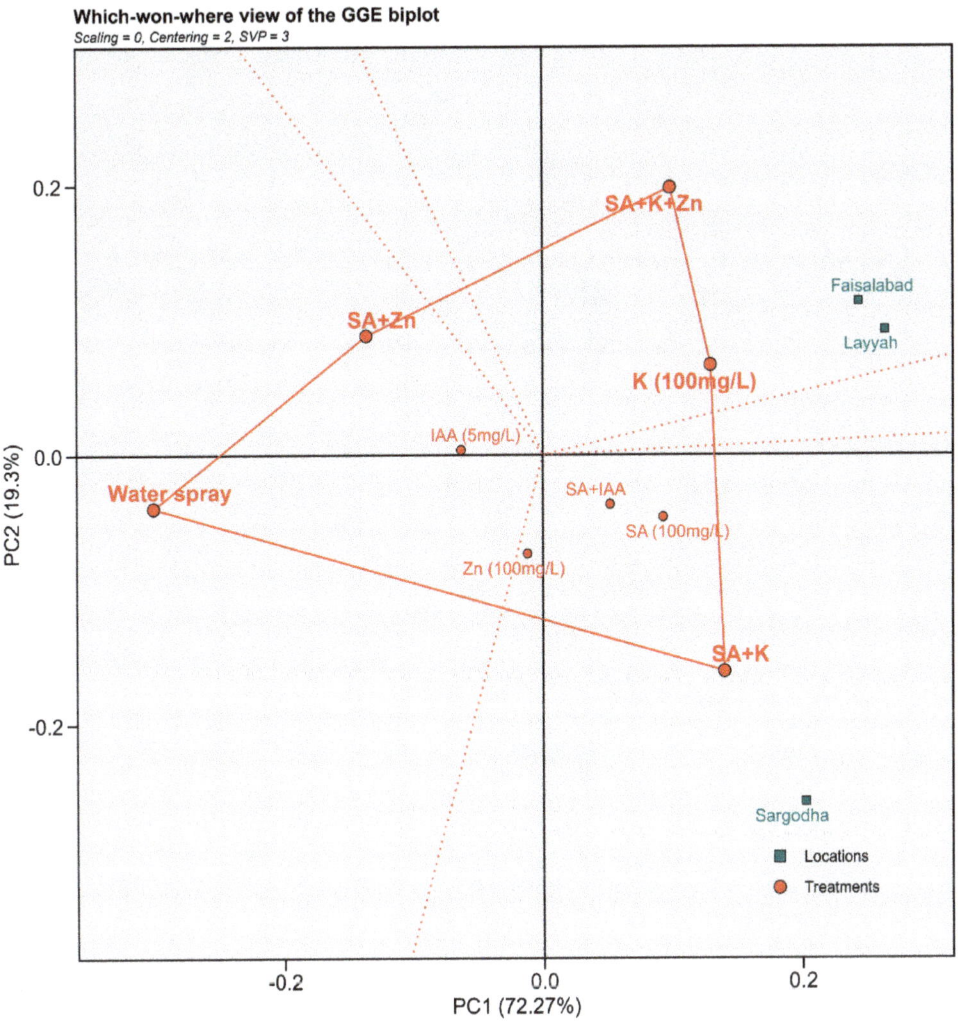

Figure 6. GGE biplot which-won-where representing the interaction between treatments and locations, unraveling the most promising treatment for the particular selected site.

3.7. Multi-Trait Genotype-Ideotype Index (MGIDI)

The MGIDI analysis was conducted separately for each location to determine the most suitable treatment for each selected site. Identifying effective treatments to achieve desired traits in citrus fruits can help farmers maximize citrus yields per hectare. According to this novel index, two treatments, SA + K and SA + K + Zn, were particularly effective in enhancing citrus fruit quality and juice quantity at the Sargodha site (Figure 7A–C). Additionally, Figure 7B indicates that fruit drop is the most consistent factor contributing to the decline in citrus productivity.

However, the Faisalabad site presented a slight variation, where SA + K + Zn and SA + IAA were the key treatments for achieving the desired outcomes. In contrast, at the Layyah site, SA + K and potassium (K) alone were crucial in attaining the desired traits in citrus fruits. Overall, our findings suggest that the Sargodha site is highly conducive to citrus cultivation due to its favorable traits, including lower fruit drop and citric acid levels compared to the other two locations. Notably, the SA + K + Zn treatment significantly reduced fruit drop at the Sargodha site compared to other treatments.

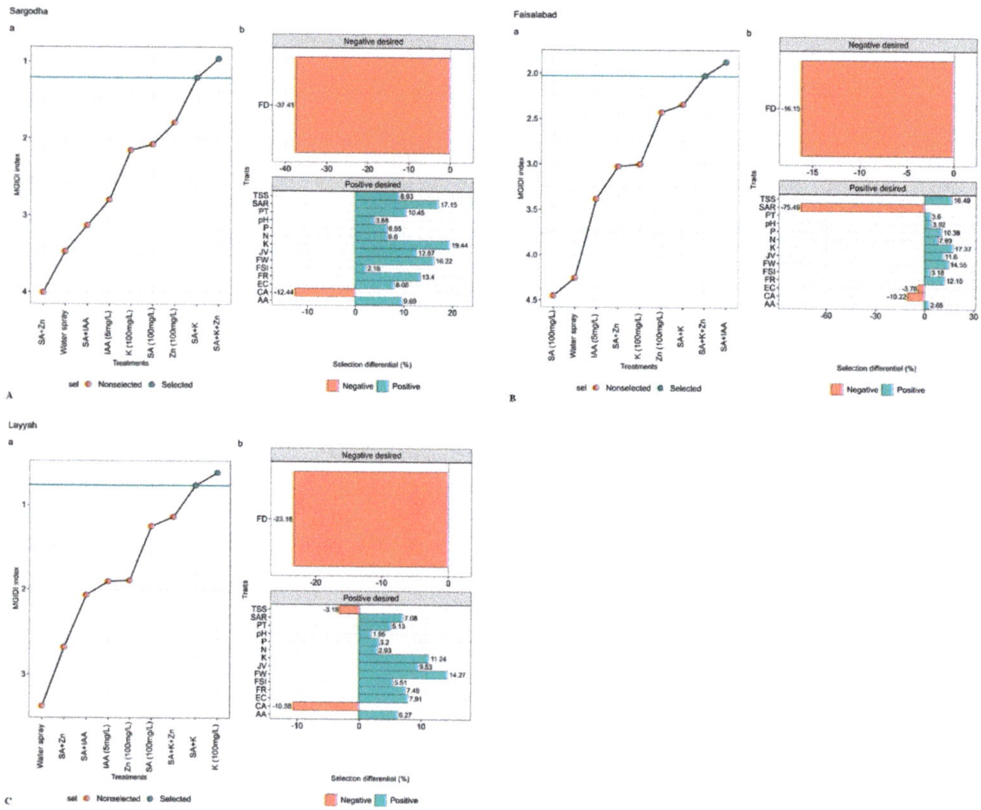

Figure 7. Multi-trait genotype ideotype index (MGIDI) for each selected location revealed promising treatments for particular sites. (**A–C**) (**a**) shows different treatments applied on the x-axis, and the y-axis indicates the MGIDI index values. Treatments below the horizontal threshold line are considered selected treatments highlighted with green color, while those above are non-selected treatments indicated with red color. Selected treatments are closer to the ideotype, revealing better performance based on the desired traits. (**A–C**) (**b**) demonstrates the selection differential for each trait. The top panel depicted the negative desired selection of FD, while the below panel demonstrated the positive and negative bars among the traits analyzed at three studied locations.

4. Discussion

The foliar application of potassium (K) and zinc (Zn) has been shown to significantly increase fruit yield by enhancing fruit weight and size in citrus plants [28]. The observed increases in fruit weight and shape index (size) with plant growth regulators (PGRs) and nutrient treatments (Figures 1A and 2A) can be attributed to the stimulation of plant growth and fruit development processes. These findings are consistent with those of Trejo et al. [29], who demonstrated that combinations of PGRs can substantially affect fruit weight and size. Regarding fruit retention (Figure 1C), the current results confirm the findings of Vani et al. [30], who reported that foliar applications of growth regulators and nutrients significantly improve fruit retention by reducing pre-harvest fruit drop (Figure 1C,D). In this study, trees treated with SA + K + Zn exhibited the minimum fruit drop, while those in control conditions had the maximum fruit drop across all orchards (Figure 1C). However, fruit drop varied by location, with the lowest levels recorded in Sargodha compared to Layyah and Faisalabad. The combination of K, Zn, IAA, and other treatments proved effective in reducing fruit drop until final harvest, ultimately boosting fruit yield. Similar results were reported by Ferrara et al. [31], who found that the foliar applications of SA, 2,4-D, K, and Zn increased fruit retention and decreased fruit drop in citrus. Additionally, the foliar supplementation of SA + K + Zn can significantly improve fruit retention per tree by reducing fruit drop [32]. Citrus fruit drop was a major issue across all orchards, primarily caused by factors such as limited water availability during flowering/fruiting, temperature fluctuations, and nutrient deficiencies, leading to hormonal imbalances [33]. Citrus is highly nutrient-sensitive, and any imbalance can negatively impact fruit quality [34], as seen in the control plants (Figure 1D). Thus, the proper care of plant health is crucial to support fruit development and reduce fruit drop, ultimately increasing yield.

The application of Zn, K, and PGRs significantly enhanced fruit weight across all orchards (Figure 2A), with the highest fruit weight observed in plants treated with SA + K + Zn. A previous report conducted by Ashraf et al. [35] also attributed increases in fruit weight to the combined use of SA, K, and Zn, while Shah et al. [36] confirmed the role of PGRs and nutrient management in this regard. These findings suggest that such treatments are essential for improving both fruit yield and quality. Peel thickness was not significantly affected by the treatments (Figure 1B), although fruits from SA + K + Zn-treated plants had thicker peels across all orchards, enhancing shelf life as thicker peels help fruits stay fresh longer. Consumers prefer such fruits for their extended freshness. Similarly, the non-significant influence of PGRs and nutrients on peel thickness was reported by Anwar et al. [6], although Yao et al. [37] noted that Zn application can increase peel thickness.

The foliar application of PGRs (SA and IAA) and nutrients (K and Zn) significantly increased juice volume across all orchards (Figure 2B). The combination of SA + K + Zn was particularly effective in enhancing juice content, as confirmed by Majeed et al. [38], who reported similar findings with Zn and K application, and Jain et al. [39] observed substantial juice volume increases with PGRs and nutrient treatments. This improvement is crucial for citrus growers, as juice factories prioritize fruits with a higher juice content, leading to better economic returns. The juice pH was only slightly affected by the treatments, with fruit from SA + K + Zn-treated plants showing a marginally higher pH (Figure 2C). However, the electrical conductivity (EC) of the juice increased significantly following the foliar application of PGRs and nutrients at all orchards (Figure 2D). Similarly, juice total soluble solids (TSS) increased across all orchards, with the highest TSS observed in fruits from trees treated with Zn, IAA, and SA + K + Zn (Figure 3A). These results are consistent with earlier studies on citrus [22], as well as research on guava [40], peach [41], pomegranate [42], and plum [43], all of which demonstrated the positive effects of PGRs and nutrients on juice quality. Citric acid concentration decreased with Zn or Zn + K application across all orchards, with the lowest levels recorded in plants treated with SA + K + Zn, while control plants had the highest citric acid concentration (Figure 3B). Additionally, the foliar application of PGRs and nutrients increased the ascorbic acid content of the juice (Figure 3C). The TSS/acid ratio also increased with the application of Zn and K, either alone or in combination, across all experimental sites (Figure 3D). These findings are supported by the literature, such as [44], who found that applying 2,4-D, GA3, and NAA at flowering improved juice ascorbic acid content, EC, TSS/acid ratio, yield, and quality in 'Kinnow' mandarin. Similarly, previous observations carried out by [45,46] demonstrated that PGR and nutrient foliar sprays enhance juice TSS, ascorbic acid content, and the TSS/acid ratio.

In the present study, juice nutrient contents (N, P, K) were significantly improved by the application of PGRs and nutrients (K and Zn). Potassium plays a vital role in several physiological processes in citrus fruit, including cell division, protein synthesis, growth, organic acid neutralization, and the formation of starch and sugars [47]. Due to its importance, citrus fruit contains relatively high potassium levels compared to other nutrients [48]. Potassium also enhances fruit size, flavor, and color [49]. These findings are consistent with previous studies, such as [50], which showed that foliar application of Zn and K increased the uptake of macronutrients (N, K, P, Ca) and micronutrients (Fe, B, Zn, Mn) in *Citrus reticulata*. The synergistic interactions between elements like P, K, N, Ca, and Zn may account for the observed increase in nutrient levels following the foliar application of plant growth regulators (SA and IAA) and nutrients (K and Zn). This is supported by previous studies, including [51], which reported an increase in juice N, P, and K content in 'Kinnow' mandarin, and [52], which found enhanced mineral content in Aonla fruit. The current findings further confirm the effectiveness of PGR and nutrient applications, resulting in significant increases in juice potassium (Figure 3A), phosphorus (Figure 3B), and nitrogen (Figure 3C). Among the orchards, the fruits from Sargodha had the highest N, P, and K content, followed by those from Layyah and Faisalabad.

Statistical analyses such as ANOVA (Table 2), PCA (Figure 4), GGE biplot (Figure 5), and the MGIDI (Figure 6) are essential in determining the most suitable treatments for citrus cultivation at the selected sites. PCA helps identify variability and interactions among traits, while the GGE biplot reveals superior treatments across different locations [53]. The novel MGIDI, in particular, assists in selecting the best treatment for achieving higher citrus productivity. Our findings related to juice quality and quantity parameters, coupled with site-specific analysis, enable the identification of the optimal location for producing superior-quality fruits, considering the variation in traits influenced by both site and treatment combinations.

The use of plant growth regulators (SA and IAA) and nutrients (K and Zn) led to significant improvements in citrus fruit characteristics, including shape index (fruit size), peel thickness, fruit retention, reduced fruit drop, fruit weight, juice volume, and juice quality parameters such as EC, TSS, ascorbic acid, and TSS/acid ratio [54]. Variations in these parameters between sites were attributed to differences in soil texture, soil nutrient status, climatic conditions, and irrigation water quality [55]. However, at all selected sites, the combined application of growth regulators and nutrients (SA, Zn, K) consistently resulted in significant enhancements across all parameters.

5. Conclusions

This study, conducted across three citrus orchards in Punjab, Pakistan (Layyah, Faisalabad, and Sargodha), concludes that the foliar application of potassium (K) and zinc (Zn) at 100 mg/L each, combined with salicylic acid (SA) at 100 mg/L and indole acetic acid (IAA) at 5 mg/L, significantly enhanced citrus tree performance regarding fruit and juice quality parameters. Applied either individually or in combination at key growth stages, these treatments reduced fruit drop, leading to improved fruit retention. Significant improvements were also observed in fruit weight, size, peel thickness, juice volume, and quality metrics such as electrical conductivity (EC), total soluble solids (TSS), ascorbic acid content, and the TSS/acid ratio. Additionally, juice nutrient levels, particularly nitrogen (N), phosphorus (P), and potassium (K), increased due to these treatments. Statistical analyses, including PCA and GGE biplot, identified the most stable and effective treatment, with the SA + K + Zn combination consistently yielding the best results across citrus orchards. While site-specific variations were influenced by differences in soil, climate, and irrigation, this treatment was consistently effective, making it a recommended strategy for boosting citrus fruit yield and juice quality, ultimately enhancing economic returns for growers.

Author Contributions: I.A., M.Y.A. and M.Y.S.; designed the experiments. I.A., M.Y.S. and G.F.; carried out the major portion of the experiments. M.Y.S.; carried out all statistical analyses. I.A. and G.F.; carried out the morphological aspects of the experiment. I.A., M.A. and M.S.; performed the nutritional analysis. I.A., M.S. and G.F.; participated in yield data collection. M.S., M.A. and M.Y.A.; supervised the experiments. M.Y.S., W.J. and G.F.; drafted the manuscript. M.S., W.J., M.Y.S. and M.A. revised the manuscript. All authors have read and agreed to the published version of the manuscript.

Funding: This research received no external funding.

Data Availability Statement: The data presented in this study are available from the corresponding authors upon reasonable request.

Acknowledgments: This research is part of PhD research conducted at NIAB, Faisalabad, Pakistan. We are grateful to the Pakistan Science Foundation (PSF), Pakistan Institute of Engineering and Applied Sciences (PIEAS), Islamabad, and the Department of Biological Sciences at the Nuclear Institute for Agriculture and Biology (NIAB), Faisalabad for providing a framework for the research program. We sincerely acknowledge the Pakistan Science Foundation for the financial support and appreciate the farmers of the Layyah, Sargodha, and Faisalabad sites who gave us their kinnow fields where we conducted our research. We sincerely thank each member of the study team as well as everyone else who helped; we are incredibly grateful for them.

Conflicts of Interest: The authors declare no conflicts of interest.

References

1. Kaur, H.; Singh, G. Recent trends in Citrus (*Citrus* spp.) peel utilization: A review. *Plant Archives (09725210)* **2021**, *21*, 88–97. [CrossRef]
2. Cheema, I.A.; Jamali, H.K. Growth of citrus fruits in Pakistan. *Amazon. Investig.* **2020**, *9*, 74–81. [CrossRef]
3. Zeng, L.; Li, P.; Yu, Z.; Nie, Y.; Li, S.; Gao, G.; Huang, D. Spatiotemporal characteristics and influencing factors of water resources' green utilization efficiency in China: Based on the EBM model with undesirable outputs and SDM model. *Water* **2022**, *14*, 2908. [CrossRef]
4. Nawaz, R.; Nadeem, M.; Abbas, T.; Shah, A.A. Mineral Characterization, fruit and leaf physiology of Citrus reticulata comparative to citrus canker stress in Sargodha, Punjab. 2024. Available online: https://assets-eu.researchsquare.com/files/rs-3912569/v1/8b6a1144-4f1b-49c8-b5a9-0024028d810a.pdf?c=1715722163 (accessed on 20 November 2024).
5. Kalcsits, L.; Lotze, E.; Tagliavini, M.; Hannam, K.D.; Mimmo, T.; Neilsen, D.; Neilsen, G.; Atkinson, D.; Casagrande Biasuz, E.; Borruso, L. Recent achievements and new research opportunities for optimizing macronutrient availability, acquisition, and distribution for perennial fruit crops. *Agronomy* **2020**, *10*, 1738. [CrossRef]
6. Anwar, S.; Ashraf, M.Y.; Saleem, M.; Shafiq, F.; Khan, N.; Khan, R.A.; Farid, G.; Ashraf, M. Integrated hormonal and nutrient management promote fruit retention and quality traits of Citrus reticulata. *J. Plant Nutr.* **2022**, *46*, 83–100. [CrossRef]
7. Gosai, S.; Adhikari, S.; Khanal, S.; Poudel, P.B. Effects of plant growth regulators on growth, flowering, fruiting and fruit yield of cucumber (*Cucumis sativus* L.): A review. *Arch. Agric. Environ. Sci.* **2020**, *5*, 268–274. [CrossRef]
8. Sardans, J.; Peñuelas, J. Potassium control of plant functions: Ecological and agricultural implications. *Plants* **2021**, *10*, 419. [CrossRef]
9. Rahman, S.; Iqbal, M.; Husen, A. Medicinal plants and abiotic stress: An overview. In *Medicinal Plants: Their Response Abiotic Stress*; Springer: Singapore, 2023; pp. 1–34.
10. Khan, A.S.; Ali, S.; Anwar, R.; Rehman, R.N.U. Preharvest Factors That Influence Postharvest Losses of Citrus Fruits. In *Citrus Production*; CRC Press: Boca Raton, FL, USA, 2022; pp. 319–343.
11. Yang, H.; Fang, R.; Luo, L.; Yang, W.; Huang, Q.; Yang, C.; Hui, W.; Gong, W.; Wang, J. Uncovering the mechanisms of salicylic acid-mediated abiotic stress tolerance in horticultural crops. *Front. Plant Sci.* **2023**, *14*, 1226041. [CrossRef]
12. Godoy, F.; Olivos-Hernández, K.; Stange, C.; Handford, M. Abiotic stress in crop species: Improving tolerance by applying plant metabolites. *Plants* **2021**, *10*, 186. [CrossRef]
13. Nage, S.; Kulkarni, U.S.; Undirwade, D.B.; Meshram, N.M.; Sant, S. The species composition, richness and diversity studies of fruit flies captured from the orchard of Nagpur mandarin. *Sci. Rep.* **2023**, *13*, 21872. [CrossRef]
14. Bacelar, E.; Pinto, T.; Anjos, R.; Morais, M.C.; Oliveira, I.; Vilela, A.; Cosme, F. Impacts of climate change and mitigation strategies for some abiotic and biotic constraints influencing fruit growth and quality. *Plants* **2024**, *13*, 1942. [CrossRef] [PubMed]
15. Dutta, S.K.; Gurung, G.; Yadav, A.; Laha, R.; Mishra, V.K. Factors associated with citrus fruit abscission and management strategies developed so far: A review. *New Zealand J. Crop Hortic. Sci.* **2023**, *51*, 467–488. [CrossRef]
16. Saini, R.K.; Ranjit, A.; Sharma, K.; Prasad, P.; Shang, X.; Gowda, K.G.M.; Keum, Y.-S. Bioactive compounds of citrus fruits: A review of composition and health benefits of carotenoids, flavonoids, limonoids, and terpenes. *Antioxidants* **2022**, *11*, 239. [CrossRef] [PubMed]
17. Olsen, S.R. *Estimation of Available Phosphorus in Soils by Extraction with Sodium Bicarbonate*; US Department of Agriculture, Circular # 939: Washington DC, USA, 1954.
18. Simard, R.R. Ammonium acetate-extractable elements. *Soil Sampl. Methods Anal.* **1993**, *1*, 39–42.
19. Soltanpour, P.N.; Schwab, A.P. A new soil test for simultaneous extraction of macro-and micro-nutrients in alkaline soils. *Commun. Soil Sci. Plant Anal.* **1977**, *8*, 195–207. [CrossRef]
20. Estefen, S.F.; Gurova, T.; Leontiev, A. Redistribution of the residual welding stresses. *Mar. Syst. Ocean Technol.* **2013**, *8*, 95–100. [CrossRef]
21. Brewer, S.; Plotto, A.; Bai, J.; Crane, J.; Chambers, A. Evaluation of 21 papaya (*Carica papaya* L.) accessions in southern Florida for fruit quality, aroma, plant height, and yield components. *Sci. Hortic.* **2021**, *288*, 110387. [CrossRef]
22. Shani, M.Y.; Ahmed, S.R.; Ashraf, M.Y.; Khan, Z.; Cocozza, C.; De Mastro, F.; Gul, N.; Pervaiz, S.; Abbas, S.; Nawaz, H. Nano-Biochar Suspension Mediated Alterations in Yield and Juice Quality of Kinnow (*Citrus reticulata* L.). *Horticulturae* **2023**, *9*, 521. [CrossRef]

23. Ashraf, M.Y.; Yaqub, M.; Akhtar, J.; Khan, M.A.; Ali-Khan, M.; Ebert, G. Control of excessive fruit drop and improvement in yield and juice quality of Kinnow (*Citrus deliciosa* × *Citrus nobilis*) through nutrient management. *Pak. J. Bot* **2012**, *44*, 259–265.
24. Yan, W.; Wu, H.X. Application of GGE biplot analysis to evaluate Genotype (G), Environment (E), and G × E interaction on *Pinus radiata*: A case study. *New Zealand J. For. Sci.* **2008**, *38*, 132–142.
25. Sankar, S.M.; Singh, S.P.; Prakash, G.; Satyavathi, C.T.; Soumya, S.L.; Yadav, Y.; Sharma, L.D.; Rao, A.R.; Singh, N.; Srivastava, R.K. Deciphering genotype-by-environment interaction for target environmental delineation and identification of stable resistant sources against foliar blast disease of pearl millet. *Front. Plant Sci.* **2021**, *12*, 656158. [CrossRef] [PubMed]
26. Ruswandi, D.; Syafii, M.; Maulana, H.; Ariyanti, M.; Indriani, N.P.; Yuwariah, Y. GGE biplot analysis for stability and adaptability of maize hybrids in western region of Indonesia. *Int. J. Agron.* **2021**, *2021*, 2166022. [CrossRef]
27. Chen, L.; Jiang, Y.; Liang, C.; Luo, Y.; Xu, Q.; Han, C.; Zhao, Q.; Sun, B. Competitive interaction with keystone taxa induced negative priming under biochar amendments. *Microbiome* **2019**, *7*, 77. [CrossRef] [PubMed]
28. Rohoma Tahir, M.A.; Bilal, H.M.; Saeed, M.S. Impact of foliar application of Zn on growth yield and quality production of citrus: A review. *Ind. J. Pure App. Biosci* **2020**, *8*, 529–534. [CrossRef]
29. Trejo, E.J.O.; Brizzolara, S.; Cardillo, V.; Ruperti, B.; Bonghi, C.; Tonutti, P. The impact of PGRs applied in the field on the postharvest behavior of fruit crops. *Sci. Hortic.* **2023**, *318*, 112103. [CrossRef]
30. Vani, N.U.; Bhagwan, A.; Kumar, A.K.; Sreedhar, M.; Sharath, S.R. Effect of Pre-Harvest Sprays of Plant Growth Regulators and Micronutrients on Fruit Set, Fruit Drop and Fruit Retention of Guava (*Psidium guajava* L.) cv. Lucknow-49. *Ind. J. Pure App. Biosci* **2020**, *8*, 254–261. [CrossRef]
31. Ferrara, G.; Palasciano, M.; Sarkhosh, A.; Cossio, F.; Babu, K.D.; Mazzeo, A. Orchard establishment and tree management. In *The Pomegranate: Botany, Production and Uses*; CAB International: Wallingford, UK, 2021; pp. 247–284.
32. Al-Saif, A.M.; Ali, M.M.; Ben Hifaa, A.B.S.; Mosa, W.F.A. Influence of spraying some biostimulants on yield, fruit quality, oil fruit content and nutritional status of olive (*Olea europaea* L.) under salinity. *Horticulturae* **2023**, *9*, 825. [CrossRef]
33. Kumar, R.; Kumar, V. Physiological disorders in perennial woody tropical and subtropical fruit crops: A review. *Indian J. Agric. Sci.* **2016**, *86*, 703–717. [CrossRef]
34. Nagdeve, S.S.; Raut, M.M.; Balpande, S.S.; Dahale, M.H.; Dhale, R.B.; Dolaskar, U.B.; Manekar, M.R. Assessment of Leaf Nutrient and Fertility Norms for Sweet Orange Under Black Soil in Nagpur District, Maharashtra, India. Available online: https://discovery.researcher.life/article/assessment-of-leaf-nutrient-and-fertility-norms-for-sweet-orange-under-black-soil-in-nagpur-district-maharashtra-india/ca699557b4e33e4d957a266e743fcac2 (accessed on 20 November 2024).
35. Ashraf, M.Y.; Ashraf, M.; Akhtar, M.; Mahmood, K.; Saleem, M. Improvement in yield, quality and reduction in fruit drop in kinnow (*Citrus reticulata Blanco*) by exogenous application of plant growth regulators, potassium and zinc. *Pak. J. Bot* **2013**, *45*, 433–440.
36. Shah, S.H.; Islam, S.; Alamri, S.; Parrey, Z.A.; Mohammad, F.; Kalaji, H.M. Plant growth regulators mediated changes in the growth, photosynthesis, nutrient acquisition and productivity of mustard. *Agriculture* **2023**, *13*, 570. [CrossRef]
37. Yao, H.-L.; Hu, X.-Z.; Li, S.-B.; Wang, H.-T.; Bai, X.-B.; Yang, C. Improvement in Tensile Strength of Mg Alloy After Immersion in SBF by Cold-Sprayed Zn Coating. *J. Therm. Spray Technol.* **2024**, *33*, 869–881. [CrossRef]
38. Majeed, A.; Rashid, I.; Niaz, A.; Ditta, A.; Sameen, A.; Al-Huqail, A.A.; Siddiqui, M.H. Balanced use of zn, Cu, fe, and b improves the yield and sucrose contents of sugarcane juice cultivated in sandy clay loam soil. *Agronomy* **2022**, *12*, 696. [CrossRef]
39. Jain, R.; Singh, A.; Giri, P.; Singh, S.P.; Chandra, A. The impact of sequential application of PGR and PGR+ nutrient combination on growth, physiological and yield attributes of sugarcane. *Plant Archives (09725210)* **2024**, *24*, 17–28. [CrossRef]
40. Mosa, W.F.A.; Sas-Paszt, L.; Gornik, K.; Ali, H.M.; Salem, M.Z.M. Vegetative growth, yield, and fruit quality of guava (*Psidium guajava* L.) cv. maamoura as affected by some biostimulants. *Bioresources* **2021**, *16*, 7379. [CrossRef]
41. Ali, I.; Abbasi, N.A.; Hafiz, I. Application of calcium chloride at different phenological stages alleviates chilling injury and delays climacteric ripening in peach fruit during low-temperature storage. *Int. J. Fruit Sci.* **2021**, *21*, 1040–1058. [CrossRef]
42. Khalil, H.A.; El-Ansary, D.O.; Ahmed, Z.F.R. Mitigation of salinity stress on pomegranate (*Punica granatum* L. cv. Wonderful) plant using salicylic acid foliar spray. *Horticulturae* **2022**, *8*, 375. [CrossRef]
43. Kaur, A.; Kaur, N.; Singh, H.; Murria, S.; Jawanda, S.K. Efficacy of plant growth regulators and mineral nutrients on fruit drop and quality attributes of plum cv. Satluj purple. *Plant Physiol. Rep.* **2021**, *26*, 541–547. [CrossRef]
44. Hiteshbhai, R.; Johar, V.; Singh, V. Effect of plant growth regulators on fruit set and quality of kinnow mandarin (*Citrus reticulata* Blanco). *Environ. Ecol.* **2023**, *41*, 7–12.
45. Moradinezhad, F.; Ranjbar, A. Foliar application of fertilizers and plant growth regulators on pomegranate fruit yield and quality: A review. *J. Plant Nutr.* **2024**, *47*, 797–821. [CrossRef]
46. Hussein, A.S.; Ibrahim, R.A.; Eissa, M.A. Exogenous pre-harvest application of abscisic and jasmonic acids improves fruit quality by enhancing sugar synthesis and reducing acidity in pomegranate (*Punica granatum* L. cv. Wonderful). *J. Soil Sci. Plant Nutr.* **2023**, *23*, 2237–2246. [CrossRef]
47. Wang, J.; Lu, Y.; Zhang, X.; Hu, W.; Lin, L.; Deng, Q.; Xia, H.; Liang, D.; Lv, X. Effects of Potassium-Containing Fertilizers on Sugar and Organic Acid Metabolism in Grape Fruits. *Int. J. Mol. Sci.* **2024**, *25*, 2828. [CrossRef] [PubMed]
48. Lu, X.; Zhao, C.; Shi, H.; Liao, Y.; Xu, F.; Du, H.; Xiao, H.; Zheng, J. Nutrients and bioactives in citrus fruits: Different citrus varieties, fruit parts, and growth stages. *Crit. Rev. Food Sci. Nutr.* **2023**, *63*, 2018–2041. [CrossRef] [PubMed]

49. Yener, H.; Altuntaş, Ö. Effects of potassium fertilization on leaf nutrient content and quality attributes of sweet cherry fruits (*Prunus avium* L.). *J. Plant Nutr.* **2021**, *44*, 946–957. [CrossRef]
50. Siddique, M.A.; Saqib, M.; Abbas, G.; Wahab, H.A.; Ahmad, N.; Khalid, M.; Akhtar, J. Foliar and soil-applied micronutrients improve yield and quality of Kinnow (*Citrus reticulata* Blanco). *Pak. J. Agric. Sci.* **2020**, *57*, 1539–1547.
51. Kumar, A.; Singh, H.; Pathania, S. Potassium fertigation improved growth, yield and quality of kinnow mandarin in potassium rich soils. *Commun. Soil Sci. Plant Anal.* **2022**, *53*, 1767–1776. [CrossRef]
52. Kumar, M.; Arya, R.K.; Kumar, M.; Gaur, R.K.; Sharma, S. Evaluation of Aonla Varieties Under Semi-Arid Conditions of Haryana. *Ekin J. Crop Breed. Genet.* **2021**, *7*, 139–144.
53. Rahmati, S.; Azizi-Nezhad, R.; Pour-Aboughadareh, A.; Etminan, A.; Shooshtari, L. Analysis of Genotype-by-Environment Interaction Effect in Barely genotypes Using AMMI and GGE biplot methods. *Heliyon* **2024**, *10*, e38131. [CrossRef]
54. Liaquat, M.; Ali, I.; Ahmad, S.; Malik, A.M.; Ashraf, H.M.Q.; Parveen, N.; Tareen, M.J.; Saeed, T.; Shah, S.H.; Zulfiqar, B. Efficiency of exogenous zinc sulfate application reduced fruit drop and improved antioxidant activity of 'Kinnow'mandarin fruit. *Braz. J. Biol.* **2021**, *83*, e244593.
55. Yüzügüllü, O.; Fajraoui, N.; Liebisch, F. Soil Texture and Ph Mapping Using Remote Sensing and Support Sampling. *IEEE J. Sel. Top. Appl. Earth Obs. Remote Sens.* **2024**, *17*, 12685–12705. [CrossRef]

horticulturae

MDPI

Article

Impact of Soil Management Practices on Soil Culturable Bacteriota and Species Diversity in Central European a Productive Vineyard under Warm and Dry Conditions

Vladimír Šimanský [1,*], Miroslava Kačániová [2], Martin Juriga [1], Natália Čmiková [2], Petra Borotová [3], Elena Aydın [4] and Elzbieta Wójcik-Gront [5]

1 Institute of Agronomic Sciences, Faculty of Agrobiology and Food Resources, Slovak University of Agriculture, 949 76 Nitra, Slovakia; martin.juriga@uniag.sk
2 Institute of Horticulture, Faculty of Horticulture and Landscape Engineering, Slovak University of Agriculture, 949 76 Nitra, Slovakia; miroslava.kacaniova@uniag.sk (M.K.); xcmikova@uniag.sk (N.Č.)
3 AgroBioTech Research Centre, 949 76 Nitra, Slovakia; petra.borotova@uniag.sk
4 Institute of Landscape Engineering, Faculty of Horticulture and Landscape Engineering, Slovak University of Agriculture, 949 76 Nitra, Slovakia; elena.aydin@uniag.sk
5 Department of Biometry, Institute of Agriculture, Warsaw University of Life Sciences—SGGW, 02-787 Warsaw, Poland; elzbieta_wojcik_gront@sggw.edu.pl
* Correspondence: vladimir.simansky@uniag.sk

Abstract: Sustainable management practices are crucial for the longevity of a monoculture vineyard, especially in the context of a changing climate. Therefore, soil management practices in a vineyard (T: tillage, T+FYM: tillage + farmyard manure, G: grass strips, G+NPK1: grass strips + rational rates of NPK, and G+NPK2: grass strips + higher rates of NPK) were tested in a temperate climate of Slovakia (Central Europe) under specific soil conditions (Rendzic Leptosol). We investigated the influence of continuous cropping on soil chemical properties and microbial communities during the dry and warm year of 2022. The results showed that the soil pH was higher by 19%, 21%, 24% and 13% in T, T+FYM, G and G+NPK1, respectively, compared to G+NPK2. The lowest soil organic matter (SOM) content was found in T, and it increased in the following order: T < T+FYM < G+NPK2 < G+NPK1 < G. Similarly, the lowest abundance of soil culturable bacteriota was found in T and it increased in the following order: T < T+FYM = G+NPK2 < G+NPK1< G. Culturable bacteriota was identified using mass spectrometry (MALDI-TOF MS Biotyper). The most numerous species group was *Bacillus*, followed by *Lactobacillus* > *Staphylococcus* > *Pseudomonas*. The most frequently isolated species were *Bacillus megaterium* (16.55%), *Bacillus cereus* (5.80%), *Bacillus thuringiensis* (4.87%), and *Bacillus simplex* (4.37%). Positive relationships between SOM and soil culturable bacteriota were found in the G and G+NPK1 treatments. Temperature also affected soil culturable bacteriota in all soil management practices, most significantly in G+NPK1. Overall, the best scenario for the sustainable management of a productive vineyard is the use of grass strips.

Keywords: *Bacillus*; fertilization; soil organic matter; grass strips; tillage

1. Introduction

The soil is a habitat for an immense variety of plant and animal life, encompassing organisms spanning a wide size range, from microscopic entities that require a powerful microscope to see, to larger forms like earthworms [1]. All of these organisms contribute to the composition of soil organic matter (SOM) [2]. SOM in its broadest sense, encompasses all of the organic materials found in soils irrespective of their origin or state of decomposition [2]. Every handful of soil contains billions of organisms, spanning nearly every living phylum [3]. Soil microorganisms, including their numbers and biodiversity, significantly influence soil quality, affecting its use and fertility [4]. Through a wide array of activities, these organisms enhance the soil's productivity. For example, they actively participate in cycles and transformations of plant-nutrient elements, facilitating their conversion from one form to another [1,5,6]. Bacteria play a significant role in the microbial communities associated with plants by colonizing the roots and rhizosphere [7]. This association helps plants overcome dry periods [8], improves soil structure [9], etc. Soil microbes are also indicators of soil quality [10], given their involvement in essential biochemical processes crucial for soil's environmental and ecological functions. This role could be pivotal in agriculture's pursuit of sustainable food security [11].

Climate change, a pressing environmental concern, has significant impacts on regional ecosystems. For instance, in Central Europe, climate change triggers alternating periods of extreme drought and heavy precipitation [12]. These fluctuations cause damage to property and human health and contribute to overall environmental degradation, including soil erosion. Soil microorganisms have their own specific requirements for life and depending on the suitability of the environmental conditions, they are able to carry out certain soil processes. For instance, for optimal mineralization of organic matter, microorganisms require a neutral soil pH, a soil temperature in the range of 25–30 °C, and a soil environment humidity of around 60% of the full water capacity [13]. Soil microorganisms facilitate the humification of organic matter under optimal conditions, requiring a neutral pH, sufficient sources of P, Mg, and N, balanced soil moisture, ample organic materials, and elevated temperatures [13,14]. Furthermore, nitrogen transformations in soil are influenced by the activity and species composition of microorganisms, organic matter quality, and environmental soil conditions [6,15]. In addition, climate change, global warming, and intensive agriculture are altering soil enzyme activities, which play a pivotal role in the decomposition of organic matter and the global cycles of carbon (C), phosphorus (P), and nitrogen (N), also serving as indicators of soil health and fertility [16]. The impact of climatic change on microbial function in soil has been well-reviewed [17–21].

Soil management practices directly influence soil microorganisms [1,4,11] by creating optimal conditions that support their growth and survival. However, the sustainability and health of terrestrial ecosystems depend heavily on soil quality. Therefore, the degree of restoration success can be evaluated based on the fertility and health of the soil. Soil physicochemical and biological indicators of vegetation restoration have been measured in earlier research [22]. Fertility and soil function depend on microorganisms [23,24]. Lundquist et al. [25], argued that microbial populations in soil break down organic matter, storing and recycling nutrients. Thus, they have an impact on plant community growth, either directly or indirectly [26]. The functional recovery of terrestrial ecosystems is significantly influenced by microbially driven soil bioprocesses [27]. For example, soil tillage reduces the communities of microorganisms and disturbs the hyphae network of microscopic fungi [28]. Monoculture reduces the diversity of microorganism species [29]. The addition of external organic matter to the soil affects qualitative and quantitative changes in microbial populations, as well as increases or decreases their activity [30]. Farmyard manure (as well as composts) is a source of a large number of diverse soil microorganisms. Several studies [31–33] have confirmed an increase in the population and diversity of microorganisms after the application of manure and compost to the soil. However, soil management practices can also influence changes in the soil environment and thus affect the soil microorganisms indirectly. As a result of different soil management practices, the soil pH, supply of nutrients, and increases/decreases in the soil organic carbon (SOC) stock can change. Different temperature and soil moisture conditions influence the abundance and diversity of microbial species composition. In addition, the supply of external organic matter or organic amendments, such as biochar, changes the chemical and physical properties of soils, and this also results in changes in the soil's biological properties, including the abundance and diversity of soil microorganisms, but also the support of specific microfauna/flora [34]. The relationships between soil management practices, soil microorganisms, and soil properties are very diverse. For example, the application of biochar or farmyard manure results in nutrients entering the soil [35], an increase in the soil's organic substances, aggregation [36], and soil water retention [37]. On the other hand, the addition of organic fertilizers can support the negative priming effect [38]. Also, the application of NPK fertilizers to the soil can be a source of available nutrients that support the abundance of microorganisms; however, their application should be in accordance with the requirements of cultivated plants and balanced with organic fertilization [4]. This is because high doses of NPK fertilizer can cause soil salinization [28] and have a negative effect on soil microorganisms [32]. Generally, growing cover crops or grass [39] is beneficial for increasing the diversity of life in the soil, both in terms of number and species.

The above-mentioned soil management practices are commonly applied to arable soils and vineyards all over the world, including the Slovak Republic. However, their effects on soil properties vary. Modification of soil management practices is more difficult in vineyards than in arable soils. Prior to planting a vineyard, a vine grower must assess the soil and climate conditions to determine the correct soil management practices from the point of view of long-term and sustainable management. Not every soil is suitable for implementing all soil management practices sustainably or for increasing production capacity, particularly given ongoing climate change and the associated rise in average air temperature. As mentioned earlier, the activity and growth of microorganisms are also influenced by temperature and soil moisture [1,3,11]. According to the Intergovernmental Panel on Climate Change [40], since the pre-industrial period, the land surface air temperature has risen nearly twice as much as the global average temperature. These changes manifest in alterations to soil processes, including those involving microorganisms. For this reason, it is necessary to optimize and implement such soil management practices that, in addition to environmental functions, will satisfy the needs of farmers and vine growers.

In the past, bacteria were identified using microbiological approaches, such as evaluating the morphological and biochemical properties of isolates, or more recently, employing molecular biology techniques. However, these methods are not suitable for rapid bacterial identification and monitoring, as they typically require skilled laboratory personnel and can take three to five days [41,42]. As an alternative to molecular identification methods and biochemical tests, MALDI-TOF MS (matrix-assisted laser desorption ionization time-of-flight mass spectrometry) has gained popularity for microbiological identification due to its speed, affordability, and labor-saving qualities. While several studies have explored its suitability for environmental microbiology, there is limited information on its effectiveness in identifying aquatic bacteria directly isolated from the environment. Furthermore, comparisons between identification methods using 16S rRNA gene sequencing and MALDI-TOF MS (Bruker Biotyper) have been conducted across various environments, including environmental mining samples, high-altitude soil samples, regular soil, and fresh vegetables [43,44]. Studies have demonstrated the vast genetic diversity of soil bacteria [45], with many bacterial species existing in soil lineages that lack known cultivated isolates [46]. Techniques such as mass spectrometry greatly facilitate the study of soil bacteriota ecology. We assumed that a significant number of these microorganisms could be cultured using simple, low-tech tools. Therefore, we evaluated the culturability of soil bacteria using a basic growth medium. Our goal was to exceed the commonly observed 5% culturability threshold, which often serves as the upper limit in cultivation research.

Based on the context provided, this study aimed to (1) quantify how soil management practices in vineyards affect soil properties, abundance, and biodiversity of culturable bacteriota, especially during dry and warm conditions; (2) assess dynamic changes in soil properties and the abundance/diversity of culturable bacteriota; and (3) examine the relationships between these factors across different soil management practices in a productive vineyard during a dry and warm year. This study tested the following hypotheses: (H1) Soil parameters, including the abundance and species diversity of culturable bacteriota, will stabilize under each soil management practice. (H2) The climate will influence changes in soil properties. (H3) Changes in soil properties, including culturable bacteriota, under NPK treatments will align with the timing and dosage of fertilizer application; higher rates of NPK are expected to negatively impact these properties, while lower rates may have a positive effect.

2. Materials and Methods

2.1. Site Description and History of Experimental Vineyard

The study was conducted at an experimental vineyard located in Dražovce, a suburb of the Nitra city (Figure 1), within the Nitra wine-growing region of Slovakia in Central Europe (48°21′6.16″ N, 18°3′37.33″ E) in 2022. Dražovce is located between the western slopes of the Tribeč mountain range (built of granitoid rocks and a packet of Mesozoic (Triasic) dolomites) that belongs to the Carpathian Mountains and the Nitra River valley, constituting a part of the Great Danubian Lowland. The slopes of the Tribeč mountain range and higher parts of the Nitra River valley are covered with colluvial deposits built of weathered Carpathian rocks and also with a widespread mantle of Quaternary silty loamy aeolian loess sediments accumulated in periglacial conditions of the last glaciation [47]. The major soil types in the proximity of the study area comprise Eutric Dolomitic Leptosol, Luvic Chernic Phaeozem, Nudiargic Luvisol, Eutric Cambisol, Haplic Calcisol, Vermic Chernozem, and Ekranic Technosol [48]. The region experiences a warm temperate climate classified as fully humid with warm summers (Cfb) according to the Köppen–Geiger classification [49]. The average annual air temperature and average annual precipitation are 10.7 °C and 559 mm, respectively. July is the hottest month with the mean air temperature ranging between 16–18 °C, and January is the coldest month with an average air temperature ranging from −2 to −4 °C. Table 1 presents the monthly precipitation and average air temperature in 2022. These meteorological properties were also compared with climatic normal 1991–2020. Overall, the year 2022 can be classified as warm and very dry.

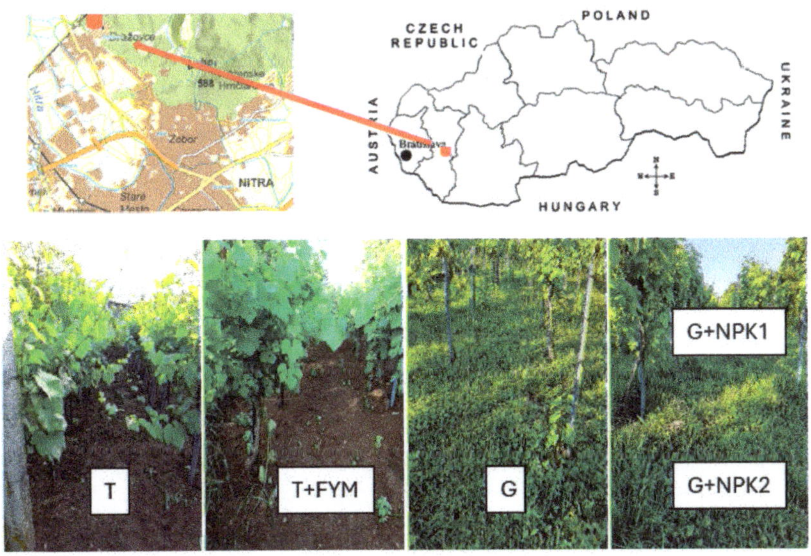

Figure 1. Location of the study area with the studied treatments.

The soil at the experimental site is classified as Rendzic Leptosol [50], which originated from Mesozoic sedimentary rocks, predominantly Cretaceous, Jurassic, and Triassic limestones. The soil is sandy loam in texture (with 57% of sand, 33% of silt, and 10% of clay), and its pH in H_2O is slightly alkaline (7.28). The initial soil properties before the establishment of the experiment were as follows: low soil organic carbon (SOC: 17 g kg^{-1}), fully saturated soil sorption complex (Bs: 99.3%), high cation exchange capacity (CEC: 47.6 cmol kg^{-1}), very low total nitrogen (Nt: 0.11 g kg^{-1}), and a good supply of available P (99 mg kg^{-1}) and K (262 mg kg^{-1}).

Table 1. Monthly precipitation and average air temperature in 2022 (evaluation of standard monthly precipitation and average air temperature are based on long-term averages for the period 1991–2020).

Month	Total Precipitation				Average Air Temperature			
	Climatic Normal (mm)	Year 2022 (mm)	Difference (%)	Classification	Climatic Normal (°C)	Year 2022 (°C)	Difference (°C)	Classification
January	32.8	0.99	3	extremely dry	−0.5	5.9	5.4	extremely warm
February	28.9	34.5	119	normal	1.3	3.6	2.3	normal
March	32.9	3.5	11	extremely dry	5.5	4.6	−0.9	normal
April	36.3	12.9	36	very dry	11.4	8.5	−2.9	cold
May	59.3	12.7	21	very dry	16.0	15.8	−0.2	normal
June	59.1	88.4	150	wet	19.6	20.7	1.1	warm
July	64.6	60.4	94	normal	21.7	21.5	−0.2	normal
August	54.6	59.9	110	normal	21.1	21.9	0.8	normal
September	58.1	6.6	11	very dry	15.9	14.4	−1.5	cold
October	46.1	27.5	60	normal	10.4	11.5	1.1	warm
November	44.9	8.7	19	very dry	5.6	5.5	−0.1	normal
December	41.6	67.2	162	wet	0.7	1.35	0.7	normal

The history of the locality where the vineyard is situated (southwestern slopes of the Tribeč Mountains) dates back to the 11th century. After the deforestation of the southern and southwestern slopes, vine cultivation has taken place here with varying intensity and has been preserved more or less to this day. Parts of this locality were reforested in the 1950s. Other neglected areas were overgrown with bushes and trees, and parts began to be used for horticulture. Before the establishment of the current vineyard, vines were grown on this site until the 1990s. Later, this vineyard was neglected and abandoned. The vineyard was covered with overgrown bushes and trees. In 1999, the site was cultivated (vegetation was removed and the area was plowed to a depth of 30 cm). In the spring of 2000, a new vineyard (*Vitis vinifera* L. cv. Chardonnay) was planted. The area between the rows of vines was plowed every autumn, and during the growing season, the vine was intensively tilled. An experiment with different soil management practices in this productive vineyard was established in 2006, with the first application of farmyard manure in autumn 2005.

2.2. Design of Experimental Vineyard

The study utilized a randomized block design, comprising five treatments, with each treatment having three replicated plots. Below are brief descriptions of the individual treatments that represent different soil management practices in a productive vineyard:

T: Tillage system—involved soil plowing up to a depth of 25 cm between the rows of vines every autumn. During the vine vegetation season, intensive tillage was performed using a cultivator to a maximum depth of 12 cm. The purpose of soil loosening was to regulate or remove weeds between the vine rows, which was done on average three times during the vegetation season. No manure or mineral fertilizers were applied.

T+FYM: A tillage system combined with the incorporation of farmyard manure by plowing in 4-year cycles. The soil tillage was the same as in the T treatment, both in autumn and during the vine vegetation season. In 2005, 2009, 2013, 2017, and 2021, farmyard manure was applied to the soil surface at a dose of 40 t ha^{-1} and incorporated to a depth of 20–25 cm. The poultry manure used had 55% organic substances in the dry matter, 2.8% total nitrogen (Nt), 1.3% phosphorus (P_2O_5), 1.2% potassium (K_2O), and a pH ranging from 6 to 8.

G: Grass between vine rows—a mixture of grasses, including *Lolium perenne* L., *Poa pratensis* L., *Festuca rubra* subsp. *commutata Gaudin*, and *Trifolium repens* L. were sown in spring 2006 at a ratio of 50:20:25:5. The aboveground grass biomass was cut down on average three times per vine vegetation season. The cut biomass was left in situ on the surface as a mulch layer. In this treatment, grass strips were not fertilized.

G+NPK1: Grass strips between vine rows and the application of NPK at the first fertilization level. Between the rows of vines, the same mix of grasses as well as the same management practice as in the case of G treatment was used. Application doses of N, P, and K were 100 kg ha^{-1}, 30 kg ha^{-1}, and 120 kg ha^{-1}, respectively. Every year, the nutrients were applied to the soil in the following ratios, 1/2 in March (bud burst) and 1/2 in May (flowering).
G+NPK2: Grass strips between vine rows and the application of NPK at the second fertilization level. Between the vine rows, the same mix of grasses and the same management practices as in the G treatment were used. Application doses of N, P, and K were 125 kg ha^{-1}, 50 kg ha^{-1}, and 185 kg ha^{-1}, respectively. Every year, the nutrients were applied to the soil in the following ratios, 2/3 in March (bud burst) and 1/3 in May (flowering).

2.3. Soil Sampling

Soil samples were collected monthly from all studied soil management treatments in the productive vineyard from February to November 2022 using a spade. Sampling was done at a depth of 0–20 cm to capture the most intensive effects of all soil management practices on soil chemical and biological properties. The samples from each treatment were then mixed into one composite sample.

2.4. Chemical Analysis

After removing the litter, roots, and stones, the soil samples were air-dried for the determination of soil pH, electrical conductivity (EC), soil organic carbon (SOC), labile carbon (C_L), and microbial parameters. Soil pH was determined potentiometrically in distilled water with a ratio of 1:2.5 (soil/water) using a pH meter (HI 2211, HANNA Instruments, Smithfield, RI, USA). EC was measured via a conductomer (DiST HANA HI 9831, HANNAInstruments, Smithfield, RI, USA), where a sample of soil was diluted with distilled water in a ratio of 1:2. SOC content was measured using the wet combustion method—oxidation of soil organic matter by a mixture of 0.07 mol L^{-1} H_2SO_4 and $K_2Cr_2O_7$, with titration using Mohr's salt [51]. The determination of C_L content was prepared by shaking of 1 g of soil sample for 2 h in 50 mL solution of 0.005 mol L^{-1} $KMnO_4$ and 0.0025 mol L^{-1} H_2SO_4. After centrifugation, C_L was determined via oxidation of 0.07 mol L^{-1} H_2SO_4 and $K_2Cr_2O_7$, with titration using 0.05 mol L^{-1} Mohr's salt. [52].

2.5. Bacterial Analysis

Bacterial populations, colony forming units (CFU), and bacterial species were evaluated in dry soil. A sample of soil weighing 1 g was diluted in 99 mL of 0.1% sterile saline solution. To acquire bacterial isolates, two culture mediums were used, tryptone soya agar (TSA, Oxoid, Merck, Austria) and plate count agar (PCA, Oxoid, Merck, Austria). PCA and TSA were preferred for cultivating soil microorganisms due to their nutrient content, solidification properties provided by agar, versatility in formulation, and standardized use in microbiological research. The bacteria community was cultivated on TSA and PCA for 48–72 h at 30 °C. The cultivated colonies were counted and the samples were created, as previously reported by Kačániová et al. [53]. The samples were transferred from a petri dish to an Eppendorf flask along with 300 mL of mixed distilled water and 900 mL of ethanol. The mixture was then centrifuged at 10,000× *g* for 2 min using an ROTOFIX 32A (Ites, Vranov, Slovakia). The supernatant was discarded, and the precipitate was allowed to dry at 20 °C. Subsequently, 30 μL of 70% formic acid and 30 μL of acetonitrile were applied to the pellet. The mixture was then centrifuged for 2 min at 10,000× *g*. The stock solution, prepared to be used as an organic reagent, contained 500 μL of pure acetonitrile, 475 μL of pure distilled water, and 25 μL of pure trifluoroacetic acid, corresponding to 50%, 47.5%, and 2.5% of the solution, respectively. "HCCA matrix portioned" was then added to 250 μL of the prepared organic solvent in an Eppendorf flask. All components for the matrix solution were obtained from Brucker (Bremen, Germany). Finally, 1 μL of supernatant from the bacterial sample was added to a MALDI plate and was allowed to dry.

A volume of 1 μL of prepared MALDI matrix covered all samples. Automatically generated mass spectra with a mass range of 2.000–20.000 Da were created using the microflex LT MALDI-TOF mass spectrometer (Bruker Daltonics, Bremen, Germany). The apparatus was calibrated using the Bruker bacterial test standard. The outcomes of the mass spectra were processed using the MALDI Biotyper 3.0 tool from Bruker Daltonics in Bremen, Germany. A score between 2.300 and 3.000 indicated highly probable identification at the species level, a score between 2.000 and 2.299 secured genus identification with probable species identification, a score between 1.700 and 1.999 indicated probable identification at the genus level, and a score of less than 1.700 was regarded as unreliable identification.

2.6. Statistical Analysis

The data were analyzed using one-way ANOVA and the means (average values of soil parameters over the whole study period) were compared using a Tukey test at $p < 0.05$. The Mann–Kendall test was used to evaluate the trends of the soil parameters during the investigated period. Multivariate clustering and principal component analyses (PCA) were performed on the coefficients determined in the multiple linear regression analysis. Cluster analysis was conducted based on Ward's method and squared Euclidean distance. The link between the microbial, physicochemical soil parameters, and climatic conditions was assessed using a correlation matrix. Spearman's correlation coefficient was used to describe the relationship between each pair of variables for each treatment separately using a monotonic function. All data were analyzed using the Statgraphics Centurion XV.I program (Statpoint Technologies, Inc., Washington, DC, USA).

3. Results

3.1. Effect of Soil Management Practices on Soil pH

The dynamics of soil pH did not exhibit a trend in any of the treatments (Table 2). In the T, T+FYM, and G treatments, the soil pH remained fairly balanced from February to November 2022. However, the soil pH responded sensitively to the application of NPK fertilizers at both levels. An increase in soil pH was observed in spring and autumn. A decrease was observed in summer and winter in the fertilized treatments. The treatment with higher NPK fertilization of the grass strips between the vine rows resulted in a more pronounced acidification effect compared to the treatment with lower fertilization. Overall, the average soil pH values from February to November 2022 were higher by 0.6 and 0.15 pH units in the T+FYM and G treatments, respectively, and lower by 0.84 and 1.68 pH units in the G+NPK1 and G+NPK2 treatments, respectively, compared to the T treatment (Table 3).

Table 2. Dynamics of soil parameters according to the results of the Mann–Kendall test.

Treatments	Abundance of Culturable Bacteriota	Soil pH	EC	C_L	SOC
	Mann-Kendall Trends				
T	Increasing	Stable/No Trend	Stable/No Trend	Decreasing	Stable/No Trend
T+FYM	Stable/No Trend	Stable/No Trend	Stable/No Trend	Decreasing	Stable/No Trend
G	Stable/No Trend	Stable/No Trend	Stable/No Trend	Stable/No Trend	Stable/No Trend
G+NPK1	Stable/No Trend	Stable/No Trend	Stable/No Trend	Stable/No Trend	Stable/No Trend
G+NPK2	Stable/No Trend	Stable/No Trend	Stable/No Trend	Stable/No Trend	Stable/No Trend

Note: EC—electrical conductivity, C_L—labile carbon, SOC—soil organic carbon, T—tillage and no fertilization, T+FYM—tillage and application of farmyard manure every 4 years at a rate of 40 t ha^{-1}, G—grass strips and no fertilization, G+NPK1—application of NPK to grass strips at 1st fertilization level, G+NPK2—application of NPK to grass strips at 2nd fertilization level.

Table 3. Results of the ANOVA for the soil parameters depending on soil management practice.

Treatments	Abundance of Culturable Bacteriota	Soil pH	EC	C_L	SOC
			m S/m	g/kg	
T	4.0 ± 0.39 [a]	6.89 ± 0.09 [bc]	277 ± 68 [a]	1.77 ± 0.24 [a]	14.4 ± 0.88 [a]
T+FYM	4.1 ± 0.30 [a]	7.01 ± 0.07 [bc]	342 ± 64 [a]	1.84 ± 0.33 [a]	18.4 ± 1.13 [b]
G	4.3 ± 0.35 [a]	7.20 ± 0.08 [c]	395 ± 65 [a]	4.02 ± 0.88 [b]	31.8 ± 6.12 [d]
G+NPK1	4.2 ± 0.46 [a]	6.51 ± 0.63 [b]	1309 ± 888 [b]	3.91 ± 0.97 [b]	27.9 ± 5.65 [c]
G+NPK2	4.1 ± 0.39 [a]	5.79 ± 0.74 [a]	2825 ± 1167 [c]	3.62 ± 0.63 [b]	25.1 ± 2.82 [c]

Note: EC—electrical conductivity, C_L—labile carbon, SOC—soil organic carbon, T—tillage and no fertilization, T+FYM—tillage and application of farmyard manure every 4 years at a rate of 40 t ha^{-1}, G—grass strips and no fertilization, G+NPK1—application of NPK to grass strips at 1st fertilization level, G+NPK2—application of NPK to grass strips at 2nd fertilization level. The values represent means and standard deviations for the treatments from February up to November 2022. Different letters within columns indicate heterogeneous groups of the means based on Tukey's test at a 0.05 significance level.

3.2. Effect of Soil Management Practices on Electrical Conductivity

Soil management practices in the productive vineyard significantly affected the average EC (Table 3) and its dynamics during the investigated period. Overall, based on the results of the Mann–Kendall analysis, the EC values remained stable across treatments (Table 2). The application of NPK fertilizers led to an increase in EC values, peaking in June, with a more pronounced effect in the G+NPK2 treatment. However, even in these treatments, no clear trend (either an increase or decrease) in EC values was observed during the warm and dry year of 2022.

3.3. Effect of Soil Management Practices on Labile Carbon Content

The dynamics of C_L for the investigated period are evaluated in Table 2. A significant linear decrease in C_L was observed in the T and T+FYM treatments by 61 and 89 mg kg^{-1} $month^{-1}$, respectively. This corresponds to a decrease of 28% and 20% C_L, respectively, over a period of 10 months. In contrast, the C_L values in the other treatments (G, G+NPK1, and G+NPK2) fluctuated considerably and were an order of magnitude higher than in the T and T+FYM treatments. In the G treatment, a steep increase in C_L was observed during the spring period, peaking in April. Then, C_L gradually decreased until the end of summer and again increased in the autumn period. A more significant sensitivity in C_L content was observed during the period from February to November 2022 in the treatments with NPK fertilization applied to the grass strips between the vine rows. Overall, the mean C_L values were higher by 4%, 127%, 120%, and 104% in the T+FYM, G, G+NPK1, and G+NPK2 treatments, respectively, compared to the T treatment (Table 3).

3.4. Effect of Soil Management Practices on Soil Organic Carbon Content

A linear decrease in C_L was observed in the T and T+FYM treatments from February to November 2022, but it did not affect the dynamics of changes in SOC content in these treatments. SOC remained relatively stable and did not show any significant trend in changes based on the Mann–Kendall analysis (Table 2). In the G treatment, two peaks of SOC increase were identified in July and November. The SOC dynamics at both levels of NPK application showed a similar trend in spring and summer but diverged during autumn. In the autumn period, a more pronounced sensitivity to changes in SOC was observed in the G+NPK1 treatment compared to the G+NPK2 treatment. However, even these changes did not have a significant effect on the SOC contents in these treatments during the investigated period (Table 2). Overall, the lowest SOC content was found in the T treatment. SOC in the other treatments increased in the following order: T < T+FYM < G+NPK2 < G+NPK1 < G treatment (Table 3).

3.5. Effect of Soil Management Practices on Abundance and Species Diversity of Soil Culturable Bacteriota

During the investigated period, the abundance of culturable bacteriota did not change significantly in the T+FYM, G, G+NPK1, and G+NPK2 treatments. Differences were observed in the peaks of the number of culturable bacteriota depending on the soil management practices. However, the overall changes in the trends of bacterial counts over the entire period (from February to November 2022) were balanced and stable. According to the Mann–Kendal test, an increase in the number of soil culturable bacteriota could be observed only in the T treatment as a result of aeration. The results showed that an intensive tillage system in the productive vineyard increased the number of culturable bacteriota at an average rate of 0.09 log CFU/g per month, which means an increase of 34% during the period of February to November 2022.

Overall, in the T treatment, 39 species, 21 genera, and 18 families of culturable bacteriota were isolated (Figure 2). The most frequently isolated species were *Bacillus megaterium* (20.17%), followed by *Bacillus thuringiensis* (6.44%), and *Bacillus cereus* (5.58%) in this treatment. Treatment with a tillage system with plowing of farmyard manure every 4 years (T+FYM) provided 40 isolated and cultivated bacterial species (Figure 3) included in 22 genera and 19 families, with the most frequently isolated species being *Bacillus megaterium* (12.99%), followed by *Bacillus thuringiensis* (9.06%) and *Bacillus simplex* (5.12%). In the G treatment, 35 species were cultivated (Figure 4), classified into 21 genera and 19 families. The most frequently isolated species in the G treatment were *Bacillus megaterium* (16.89%), *Bacillus* spp. (10.50%), *Bacillus cereus* (8.68), *Bacillus thuringiensis* (4.57%), and *Bacillus licheniformis* (4.57%). In the G+NPK1 treatment, 45 species of culturable bacteriota, 26 genera, and 21 families were isolated (Figure 5). The most frequently isolated species were *Bacillus megaterium* (15.02%), followed by *Bacillus simplex* (8.06%), *Bacillus cereus* (6.96%), and *Bacillus licheniformis* (5.13%). In contrast to the G+NPK1 treatment, which provided the highest variability of analyzed species, the G+NPK2 treatment had the lowest number of culturable bacteriota with 33 species, 15 genera, and 14 families (Figure 6). The most frequent species in this treatment were *Bacillus megaterium* (18.48%), followed by *Bacillus amyloliquefaciens* (6.64%), *Bacillus endophyticus* (6.16%), and *Bacillus simplex* (6.16%). In summary, our study isolated 140 of soil culturable bacteriota species, 59 genera, and 40 families (Figure 7). The most represented species were *Bacillus megaterium* (16.55%), *Bacillus cereus* (5.80%), *Bacillus thuringiensis* (4.87%), and *Bacillus simplex* (4.37%).

3.6. Relationships between Soil Parameters and Soil Culturable Bacteriota

The PCA analysis (Figure 8A) revealed a correlation between the averages of C_L, SOC, and the number of culturable bacteriota throughout the entire study period (February to November 2022). The average values of these variables were highest in the G and G+NPK1 treatments. EC was negatively correlated with soil pH. The average value of EC was highest in the G+NPK2 treatment, whereas soil pH was highest in the T and T+FYM treatments. The soil management practices in the productive vineyard in the T and T+FYM treatments were significantly different from those under other management practices (Figure 8B).

Table 4 presents the correlation coefficients between the number of culturable bacteriota and other soil properties and climatic characteristics (variables) under different soil management practices in the productive vineyard from February to November 2022. During the investigated period, average air temperature was the most significant variable among all properties, positively influencing the number of culturable bacteriota, not only when all treatments were assessed together, but also separately, especially in the treatments with grass strips between vine rows (G, G+NPK1, and G+NPK2). The strongest correlation was found in the G+NPK1 treatment compared to the G and G+NPK2 treatments. In the T treatment, the number of culturable bacteriota increased as a result of greater nutrient availability, reflected by higher EC values. No correlations between culturable bacteriota abundance and other variables were found in the T+FYM treatment. Notably, culturable bacteriota abundance had a significant influence on SOC, but only in the G+NPK2 treatment, where lower SOC values corresponded to higher culturable bacteriota abundance. Furthermore, the correlation analysis revealed a significant positive correlation between air temperature and precipitation (r = 0.71) in our experiment.

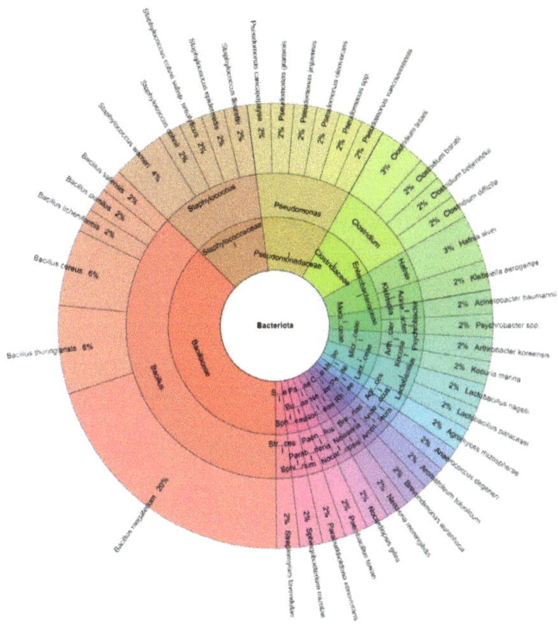

Figure 2. Krona chart: Isolated species of culturable bacteriota from the T treatment.

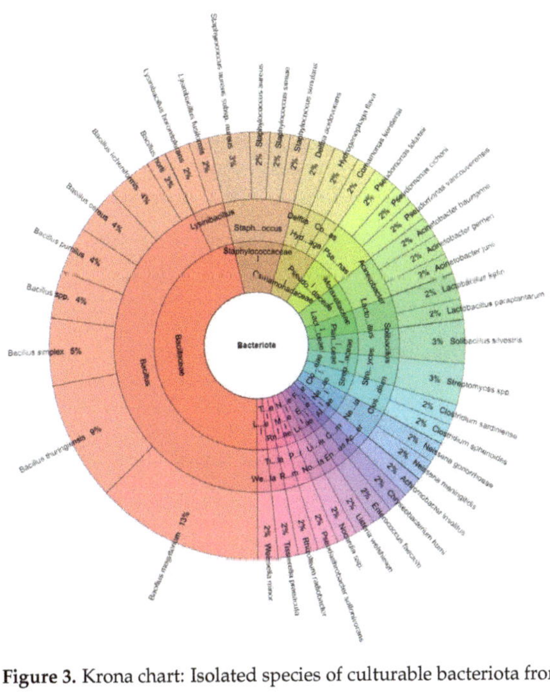

Figure 3. Krona chart: Isolated species of culturable bacteriota from the T+FYM treatment.

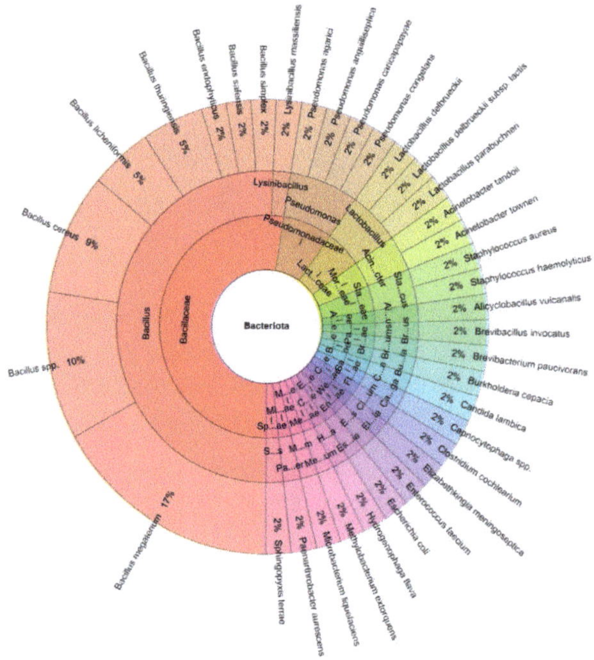

Figure 4. Krona chart: Isolated species of culturable bacteriota from the G treatment.

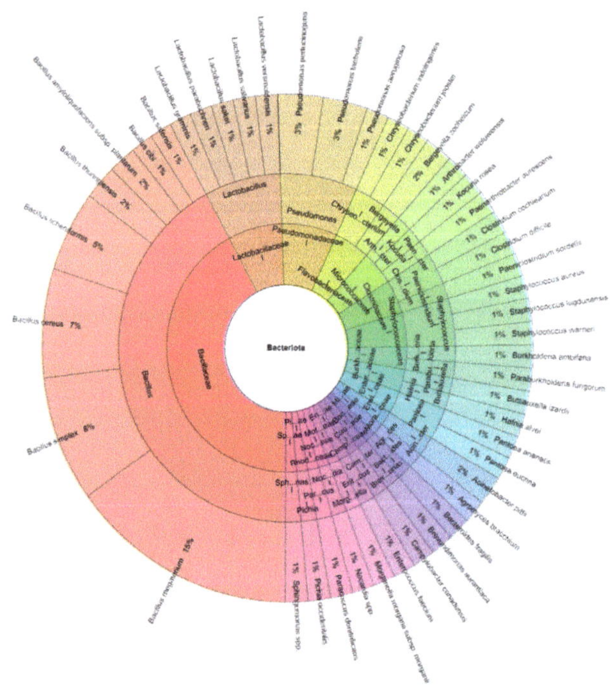

Figure 5. Krona chart: Isolated species of culturable bacteriota from the G+NPK1 treatment.

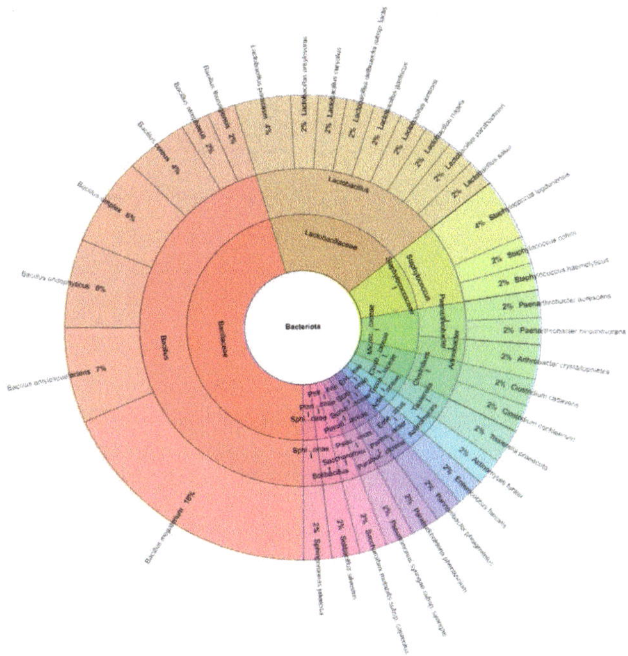

Figure 6. Krona chart: Isolated species of culturable bacteriota from the G+NPK2 treatment.

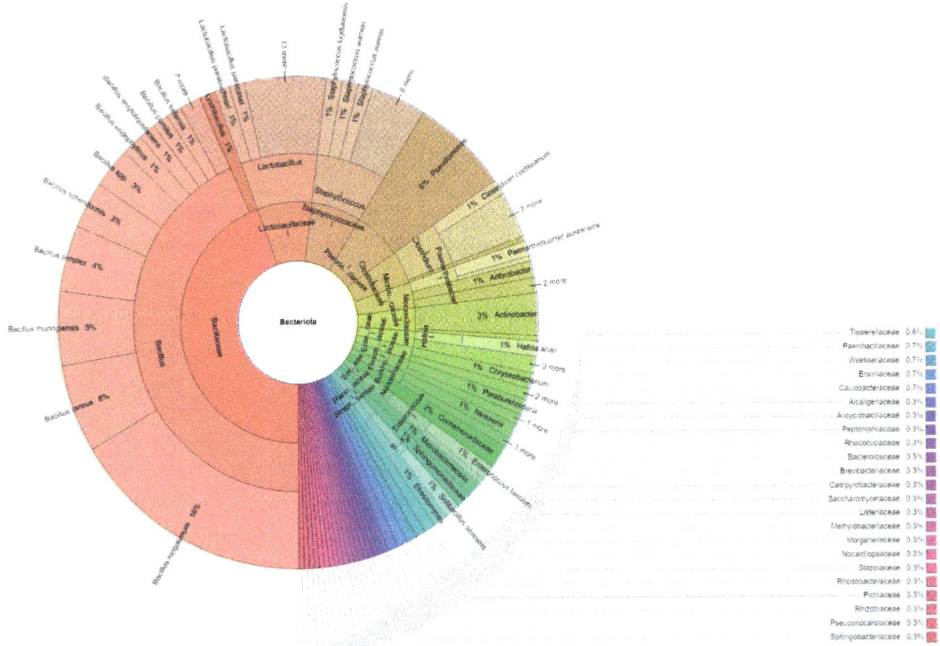

Figure 7. Krona chart: Isolated species of culturable bacteriota from all treatments.

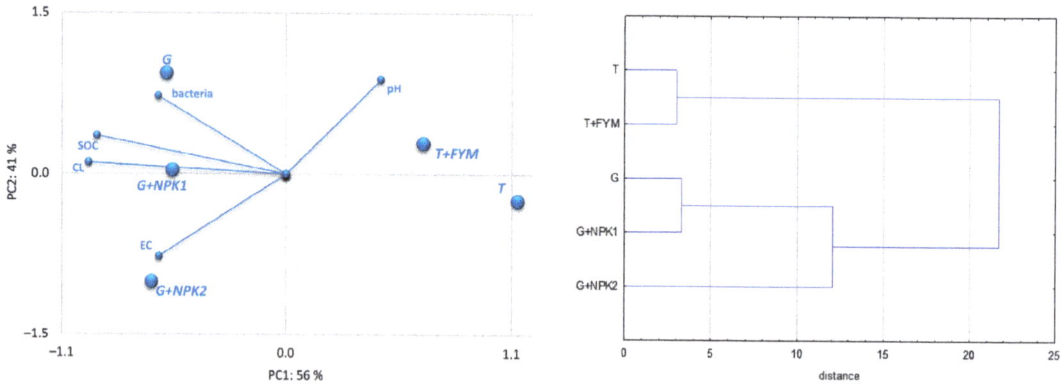

(A) (B)

Figure 8. (**A**) Relationships between objects (T, T+FYM, G, G+NPK1, and G+NPK2) and averages of the describing variables (culturable bacteriota, pH, EC, C_L, and SOC) over the entire study period, and (**B**) Similarities between objects T, T+FYM, G, G+NPK1, and G+NPK2 based on the describing variables (culturable bacteriota, pH, EC, C_L, and SOC) over the entire study period.

Table 4. Correlation coefficients between culturable bacteriota, soil pH, EC, C_L, SOC, air temperature, and precipitation for the T, T+FYM, G, G+NPK1, and G+NPK2 treatments. Each treatment is referred to with a different color: T (light blue)—tillage with no fertilization, T+FYM (light green)—tillage and the application of farmyard manure every 4 years at a rate of 40 t ha^{-1}, G (light gray)—grass strips with no fertilization, G+NPK1 (light red)—application of NPK fertilizers to grass strips at the first fertilization level, G+NPK2 (yellow)—application of NPK fertilizers to grass strips at the second fertilization level. Numbers in red font present significant correlations at a significance level of 0.05.

	Culturable Bacteriota	Soil pH	EC	C_L	SOC	Air Temperature	Precipitation
Culturable bacteriota		−0.09	0.69	−0.26	−0.26	0.35	−0.08
Soil pH	−0.07		0.16	0.40	−0.25	−0.23	−0.20
EC	0.58	0.13		−0.25	−0.13	0.62	0.22
C_L	−0.19	0.22	−0.14		−0.14	−0.07	−0.02
SOC	−0.31	0.25	−0.27	0.20		0.31	0.11
Air temperature	0.17	−0.34	0.68	−0.38	−0.39		0.71
Precipitation	−0.13	−0.36	0.32	−0.01	−0.37	0.71	
Culturable bacteriota		−0.20	0.52	0.35	0.43	0.68	0.17
Soil pH			−0.44	−0.73	−0.92	−0.45	−0.39
EC				0.23	0.53	0.69	0.43
C_L					0.87	0.35	0.17
SOC						0.52	0.32
Air temperature							0.71
Culturable bacteriota		0.09	0.54	−0.08	0.08	0.79	0.32
Soil pH	0.17		−0.51	−0.12	−0.26	0.00	−0.49
EC	0.37	−0.43		−0.01	−0.13	0.78	0.90
C_L	−0.07	0.47	−0.03		0.74	−0.07	−0.02
SOC	−0.65	0.01	−0.11	0.70		−0.18	−0.26
Air temperature	0.63	−0.02	0.64	0.25	−0.01		0.71
Precipitation	0.05	−0.44	0.81	0.19	0.32	0.71	

4. Discussion

4.1. Effect of Soil Management Practices on Soil Properties and Soil Culturable Bacteriota

In the experimental vineyard, the soil type is Rendzic Leptosol, which represents about 4.7% (115 thousand ha) of the total agricultural land in Slovakia [54]. Leptosols are the most extensive reference soil group on earth, extending over about 1655 million ha. They are found from the tropics to the polar regions and from sea level to the highest mountains [50]. Therefore, this experiment is informative for the Slovak territory and significant for understanding a substantial proportion of the world's soils. The vine is generally not demanding in terms of available nutrient content, as it is able to efficiently utilize them very well from the overall soil supply [55]. However, this does not imply that the vine does not require optimal soil conditions for its growth and development. White [56] identified several crucial factors for successful vine growth: soil depth, soil structure, water content, soil cohesion, chemical properties, nutrient supply, and soil organisms. Most of these soil parameters can be effectively managed by vine growers through soil management practices in vineyards [57,58], or they can be minimized even before planting the vines [59]. Among the physico-chemical properties, soil pH is particularly crucial, as it directly affects the abundance and species diversity of soil organisms, soil reactions and processes, the availability and mobility of nutrients in the soil, and their transfer to plants, including vines [55,56]. Fertile soils can exhibit a relatively broad range of soil pH, while most cultivated plants thrive within a pH range of 5.5 to 6.5 [15]. This range is also suitable for most soil microorganisms, as plants grow well and produce more root exudates, providing a carbon source for the survival and multiplication of microbes [60]. However, most bacteria thrive around a neutral pH [61]. In this study, the soil pH ranged from 5.7 to 7.2 (slightly acid to neutral) and overall soil management practices in the productive vineyard significantly affected its values (Table 3). Different effects of soil management practices on pH changes were also observed throughout the warm and dry year of 2022 (Table 2). Soil pH was affected mainly by the addition of NPK fertilizers, where a higher rate of NPK markedly influenced the acidification effect. However, in terms of nutrient regime and plant nutrient uptake, altered soil pH values may appear to be beneficial [62] in the case of the G+NPK1 treatment in this study. From a geological perspective, the soil in the vineyard was formed from Mesozoic sedimentary rocks, predominantly Cretaceous, Jurassic, and Triassic limestones [48], and had a medium content of organic C [20], which ensured a functioning buffering system [63], especially in the T, T+FYM, and G treatments. These soil management practices in the productive vineyard did not cause significant fluctuations in soil pH during the warm and dry year of 2022 (Table 2). Moreover, in the T+FYM treatment, poultry manure with 55% organic substances in the dry matter was applied, and its pH ranged from 6 to 8. In terms of soil pH, a stressful environment for the soil fauna was not created compared to the G+NPK2 treatment (Tables 2 and 3).

The total nutrient supply (Table 3) and the dynamics of changes (Table 2) during the warm and dry year of 2022 were dependent on soil management in the vineyard. A high salt concentration above 1600 m S m^{-1} represents strong salinization [63], which is a stressful condition for cultivated plants and soil microorganisms [62]. In the G+NPK2 treatment, EC values were several times higher than in the T, T+FYM, and G treatments. In the G+NPK1 treatment, EC values were relatively high, but below the critical level. From an economic perspective, as well as from the perspective of minimizing stress for soil culturable bacteriota (ecological perspective), the EC values were the most optimal in the G treatment (Tables 2 and 3).

Soil management plays a key role in changes in SOM [64]. In the productive vineyard, the implementation of various soil management practices had a statistically significant impact on the overall SOM (Table 3). Additionally, significant changes in the dynamics of the labile fraction (C_L) were observed in the T and T+FYM treatments during the warm and dry year of 2022 (Table 2). The SOM in treatments increased in the following order: T < T+FYM < G+NPK2 < G+NPK1 < G. In the T treatment, the lowest SOM was linked to intensive SOM mineralization. Intensive tillage systems promote soil aeration and support aerobic microflora responsible for increased SOM mineralization, resulting in a decrease in C and an overall low C stock in the soil [65]. Although C_L significantly decreased, the stable SOM fraction did not show any increase or decrease in trend during the period of 2022. This indicates that the SOM is in an equilibrium state in terms of its balance [66]. Trends in other treatments showed stability in SOC over the observed period (Table 2). The application of FYM results in an increase in SOM [67] and its supply is higher in such soils. Compared to the T treatment, SOC increased from 14.4 to 18.4 g kg^{-1} in the T+FYM treatment. Permanent grassing and grass strips in vineyards have a positive effect on increasing SOM [58] and this is also evidenced by the results of our study. In the G treatment, the trend in SOM was balanced during the warm and dry year of 2022 (Table 2). It is noteworthy that in the long term, SOM in this treatment (grass strips) is constantly growing, as reported by Šimanský et al. [68]. This is typically the regeneration scenario of previously intensively cultivated land [69]. In the G treatment, the observed SOC was higher by 55, 42, 22, and 21% when compared to T, T+FYM, G+NPK1, and G+NPK2 treatments, respectively. NPK fertilization can increase crop yields and plant and root residue input into the soil system and eventually result in an increase in the soil's organic carbon content [70]. On the other hand, fertilizers can also decrease the C content compared to unfertilized soil [71], which can be reflected in a decline in SOM parameters. However, application dose is critical, as evidenced by the findings of this study. In both treatments with NPK application, SOC was lower than in the G treatment and SOM did not show any significant changes during the period of the warm dry year (Table 2).

Interestingly, the abundance of culturable bacteriota in all treatments was balanced (Table 3) and did not change significantly in T+FYM, G, G+NPK1, and G+NPK2 treatments during the warm and dry year of 2022. A significant change occurred only in the case of the T treatment, where we observed a significant increase in counts of culturable bacteriota during the studied period. This may be related to the formation of optimal aerobic conditions for the variance of bacterial species and the subsequently more intensive mineralization of SOM [54], as shown in the results of this study (Table 2—decreasing trend in C_L values in the T treatment, and Table 3—the lowest values of SOC and C_L in the T treatment). Soil microorganisms are regarded as reliable indicators of soil quality [32]. Persistent monoculture and long-term fertilization impact the microbial community compositions in soil [72], as confirmed by the findings of our study (Figures 2–7).

The representation of individual species, genera, and families considerably differed depending on the soil management practices in the productive vineyard. In the T+FYM treatment, the second-greatest overall biodiversity of microorganisms was observed (Figure 3). As stated by Iqbal et al. [33], the incorporation of FYM provides the soil with a whole range of soil microorganisms living in the manure, in addition to nutrients. FYM is usually plowed/incorporated into the soil, which promotes the development of aerobic microflora, so the fact that the T treatment (Figure 2) reached the third-highest biodiversity is not a surprising result. According to Natywa et al. [73], agrotechnical treatments such as fertilization have a significant impact on the activity of soil microorganisms. This is also confirmed by the results of Wolny-Koładka et al. [4]. However, the abundance and species of culturable bacteriota were highly differentiated and changed depending on the type and amount of applied fertilization as well as the plant species cultivated. Overall, the greatest microbial diversity was found in the G+NPK1 treatment (Figure 5). On the other hand, a higher rate of NPK (Figure 6) resulted in a decrease in bacterial biodiversity. The main reason may be due to the higher concentration of salts in the soil solution, as reported by Zhao et al. [29], which is also supported by the results of our study (Table 3). High EC values come from higher rates of NPK fertilizers, which means a negative effect on the biodiversity of soil microorganisms [32]. In the case of this study, *Bacillus* was the most numerous genus analyzed in our soil samples (Figure 7). *Bacillus* is a bacterium normally present as part of the soil microbiome. The ability to form spores also helps this bacterium to survive even in less favorable conditions. Among the aerobic bacteria, there were also more represented bacteria such as the genera *Lactobacillus*, *Staphylococcus*, and *Pseudomonas*, which are commonly found as part of the soil microbiome [74].

The results of this study showed that hypothesis H1 was only partially confirmed, because not all soil parameters, including abundance and species variability of culturable bacteriota, were stabilized during the warm and dry year of 2022. Overall, the application of lower NPK fertilizers to grass strips between vine rows had the most favorable effect on the overall diversity of soil life. The management of grass strips also confirmed its validity in a productive vineyard. However, the high levels of NPK fertilization in the grass strips caused a decline or had no effect on the changes in soil properties when compared to the G treatment. Higher rates of NPK fertilizers had more negative effects than lower levels of NPK, therefore H3 was confirmed.

4.2. Relationships between Soil Properties under Different Soil Management Practices

Vineyard systems differ qualitatively and quantitatively in their soil microbiomes [75], as confirmed by the findings of this study (Figures 2–7). Soil microbial communities are impacted by a variety of environmental variables, including climate, soil characteristics, cultivars, and agricultural management [76,77]. In this study, PCA analysis showed a correlation between SOM and soil culturable bacteriota. The average values of these variables were found to be high in the G and G+NPK1 treatments (Figure 8A). This observation suggests that these soil management practices, implemented in a productive vineyard, are not only associated with the relationship between SOM and soil culturable bacteriota but are also deemed most suitable in terms of the considered parameters. SOM is an important source of C and nutrients for microorganism activity [2] and is also transformed by their activity [13]. The fact that these positive relationships are registered in the G and G+NPK1 treatments was primarily related to their high contents of SOC, C_L, and abundance of culturable bacteriota (Table 3). According to Zarraonaindia et al. [78], soil physicochemical factors such as soil pH, temperature, moisture, carbon, and nitrogen pools have a major impact on soil bacterial diversity and composition. Additionally, human activities, including different soil management practices, play a crucial role [4,33]. The PCA analysis also showed that EC negatively correlated with soil pH. The average EC value was high in the G+NPK2 treatment, while soil pH was high in the T and T+FYM treatments. In the G+NPK2 treatment, the high concentration of nutrients could be the result of their excessive application to the soil and insufficient uptake by vine plants and grass strips in the vineyard. High concentrations of nutrients can cause acidification. In addition to monoculture cultivation, which also has a negative impact on soil pH and EC [29], the type of fertilizer itself has a significant effect. Physiologically acidic fertilizers cause soil acidification after their application [15]. Sulfate or nitrate forms of N fertilizers reduce the soil pH at high concentrations of nutrients [55]. Our results indicated that the interaction between many factors had a great effect on the chemical and microbiological properties of the soil. The addition of mineral–organic mixtures produces changes in the chemical properties of the soil and hence affects its microbial characteristics [4]. Management in the T and T+FYM treatments was significantly different from other managements, as confirmed by cluster analysis (Figure 8B).

The diversity and composition of microbial communities are significantly influenced by soil characteristics [79,80] in individual soil management practices. The climate, particularly temperature and rainfall patterns, also have an impact on microbial communities through their effects on the soil [81], as our study's findings partially suggest. Average air temperature was the most significant variable among all investigated variables. It positively affected the number of culturable bacteriota, not only when all treatments were considered together, but also separately, especially in the G, G+NPK1, and G+NPK2 treatments. Rainfall increased the concentration of salts in the soil solution (high EC values) in the G+NPK2 treatment (Table 4). This could be related to the solubility of fertilizers after an intense rainfall event in June 2022 (Table 1). Additionally, the lowest soil pH was found during June, which was evaluated as a wet month compared to the 30-year climatic normal. Overall, however, the timing of fertilization (March and May) did not affect the dynamics of the soil properties (SOC, soil pH, EC—Table 2). In the other soil parameters, including the abundance of culturable bacteriota, there were no changes during the months, thus partially confirming H2

5. Conclusions

Overall, soil parameters, including the biodiversity of soil culturable bacteriota, depended on the management of moderately coarse soil in the productive vineyard. However, the dynamics of the changes in soil properties were largely balanced during the warm and dry year of 2022. Grass strips in the vine rows without tillage, as well as applying NPK fertilizer to grass strips at the 1st level (low rate) had the highest SOM content and culturable bacteriota abundance overall. A positive relationship between SOM and culturable bacteriota abundance was also noted in both of these treatments. The greatest biodiversity was found in grass strips fertilized with a low dose of NPK. In contrast, acidification and a high concentration of salts in the soil solution were observed in the treatment with grass strips and application of NPK fertilizer at a high dose. During the warm and dry year, the average air temperature was the most significant variable in almost all the soil management practices in the productive vineyard, with its most pronounced effect found in the case of grass strips and grass strips combined with NPK fertilization at a lower dose.

Based on the results of this study and considering the evaluated parameters, grassing with NPK application at a low rate appears to be the best scenario for sustainable management of vineyards in a changing climate. However, from an economic and eco-friendly perspective (no costs for NPK fertilization), the best soil management practice in vineyards on sandy loam soils in Central Europe is grass striping between vine rows. Since other biological, chemical, and physical properties were not assessed, further research activities (focused on their determination and evaluation) are necessary to draw completely comprehensive recommendations.

Author Contributions: Conceptualization, V.Š.; methodology, V.Š., M.K., P.B. and E.W.-G.; software, M.K., V.Š. and E.W.-G.; validation, M.J., E.A. and N.Č.; formal analysis, V.Š., M.K. and E.W.-G.; investigation, V.Š., M.J., N.Č. and P.B.; resources, V.Š.; data curation, V.Š., M.K., P.B. and E.W.-G.; writing—original draft preparation, V.Š., M.K. and E.W.-G.; writing—review and editing, M.J., N.Č., E.A. and P.B.; visualization, V.Š., M.K. and E.W.-G.; project administration, V.Š.; funding acquisition, V.Š. and E.A. All authors have read and agreed to the published version of the manuscript.

Funding: This study was supported by the Slovak Research and Development Agency under the contract No. APVV-21-0089 and Cultural and Educational Grant Agency of the Ministry of Education, Youth and Sports of the Slovak Republic, project No. KEGA 006SPU-4/2024.

Data Availability Statement: The raw data supporting the conclusions of this article will be made available by the authors on request.

Acknowledgments: The authors express their gratitude to the editor and the reviewers for their constructive comments.

Conflicts of Interest: The authors declare no conflicts of interest.

References

1. Millar, C.E.; Turk, L.M.; Foth, H.D. *Fundamentals of Soil Science*; John Wiley and Son, Inc.: New York, NY, USA, 1962; p. 526.
2. Murphy, B.W. *Soil Organic Matter and Soil Function—Review of the Literature and Underlying Data*; Department of the Environment: Canberra, Australia, 2014; p. 155.
3. Weil, R.R.; Brady, N.C. *The Nature and Properties of Soils*; Pearson Education Limited: London, UK, 2017; p. 1104.
4. Wolny-Koładka, K.; Jarosz, R.; Marcińska-Mazur, L.; Lošák, T.; Mierzwa-Hersztek, M. Effect of mineral and organic additions on soil microbial composition. *Int. Agrophys.* **2022**, *36*, 131–138. [CrossRef]
5. Jia, Y.; Liu, Z.; Zhou, L.; Liu, X.; Ma, K.; Feng, X. Soil organic carbon sourcing variance in the rhizosphere vs. non-rhizosphere of two mycorrhizal tree species. *Soil Biol. Biochem.* **2023**, *176*, 108884. [CrossRef]
6. Zhang, Y.; Shangguan, Z. Long-term N addition accelerated organic carbon mineralization in aggregates by shifting microbial community composition. *Agric. Ecosyst. Environ.* **2023**, *342*, 108249. [CrossRef]
7. Hardoim, P.R.; van Overbeek, L.S.; van Elsas, J.D. Properties of bacterial endophytes and their proposed role in plant growth. *Trends Microbiol.* **2008**, *16*, 463–471. [CrossRef] [PubMed]
8. Krause, S.M.B.; Dohrmann, A.B.; Gillor, O.; Christensen, B.T.; Merbach, I.; Tebbe, C.C. Soil properties and habitats determine the response of bacterial communities to agricultural wastewater irrigation. *Pedosphere* **2020**, *30*, 146–158. [CrossRef]
9. Bronick, C.J.; Lal, R. Soil structure and management: A review. *Geoderma* **2005**, *124*, 3–22. [CrossRef]
10. Schloter, M.; Nannipieri, P.; Sørensen, S.J.; van Elsas, J.D. Microbial indicators for soil quality. *Biol. Fertil. Soils* **2018**, *54*, 1–10. [CrossRef]
11. Meetei, T.T.; Devi, Y.B.; Thounaojam, T.C. Role of Soil Organisms in Maintaining Soil Health. In *Microbial Based Land Restoration Handbook, Volume 2 Soil and Plant Health Development*; Pandey, V., Pankaj, U., Eds.; Taylor and Francis Group: Abingdon, UK; CRC Press: Boca Raton, FL, USA, 2022. [CrossRef]

12. Climate Change Adaptation. Available online: https://climate-adapt.eea.europa.eu/en/countries-regions/countries/slovakia (accessed on 23 April 2024).
13. Váchalová, R.; Kolář, L.; Muchová, Z. *Primární Organická Pudní Hmota a Humus, Dvě Složky Pudní Organické Hmoty [Primary Soil Organic Matter and Humus, Two Components of Soil Organic Matter]*; SUA: Nitra, Slovakia, 2016; p. 122. (In Czech)
14. Bereza-Boruta, B. Selected enzymatic activities of actinomycetes of Streptomyces genus isolated from potato plantation. *Agricultura* **2002**, *1*, 27–36.
15. Kováčik, P.; Ryant, P. *Agrochémia, Princípy A Prax [Agrochemistry, Principles and Practice]*; SPU: Nitra, Slovakia, 2019; p. 358. (In Slovak)
16. Daunoras, J.; Kačergius, A.; Gudiukaité, R. Role of soil microbiota enzymes in soil health and activity changes depending on climate change and the type of soil ecosystem. *Biology* **2024**, *13*, 85. [CrossRef]
17. Blankinship, J.C.; Niklaus, P.A.; Hungate, B.A. A meta-analysis of responses of soil biota to global change. *Oecologia* **2011**, *165*, 553–565. [CrossRef]
18. Henry, H.A.L. Soil extracellular enzyme dynamics in a changing climate. *Soil Biol. Biochem.* **2012**, *47*, 53–59. [CrossRef]
19. Manzoni, S.; Schimel, J.P.; Porporato, A. Responses of soil microbial communities to water stress: Results from a meta-analysis. *Ecology* **2012**, *93*, 930–938. [CrossRef]
20. A'Bear, A.D.; Jones, T.H.; Boddy, L. Potential impacts of climate change on interactions among saprotrophic cord-forming fungal mycelia and grazing soil invertebrates. *Fungal Ecol.* **2014**, *10*, 34–43. [CrossRef]
21. Chen, S.; Zou, J.; Hu, Z.; Chen, H.; Lu, Y. Global annual soil respiration in relation to climate, soil properties and vegetation characteristics: Summary of available data. *Agric. For. Meteorol.* **2014**, *198–199*, 335–346. [CrossRef]
22. Zhang, N.N.; Sun, G.; Zhong, B.; Wang, E.T.; Zhao, C.Z.; Wang, Y.J.; Cheng, W.; Wu, N. Impacts of wise grazing on physicochemical and biological features of soil in a sandy grassland on the Tibetan Plateau. *Land Degrad. Dev.* **2019**, *30*, 719–729. [CrossRef]
23. Pereira e Silva, M.C.; Semenov, A.V.; Schmitt, H.; van Elsas, J.D.; Salles, J.F. Microbe-mediated processes as indicators to establish the normal operating range of soil functioning. *Soil Biol. Biochem.* **2013**, *57*, 995–1002. [CrossRef]
24. Tibbett, M.; Gil-Martínez, M.; Fraser, T.; Green, I.D.; Duddigan, S.; De Oliveira, V.H.; Raulund-Rasmussen, K.; Sizmur, T.; Diaz, A. Long-term acidification of pH neutral grasslands affects soil biodiversity, fertility and function in a heathland restoration. *Catena* **2019**, *180*, 401–415. [CrossRef]
25. Lundquist, E.J.; Jackson, L.E.; Scow, K.M. Wet–dry cycles affect dissolved organic carbon in two California agricultural soils. *Soil Biol. Biochem.* **1999**, *31*, 1031–1038. [CrossRef]
26. Kardol, P.; Wardle, D.A. How understanding aboveground–belowground linkages can assist restoration ecology. *Trends Ecol. Evol.* **2010**, *25*, 670–679. [CrossRef]
27. Smith, R.S.; Shiel, R.S.; Bardgett, R.D.; Millward, D.; Corkhill, P.; Rolph, G.; Hobbs, P.J.; Peacock, S. Soil microbial community, fertility, vegetation and diversity as targets in the restoration management of a meadow grassland. *J. Appl. Ecol.* **2003**, *40*, 51–64. [CrossRef]
28. Wang, Y.; Li, C.; Tu, C.; Hoyt, G.D.; DeForest, J.L.; Hu, S. Long-term no-tillage and organic input management enhanced the diversity and stability of soil microbial community. *Sci. Total Environ.* **2017**, *609*, 341–347. [CrossRef]
29. Zhao, Q.; Xiong, W.; Xing, Y.; Sun, Y.; Lin, X.; Dong, Y. Long-term cofee monoculture alters soil chemical properties and microbial communities. *Sci. Rep.* **2018**, *8*, 6116. [CrossRef]
30. Tian, J.; Wang, J.; Dippold, M.; Gao, Y.; Blagodatskaya, E.; Kuzyakov, Y. Biochar affects soil organic matter cycling and microbial functions but does not alter microbial community structure in a paddy soil. *Sci. Total Environ.* **2016**, *556*, 89–97. [CrossRef]
31. Toyota, K.; Kuninaga, S. Comparison of soil microbial community between soils amended with or without farmyard manure. *Appl. Soil Ecol.* **2006**, *33*, 39–48. [CrossRef]
32. Zhen, Z.; Liu, H.; Wang, N.; Guo, L.; Meng, J.; Ding, N.; Wu, G.; Jiang, G. Effects of manure compost application on soil microbial community diversity and soil microenvironments in a temperate cropland in China. *PLoS ONE* **2014**, *9*, e108555. [CrossRef]
33. Iqbal, A.; Tang, X.; Ali, I.; Yuan, P.; Khan, R.; Khan, Z.; Adnan, M.; Wei, S.; Jiang, L. Integrating low levels of organic fertilizer improves soil fertility and rice yields in paddy fields by influencing microbial communities without increasing CH_4 emissions. *Appl. Soil Ecol.* **2023**, *189*, 104951. [CrossRef]
34. Lehmann, J.; Rillig, M.C.; Thies, J.; Masiell, C.A.; Hockaday, W.C.; Crowley, D. Biochar effects on soil biota, A review. *Soil Biol. Biochem.* **2011**, *43*, 1812–1836. [CrossRef]
35. Šimanský, V.; Aydın, E.; Horák, J. Is it possible to control the nutrient regime of soils with different texture through biochar substrates? *Agronomy* **2022**, *12*, 51. [CrossRef]
36. Šimanský, V.; Wójcik-Gront, E.; Horváthová, J.; Pikuła, D.; Lošák, T.; Parzych, A.; Lukac, M.; Aydın, E. Changes in relationships between humic substances and soil structure following different mineral fertilization of *Vitis vinifera* L. in Slovakia. *Agronomy* **2022**, *12*, 1460. [CrossRef]
37. Igaz, D.; Šimanský, V.; Horák, J.; Kondrlová, E.; Domanová, J.; Rodný, M.; Buchkina, N.P. Can a single dose of biochar affect selected soil physical and chemical characteristics? *J. Hydrol. Hydromech.* **2018**, *66*, 421–428. [CrossRef]
38. Kuzyakov, Y.; Chang, W. Photosynthesis controls of rhizosphere respiration and organic matter decomposition. *Soil Biol. Biochem.* **2001**, *33*, 1915–1925. [CrossRef]

39. Burns, K.N.; Bokulich, N.A.; Cantu, D.; Greenhut, R.F.; Kluepfel, D.A.; O'Geen, A.T.; Strauss, S.L.; Steenwerth, K.L. Vineyard soil bacterial diversity and composition revealed by 16S rRNA genes: Differentiation by vineyard management. *Soil Biol. Biochem.* **2016**, *103*, 337–348. [CrossRef]
40. Intergovernmental Panel on Climate Change (IPCC). *Climate Change and Land: An IPCC Special Report on Climate Change, Desertification, Land Degradation, Sustainable Land Management, Food Security, and Greenhouse Gas Fluxes in Terrestrial Ecosystems*; IPCC: Geneva, Switzerland, 2019.
41. Barreiro, J.R.; Ferreira, C.R.; Sanvido, G.B.; Kostrzewa, M.; Maier, T.; Wegemann, B.; Böttcher, V.; Eberlin, M.N.; dos Santos, M.V. Short communication: Identification of subclinical cow mastitis pathogens in milk by matrix-assisted laser desorption/ionization time-of-flight mass spectrometry. *J. Dairy Sci.* **2010**, *93*, 5661–5667. [CrossRef]
42. Böhme, K.; Fernández-No, I.C.; Barros-Velázquez, J.; Gallardo, J.M.; Cañas, B.; Calo-Mata, P. Rapid species identification of seafood spoilage and pathogenic Gram-positive bacteria by MALDI-TOF mass fingerprinting. *Electrophoresis* **2011**, *32*, 2951–2965. [CrossRef]
43. Pandey, A.; Jain, R.; Sharma, A.; Dhakar, K.; Kaira, G.S.; Rahi, P.; Dhyani, A.; Pandey, N.; Adhikari, P.; Shouche, Y.S. 16S rRNA gene sequencing and MALDI-TOF mass spectrometry based comparative assessment and bioprospection of psychrotolerant bacteria isolated from high altitudes under mountain ecosystem. *SN Appl. Sci.* **2019**, *1*, 278. [CrossRef]
44. El-Nemr, I.M.; Mushtaha, M.; Sundararaju, S.; Fontejon, C.; Suleiman, M.; Tang, P.; Goktepe, I.; Hasan, M.R. Application of MALDI biotyper system for rapid identification of bacteria isolated from a fresh produce market. *Curr. Microbiol.* **2019**, *76*, 290–296. [CrossRef]
45. Surányi, B.B.; Zwirzitz, B.; Mohácsi-Farkas, C.; Engelhardt, T.; Domig, K.J. Comparing the efficacy of MALDI-TOF MS and sequencing-based identification techniques (Sanger and NGS) to monitor the microbial community of irrigation water. *Microorganisms* **2023**, *11*, 287. [CrossRef] [PubMed]
46. Jesser, K.J.; Noble, R.T. Vibrio ecology in the neuse river estuary, North Carolina, characterized by next-generation amplicon sequencing of the gene encoding heat shock protein 60 (hsp60). *Appl. Environ. Microbiol.* **2018**, *84*, e00333-18. [CrossRef] [PubMed]
47. Hreško, J.; Pucherová, Z.; Baláž, I.; Ambróz, M.; Bezák, P. *Krajina Nitry a jej Okolia [The Landscape of Nitra and its Surroundings]*; UKF: Nitra, Slovakia, 2006; p. 254. (In Slovak)
48. Jankowski, M.; Šimanský, V.; Markiewicz, M.; Pilichowska, A.; Michalak, J. Differently Used Soils of the Tribeč Mountain Range and Nitra Valley Slope. In *Soil Sequences Atlas IV*; Switoniak, M., Charzyński, P., Eds.; Nicolaus Copernicus University: Toruń, Poland, 2018; pp. 139–158.
49. Kottek, M.; Grieser, J.; Beck, C.H.; Rudolf, B.; Rubel, F. World map of the Köppen-Geiger climate classification updated. *Meteorol. Z.* **2006**, *15*, 259–263. [CrossRef]
50. World Reference Base for Soil Resources 2014. In *International Soil Classification System for Naming Soils and Creating Legends for Soil Maps. Update 2015*; World Soil Resources Reports No. 106; FAO: Rome, Italy, 2015. Available online: http://www.fao.org/3/i3794en/I3794en.pdf (accessed on 23 April 2024).
51. Dziadowiec, H.; Gonet, S. *Przewodnik Metodyczny do Badań Materii Organicznej Gleb [Methodological Guidebook for The Organic Matter Researches]*; PTG: Warszawa, Poland, 1999; pp. 31–34. (In Polish)
52. Loginow, W.; Wisniewski, W.; Gonet, S.S.; Ciescinska, B. Fractionation of organic carbon based on susceptibility to oxidation. *Pol. J. Soil Sci.* **1987**, *20*, 47–52.
53. Kačániová, M.; Klūga, A.; Kántor, A.; Medo, J.; Žiarovská, J.; Puchalski, C.; Terentjeva, M. Comparison of MALDI-TOF MS Biotyper and 16S rDNA sequencing for the identification of Pseudomonas species isolated from fish. *Microb. Pathog.* **2019**, *132*, 313–318. [CrossRef] [PubMed]
54. Šimanský, V.; Polláková, N.; Chlpík, J.; Kolenčík, M. *Pôdoznalectvo [Soil Science]*; SPU: Nitra, Slovakia, 2018; p. 398. (In Slovak)
55. Vaněk, V.; Ložek, O.; Balík, J.; Pavlíková, D.; Tlustoš, P. *Výživa Pol'ných a Záhradných Plodín [Nutrition of Field and Garden Crops]*; Profi Press: Praha, Czech Republic, 2013; p. 175. (In Slovak)
56. White, R.E. *Understanding Vineyard Soils*; Oxford University Press: New York, NY, USA, 2015; p. 280.
57. Torre, C.M.; Morano, P.; Tajani, F. Saving soil for sustainable land use. *Sustainability* **2017**, *9*, 350. [CrossRef]
58. Lazcano, C.; Decock, C.; Wilson, S.G. Defining and managing for healthy vineyard soils, intersections with the concept of terroir. *Front. Environ. Sci.* **2020**, *8*, 68. [CrossRef]
59. Šimanský, V.; Bajčan, D. The response of different soil management practices in a vineyard to water availability. *Acta Fytotechn. Zootechn.* **2013**, *16*, 53–57.
60. Graham, P.H.; Draeger, K.J.; Ferrey, M.L.; Conroy, M.J.; Hammer, B.E.; Martinez, E.; Aarons, S.R.; Quinto, C. Acid pH tolerance in strains of Rhizobium and Bradyrhizobium, and initial studies on the basis for acid tolerance of Rhizobium tropici UMR1899. *Can. J. Microbiol.* **1994**, *40*, 198–207. [CrossRef]
61. Zifcakova, L. Factors Affecting Soil Microbial Processes. In *Carbon and Nitrogen Cycling in Soil*; Datta, R., Meena, R.S., Pathan, S.I., Ceccherini, M.T., Eds.; Springer: Berlin/Heidelberg, Germany, 2020; pp. 439–461.
62. Ratzke, C.; Gore, J. Modifying and reacting to the environmental pH can drive bacterial interactions. *PLoS Biol.* **2019**, *16*, e2004248. [CrossRef] [PubMed]
63. Hanes, J. *Analýza Sorpčných Vlastností Pôd [Analyzes of Soil Sorptive Characteristics]*; SSCRI: Bratislava, Slovakia, 1999; p. 138. (In Slovak)

64. Das, S.; Deb, S.; Sahoo, S.S.; Sahoo, U.K. Soil microbial biomass carbon stock and its relation with climatic and other environmental factors in forest ecosystems: A review. *Acta Ecol. Sin.* **2023**, *43*, 933–945. [CrossRef]
65. Jakab, G.; Madarász, B.; Masoudi, M.; Karlik, M.; Király, C.; Zacháry, D.; Filep, T.; Dekemati, I.; Centeri, C.; Al-Graiti, T.; et al. Soil organic matter gain by reduced tillage intensity: Storage, pools, and chemical composition. *Soil Tillage Res.* **2023**, *226*, 105584. [CrossRef]
66. Lal, R. Soil carbon sequestration impacts on global climate change and food security. *Science* **2004**, *304*, 1623–1627. [CrossRef] [PubMed]
67. Masood, S.; Naz, T.; Javed, M.T.; Ahmed, I.; Ullah, H.; Iqbal, M. Effect of short-term supply of farmyard manure on maize growth and soil parameters in pot culture. *Arch. Agron. Soil Sci.* **2013**, *60*, 337–347. [CrossRef]
68. Šimanský, V.; Jonczak, J.; Pikuła, D.; Lukac, M. Grass sward cover improves soil organic carbon and nitrogen in a vineyard. *Soil Sci. Plant Nutr.* **2023**, *69*, 240–249. [CrossRef]
69. Conant, R.T.; Cerri, C.E.P.; Osborne, B.B.; Paustian, K. Grassland management impacts on soil carbon stocks: A new synthesis. *Ecol. Appl.* **2017**, *27*, 662–668. [CrossRef]
70. Tian, K.; Zhao, Y.; Xu, X.; Hai, N.; Huang, B.; Deng, W. Effects of long-term fertilization and residue management on soil organic carbon changes in paddy soils of China: A meta-analysis. *Agric. Ecosyst. Environ.* **2015**, *204*, 40–50. [CrossRef]
71. Shimizu, M.M.; Marutani, S.; Desyatkin, A.R.; Jin, T.; Hata, H.; Hatano, R. The effect of manure application on carbon dynamics and budgets in a managed grassland of Southern Hokkaido, Japan. *Agric. Ecosyst. Environ.* **2009**, *130*, 31–40. [CrossRef]
72. Zhang, J.; Wang, P.; Tian, H.; Xiao, Q.; Jiang, H. Pyrosequencing-based assessment of soil microbial community structure and analysis of soil properties with vegetable planted at different years under greenhouse conditions. *Soil Tillage Res.* **2019**, *187*, 1–10. [CrossRef]
73. Natywa, M.; Sawicka, A.; Wolna-Maruwka, A. Microbial and enzymatic activity in the soil under maize crop in relation to differentiated nitrogen fertilisation. *Water-Environ. Rural Areas* **2010**, *10*, 111–120.
74. Saxena, A.K.; Kumar, M.; Chakdar, H.; Anuroopa, N.; Bagyaraj, D.J. Bacillus species in soil as a natural resource for plant health and nutrition. *J. Appl. Microbiol.* **2020**, *128*, 1583–1594. [CrossRef]
75. Vega-Avila, A.D.; Gumiere, T.; Andrade, P.A.M.; Lima-Perim, J.E.; Durrer, A.; Baigori, M.; Vazquez, F.; Andreote, F.D. Bacterial communities in the rhizosphere of Vitis vinifera L. cultivated under distinct agricultural practices in Argentina. *Anton. Leeuw.* **2015**, *107*, 575–588. [CrossRef]
76. Reeve, J.R.; Schadt, C.W.; Carpenter-Boggs, L.; Kang, S.; Zhou, J.; Reganold, J.P. Effects of soil type and farm management on soil ecological functional genes and microbial activities. *ISME J.* **2010**, *4*, 1099–1107. [CrossRef]
77. Lamb, E.G.; Kennedy, N.; Siciliano, S.D. Effects of plant species richness and evenness on soil microbial community diversity and function. *Plant Soil.* **2011**, *338*, 483–495. [CrossRef]
78. Zarraonaindia, I.; Owens, S.M.; Weisenhorn, P.; West, K.; Hampton-Marcell, J.; Lax, S.; Bokulich, N.A.; Mills, D.A.; Martin, G.; Taghavi, S.; et al. The soil microbiome influences grapevine-associated microbiota. *mBio* **2015**, *6*, e02527-14. [CrossRef] [PubMed]
79. Rousk, J.; Bååth, E.; Brookes, P.C.; Lauber, C.L.; Lozupone, C.; Caporaso, J.G.; Knight, R.; Fierer, N. Soil bacterial and fungal communities across a pH gradient in an arable soil. *ISME J.* **2010**, *4*, 1340–1351. [CrossRef] [PubMed]
80. Fierer, N.; Jackson, R.B. The diversity and biogeography of soil bacterial communities. *Proc. Natl. Acad. Sci. USA* **2006**, *103*, 626–631. [CrossRef]
81. Burns, K.N.; Kluepfel, D.A.; Strauss, S.L.; Bokulich, N.A.; Cantu, D.; Steenwerth, K.L. Vineyard soil bacterial diversity and composition revealed by 16S rRNA genes: Differentiation by geographic features. *Soil Biol. Biochem.* **2015**, *91*, 232–247. [CrossRef]

horticulturae

MDPI

Article

Effect of Transplanting Time and Nitrogen–Potassium Ratio on Yield, Growth, and Quality of Cauliflower Landrace Gigante di Napoli in Southern Italy

Alessio Vincenzo Tallarita [1], Eugenio Cozzolino [2,*], Antonio Salluzzo [3], Agnieszka Sekara [4], Robert Pokluda [5], Otilia Cristina Murariu [6,*], Lorenzo Vecchietti [7], Luisa del Piano [2], Pasquale Lombardi [8], Antonio Cuciniello [2] and Gianluca Caruso [1]

1 Department of Agricultural Sciences, University of Naples Federico II, Portici, 80055 Naples, Italy; alessiovincenzo.tallarita@unina.it (A.V.T.); gcaruso@unina.it (G.C.)
2 Council for Agricultural Research and Economics (CREA)—Research Center for Cereal and Industrial Crops, 81100 Caserta, Italy; luisa.delpiano@crea.gov.it (L.d.P.); antonio.cuciniello@crea.gov.it (A.C.)
3 Territorial and Production Systems Sustainability Department—Research Centre Portici, Italian National Agency for New Technologies, Energy and Sustainable Economic Development (ENEA), 80055 Portici, Italy; antonio.salluzzo@enea.it
4 Department of Horticulture, Faculty of Biotechnology and Horticulture, University of Agriculture, 31-120 Krakow, Poland; agnieszka.sekara@urk.edu.pl
5 Department of Vegetable Growing and Floriculture, Faculty of Horticulture, Mendel University, 613 00 Brno, Czech Republic; robert.pokluda@mendelu.cz
6 Department of Food Technology, University 'Ion Ionescu de la Brad' of Life Sciences of Iasi, 700490 Iasi, Romania
7 Hydro Fert S.r.l., 76121 Barletta, Italy; l.vecchietti@hydrofert.it
8 Research Center for Vegetable and Ornamental Crops, 84098 Pontecagnano Faiano, Italy; pasquale.lombardi@crea.gov.it
* Correspondence: eugenio.cozzolino@crea.gov.it (E.C.); otilia.murariu@iuls.ro (O.C.M.)

Abstract: Research has been increasingly focusing on the preservation of the biodiversity of vegetable crops under sustainable farming management. An experiment was carried out in southern Italy on *Brassica oleracea* L. var. botrytis, landrace Gigante di Napoli, to assess the effects of two transplanting times (9 September and 7 October), in factorial combination with five nitrogen–potassium ratios (0.6; 0.8; 1.0; 1.2; and 1.4) on plant growth, yield, and quality of cauliflower heads. A split-plot design was used for the treatment distribution in the field, with three replications. The earlier transplant and the 1.2 N:K ratio led to the highest yield, mean weight, and firmness of cauliflower heads which were not significantly affected by both transplanting time and N:K ratio in terms of colour components. The 1.2 N:K ratio led to the highest head diameter with the earlier transplant, whereas the 1.0 ratio was the most effective on this parameter in the later crop cycle. The highest nitrate, nitrogen, and potassium concentrations in the heads were recorded with the earlier transplanting time. Antioxidant activity, ascorbic acid, and polyphenol content increased with the rise of the N:K ratio. The element use efficiency was constantly negative with the N:K increase for nitrogen and was augmented until the 1.2 ratio for potassium. The results of our investigation showed that the optimal combination between transplanting time and N:K ratio is a key aspect to improve head yield and quality of the cauliflower landrace Gigante di Napoli, under the perspective of biodiversity safeguarding and valorisation.

Keywords: local variety; *Brassica oleracea* var. botrytis; antioxidant activity; ascorbic acid; polyphenols

1. Introduction

The preservation of agricultural biodiversity encompasses a multitude of practices aimed at sustaining the genetic variability of plants, especially those utilised in food production [1]. Within the latter context, the cauliflower landrace (*Brassica oleracea* L. var. botrytis) Gigante di Napoli is a traditional crop in the Campania region (Italy), with an important cultural heritage and cultivation practices traditionally rooted in local agriculture. It is appreciated for its organoleptic quality, high texture, colour ranging from white to cream of the heads, delicate taste, and versatility in cooking [2,3].

From the agronomic point of view, Gigante di Napoli requires specific choices to express its full development potential, including the most appropriate transplanting time and adequate fertilisation management, particularly referring to the nitrogen/potassium ratio. The latter aspects influence both the yield and quality of the final product and the preservation of the global characteristics of this landrace [4,5].

The cultivation and promotion of Gigante di Napoli are essential either to keep alive a part of Italian agricultural and cultural heritage or to offer farmers an opportunity to diversify and sustainably manage the crop system. In order to make it easier and feasible for farmers to choose and cultivate local varieties, it is crucial to help them identify, under the current climate change conditions, the best fitting transplanting time and fertilisation methods.

Transplanting time is a key factor that significantly influences plant growth, and crop yield and quality [6,7]. The choice of the optimal transplanting time is affected by several factors, including genotype, climatic and cultivation requirements [8,9]. Based on cauliflower characteristics for germination, growth, and ripening [10], temperatures under 10 °C can slow down its development, while temperatures higher than 25 °C may induce the premature 'sprout' of the heads, compromising the quality of the final product [11]. Moreover, the choice of the appropriate transplanting time allows for a more efficient use of water and nutrient resources [12], both by reducing the need for supplementary irrigation and identifying the phases of the most efficient plant nutrient uptake, thus optimising the fertilisation.

The ratio between nitrogen (N) and potassium (K) in the soil plays a crucial role in plant nutrition and directly influences crop yield and quality [13]. The two mentioned elements are among the primary nutrients needed by plants and have complementary but distinct functions that, when properly balanced, can lead to optimal results in terms of yield and quality. In fact, nitrogen is essential for the synthesis of proteins, nucleic acids, and chlorophyll, promoting leaf development and plant tissue formation through its key function in photosynthesis. Adequate nitrogen availability in the soil supports rapid and vigorous growth, but an excess of this mineral element may lead to excessive leaf growth which slows and reduces flowering and/or fruiting, as well as potential negative impacts on the environment, such as the eutrophication of surface waters [14,15].

Potassium is vital for osmotic regulation, enzyme activation and sugar transport, contributing to plant resistance to stress, improving their ability to withstand adverse conditions such as drought, cold, and disease [16]. This element is also crucial to fruit quality, influencing attributes such as colour, flavour, firmness, and storability [17]. The most appropriate N:K ratio is therefore essential for balancing vegetative growth with the reproduction phase and stress resistance, with consequent positive effects on yield and crop quality. Determining the optimal ratio depends on the genotype, and the specific environmental conditions of the cultivation area [18,19].

This study aimed to investigate the effects of transplanting time and nitrogen/potassium ratio on plant growth, yield, and quality of cauliflower landrace Gigante di Napoli in the Campania region (southern Italy), to highlight the importance of tailored agricultural practices in maintaining crop diversity, thus supporting sustainable agriculture and food security in the face of climate change.

2. Materials and Methods

2.1. Plant Material, Growing Conditions, and Experimental Protocol

Research was conducted at the Department of Agricultural Sciences, University of Naples Federico II, Naples, Italy (40°500 N, 14°150 E, 17 m a.s.l., in a Mediterranean or Csa climate according to the Köppen classification scheme) [20] on cauliflower (*Brassica oleracea* (L.) var. botrytis, landrace Gigante di Napoli), in open fields, in 2021 and 2022. Cauliflower plants were grown in a sandy loam soil and the roots developed in the superficial 40 cm profile having 77% sand, 16% silt, and 7% clay, and 2.3% organic matter content (*w*/*w*), 0.13% total nitrogen, 33.7 mg kg^{-1} P, and 1369 mg kg^{-1} exchangeable K. The average monthly total rainfall and temperature were the following: 10 mm and 25 °C in September; 20 mm and 22 °C in October; 90 mm and 18 °C in November; 70 mm and 14 °C in December; and 60 mm and 10 °C in January. The plants were spaced 50 cm along the rows which were 70 cm apart, with a density of 2.9 plants per m^2.

The experimental protocol was based on the factorial combination of two transplanting times (9 September, 7 October) and five nitrogen/potassium ratios (0.6; 0.8; 1.0; 1.2; and 1.4), using a split-plot design for the treatment distribution in the field, with three repetitions, with the experimental unit covering a 10.5 m^2 (3.5 × 3.0 m) surface area. The N:K ratios were applied by keeping constant the potassium dose and changing the nitrogen one, based on the following average cauliflower requirements: 250 kg ha^{-1} N, 60–100 kg ha^{-1} P_2O_5, and 250 kg ha^{-1} K_2O. The mentioned fertilisations were achieved by supplying calcium nitrate (15.5 N, 23.0 CaO), potassium nitrate (13 N, 46 K_2O), potassium sulphate (50 K_2O, 45 SO_3), and phosphoric acid (54% P_2O_5), partly during pre-transplanting (35%) and the remainder during the crop cycles (65%).

Drip irrigation was activated when the soil available water capacity (AWC) decreased to 70%. The harvests were carried out when the cauliflower heads completed their growth, cutting the whole plants of the central rows at 2 cm above soil level, excluding the border ones, on 21 November and 27 December, for the earlier and the later transplanting times, respectively, as an average of 2021 and 2022.

At harvest, 15-plant samples were randomly collected in each plot to assess the fresh and dry weight (in an oven at 70 °C under vacuum, at 15 kPa pressure, and until they reached constant weight) of the cauliflower heads.

For each sample, firmness (digital penetrometer with 8 mm tip T.R. Turoni s.r.l., Forlì, Italy) and colour (L*, a*, b*—CIElab), using a Minolta CR-300 Chroma Meter (Minolta Camera Co. Ltd., Osaka, Japan), were measured.

2.2. Nitrate and N, P, and K Determinations

Head samples of 0.5 g per treatment were ground using an IKA mill (IKA-Werke, Staufen, Germany), sieved through a 2 mm screen, to measure nitrate content by a Foss continuous flow Analyzer—FIAstar 5000 (FOSS Italia S.r.l., Padova, Italy), as reported by Di Mola et al. [21]. An aliquot of the samples was washed with ultrapure water, dried, and stored in the freezer (−80 °C) to perform the following analyses: ascorbic acid, total phenols, and antioxidant activity. Frozen 'curds' were roughly cut with a ceramic blade knife, brought back to −20 °C, and homogenized in a knife mill Grindomix GM300 Retsch (Haan, Germany). The grinding program was in two steps: 1000 rpm revolution speed, direction impact, for 20 s (pre-grinding) and 2000 rpm revolution speed, direction cut, for 30 s (fine grinding). The homogenate obtained was divided into sub-samples and stored at −80 °C and used for the determination of ascorbic acid, total phenols, and antioxidant activity.

2.3. Ascorbic Acid

Ascorbic acid extraction procedure (in duplicate for each sample) was conducted according to the AOAC's *Official Methods of Analysis*. Briefly, 10 mL of extraction solution (30 g L^{-1} MPA—80 mL L^{-1} acetic acid—1 mmol L^{-1} EDTA) were added to 3–4 g of homogenate and the mixture was shaken for 3 min with a disperser (Ultra-TuraxT25 IKA-Werke GmbH & Co. KG, Staufen, Germany). The resulting extract after centrifugation (10,000 rpm; 20 min; 4 °C) was immediately analysed by the colorimetric method using the Folin Phenol Reagent according to Jagota and Dani [22]. To 400 µL of the sample, 200 µL of extraction solution, 1.2 mL of ultrapure water, and 200 µL of Folin–Ciocalteu's phenol reagent 2N (diluted 1:5 with ultrapure water) were added. After 30 min in darkness and at room temperature, the absorbance of the blue colour developed was measured at 760 nm.

This procedure was repeated in triplicate. For quantification, six standard solutions of ascorbic acid (range 25–150 μg mL^{-1}) were prepared and estimated as above ($R^2 \geq 0.996$).

2.4. Total Phenols

Total phenol extraction was performed following the procedure as described by Kaur et al. [23] with minor modifications. Approximately 5 g of sample (in triplicate) were homogenized in 25 mL of 80:20 CH_3OH/H_2O mixture and sonicated in an ultrasonic bath (Elmasonic P Elma, Singen, Germany) at 40 °C for 3 h in darkness. Afterwards, the extracts were centrifuged for 20 min at 10,000 rpm (HF 14.94 rotor, radius 9.7 1086 RCF) at 4 °C (Centifuge 17R Heraeus Sepatech, Waltham, MA, USA) and the upper layer was separated and kept at −20 °C until analyses were performed. The total phenolic content was determined according to Folin–Ciocalteu's method [24] with small modifications. In the test tube, the following were added in sequence: 1 mL H_2O, 100 μL of cauliflower extract, 100 μL of Folin–Ciocalteu's phenol reagent 2 N solution, and, after 10 min, 800 μL of Na_2CO_3 75 g L^{-1} solution, so that the final pH was ≥10 (test final volume 2 mL). The mixture was allowed to stand for 120 min at room temperature and in darkness, and absorption was measured at 765 nm against a reagent blank. Measurements were performed in duplicate, and results were expressed as mg of gallic acid equivalent per 100 mg of fresh weight (GAE 100 g^{-1} f.w.). For calibration, six standard solutions of gallic acid (range 0–150 μg mL^{-1}) were prepared and estimated as above ($R^2 \geq 0.998$).

2.5. Antioxidant Activity

The antioxidant activity was measured on the cauliflower extract prepared as for total phenolic determination, using the DPPH assay based on the measurement of the scavenging ability of the antioxidant toward the stable radical DPPH. A 3.9 mL aliquot of a 0.0634 mM of DPPH solution in methanol was added to 0.1 mL of each extract and shaken vigorously. The change in the absorbance of the mixture at 515 nm was observed and the antiradical activity of the sample was evaluated from the percentage of DPPH remaining when the kinetics reached a steady state. It has been found experimentally that for cauliflower fruit, this occurs after 150 min. A stock solution of DPPH 634 μM (12.5 mg in 50 mL of degassed methanol) was prepared at the assay moment. For the test, instead, a 10-fold fresh diluted solution is used. Dark glass test tubes were used with Teflon-coated rubber septa and aluminium caps (Supelco Inc., Bellefonte, PA, USA), and the DPPH solution was degassed and flushed with helium gas for 3 min. Trolox equivalency is used as a benchmark for the oxidant capacity of the extracts. Therefore, the calibration curve was constructed using nine standard solutions containing this antioxidant (range 0.15–20 μM) and the results are expressed as μmol of Trolox equivalent per 100 g of fresh weight (μmol TE 100 g^{-1} f.w.).

2.6. Nitrogen Use Efficiency (NUE) and Potassium Use Efficiency (KUE) on Yield Calculation

For nitrogen use efficiency (NUE) and potassium use efficiency (KUE) on yield, the following formulas, based on average values referred to the variable examined per level of N:K treatment, were adopted:

$$NUE = \frac{\text{Dry weight of crop}\left(\text{kg ha}^{-1}\right)}{\text{Amount of nitrogen applied}\left(\text{kg ha}^{-1}\right)} \qquad KUE = \frac{\text{Dry weight of crop}\left(\text{kg ha}^{-1}\right)}{\text{Amount of potassium applied}\left(\text{kg ha}^{-1}\right)}$$

2.7. Statistical Analysis

Data were processed by analysis of variance (two-way ANOVA) and mean separations were performed through the Tukey's test, with reference to 0.05 probability level, using the SPSS software version 29.

3. Results and Discussion

3.1. Yield and Growth Parameters

No significant differences arose between the two research years (2021 and 2022) regarding the variables examined and, therefore, the average values were reported.

The later transplanted crops exhibited a longer growing period, compared to the earlier ones (81.4 vs. 73.1 days), resulting in significantly higher yields (11.2 vs. 9.0 t ha^{-1}), as observed in Table 1.

Table 1. Effect of transplanting time and N:K ratio on yield and plant growth parameters of cauliflower.

Experimental Treatment	Crop Cycle Duration (Days)	Yield (t ha^{-1})	Head Fresh Weight (kg)	Head Dry Weight (kg)	Firmness (N)	Colour L*	Colour a*	Colour b*
Transplanting time								
9 September	73.1 b	11.1 a	0.83 a	0.17	6.65 a	85.9	0.05	0.65
7 October	81.4 a	9.0 b	0.61 b	0.16	5.44 b	85.8	0.54	0.04
				n.s.		n.s.	n.s.	n.s.
N:K ratio								
0.6	74.8 b	9.2 d	0.41 d	0.14 c	4.71 d	85.5	0.44	−0.10
0.8	75.1 b	9.1 d	0.47 c	0.15 c	5.81 c	86.1	−1.04	0.85
1.0	76.9 ab	11.0 b	0.76 b	0.17 b	6.45 b	87.0	0.62	0.83
1.2	79.7 a	11.3 a	1.10 a	0.19 a	6.96 a	85.6	0.81	−0.17
1.4	79.8 a	9.8 c	0.86 b	0.17 b	6.31 b	84.8	0.40	0.31
						n.s.	n.s.	n.s.

n.s. not significant; within each column, mean values followed by different letters are significantly different at $p \leq 0.05$ according to Tukey's test.

No significant interactions between the two experimental factors on the variables examined arose, with the exception of the diameter of cauliflower heads (Figure 1), which was the highest corresponding to 1.2 N:K with the earlier transplant and to 1.0 with the later one. The earlier transplanting time also led to a higher mean head fresh weight (0.83 kg), compared to the later one (0.61 kg). However, no significant differences were detected in the dry weight, indicating that the increased biomass in the earlier crops was due to the higher water content. Furthermore, the head firmness associated with the 9 September transplant was higher (6.65 N) as compared to the later crop cycle (5.44 N). The colour parameters (L*, a*, b*) were not significantly affected by the transplanting time.

The 1.2 and 1.4 N:K ratios led to a longer crop cycle, compared to 0.6 and 0.8. The highest yield was recorded at 1.2 N:K (11.3 t ha^{-1}), due to the highest head fresh weight (1.10 kg), and the lowest at 0.6 ratio (0.41 kg).

Both the head dry weight and firmness reached the highest value at 1.2 N:K (0.19 kg and 6.96 N, respectively) and the lowest at 0.6 (0.14 kg and 4.71 N, respectively).

The colour components (L*, a*, b*) were not significantly affected by the N:K ratio.

In a previous investigation [25], the transplanting time significantly influenced the growth and yield of cauliflower, with the transplant practiced at the end of November enhancing biometrical parameters and yield, particularly plant height, leaf number, stem diameter, and head size.

Furthermore, Kartika et al. [26] reported the detrimental effects of delayed transplanting time on yield and plant growth indicators, such as leaf number, canopy area, and curd characteristics. Moreover, Sultana et al. [27] found that the optimal transplanting time to maximize growth and yield also depends on cultivar and latitude.

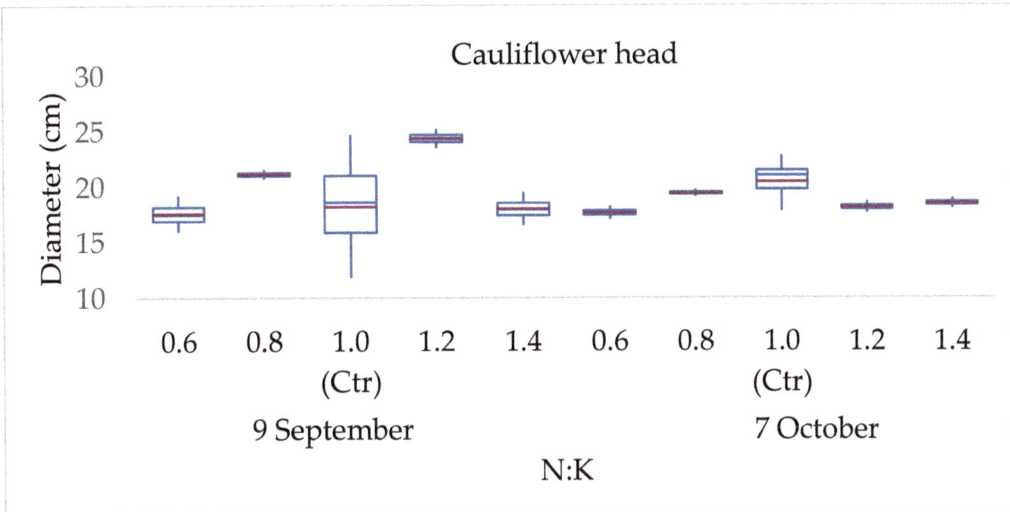

Figure 1. Interaction between transplanting time and N:K ratio on cauliflower head diameter.

Cauliflower plants experience physiological and biochemical reactions when subjected to cold temperatures, encompassing acclimatisation which allows them to progressively adapt to the modified environment, thus enhancing their ability to benefit from the mentioned stress [28] resulting in a remarkable rise in marketable yields. In this respect, the adaptation to low temperatures involves several phenomena, including alterations in protein synthesis, variations in membrane composition, and activation of antioxidant defence systems.

Rahman et al. [29] highlighted the relationship between planting time and plant hormonal status, which affects cauliflower production. Wadan et al. [30] investigated the effects of nitrogen/potassium ratios on yield, revealing that a combination of 120 kg N and 90 kg K per hectare significantly enhanced yield and growth parameters, including head weight and leaf area.

3.2. Nitrate, Nitrogen, Phosphorus, and Potassium Concentrations

In a study conducted by Yildirim [31], it was shown that raising the amount of nitrogen fertiliser in the rhizosphere led to an increase in the availability of other minerals in the soil solution which, in turn, resulted in improved growth and higher yield and weight of cauliflower heads. Further, the amount of nitrogen supply is reportedly crucial for plant development as well as nitrate absorption and accumulation [32].

Marschner [33] emphasized the significance of the nitrogen–potassium relationship in plant metabolism and the potential consequences of imbalances, i.e., a decreased rate of photosynthesis and nitrogen absorption, as well as the influence of the ratio between the two elements in the soil on their absorption.

Nitrogen has a crucial role in promoting plant growth as it is a vital constituent of chlorophyll. Probably, the higher nitrogen availability resulted in higher chlorophyll levels, which improves the plant capacity to convert light into energy for its growth and development. This hypothesis is consistent with the reports of Katuwal et al. [34] who found that cauliflower plants subjected to different nitrogen levels showed an increase in head fresh weight, leaf area, and plant height, upon the augmentation of nitrogen supply. The latter is notoriously essential for the synthesis of proteins, which have a crucial role in the organization and functioning of plant cells, including the enzyme enhancement of all metabolic processes.

The influence of transplant nutrient conditioning (TNC) and the nitrogen–potassium (N:K) ratio on the growth characteristics, and on cauliflower diameter, was also reported in previous study [31], elucidating that TNC at different levels of N, P, and K significantly impacted early growth traits such as stem diameter in cauliflower seedlings.

An increase in TNC levels of N enhanced growth but also induced greater transplant shock, which was compensated by faster growth in later stages, suggesting a critical balance between nutrient provision and the plant stress response.

Further confirming the pivotal role of nitrogen, Katuwal et al. [34] demonstrated that the application of 130 kg ha^{-1} N improved yield and biometrical parameters, including stem diameter and head girth.

Similarly, Cutcliffe and Munro [35] reported that cauliflower yield was significantly augmented by nitrogen and phosphorus applications, optimally at N applications ranging from 112 to 224 kg ha^{-1}.

Wadan et al. [30] further reported that fertilizer levels of 120 kg ha^{-1} N plus 90 K kg ha^{-1} substantially enhanced yield performance in terms of head weight and diameter and leaf area. The latter finding indicates that the appropriate N:K ratio is crucial to achieve optimal growth and head size in cauliflower production.

As shown in Table 2, the earlier transplanting time led to higher concentrations of nitrate and nitrogen content in cauliflower heads (341 and 3728 ppm, respectively), compared to the later transplant (278.9 and 3050 ppm, respectively). Potassium accumulated at a significantly higher extent in the heads for the earlier transplanting time compared to the later one.

Table 2. Effect of transplanting time and N:K ratio on quality parameters of cauliflower heads.

Experimental Treatment	Nitrate (mg kg^{-1})	N (mg kg^{-1})	P (mg kg^{-1})	K (mg kg^{-1})
Transplanting time				
9 September	341.00 a	3728.27 a	1081.25	4454.17 a
7 October	278.90 b	3050.41 b	884.62	3644.32 b
			n.s.	
N:K ratio				
0.6	393.90 a	3706.25 a	1016.21 a	5816.14 a
0.8	342.23 b	3483.88 b	1036.50 a	4942.54 b
1.0	303.21 c	3449.48 b	999.52 a	3105.98 c
1.2	273.28 d	3157.52 c	958.75 ab	3378.85 c
1.4	237.36 e	3149.57 c	903.59 b	3002.73 d

n.s. not significant; within each column, mean values followed by different letters are significantly different at $p \leq 0.05$, according to Tukey's test.

Increasing the nitrogen–potassium ratio, a decreasing trend of nitrate and nitrogen concentrations in cauliflower heads was recorded. The lowest nitrogen–potassium ratio (0.6) resulted in the highest nitrate content (393.9 ppm), which gradually decreased up to the 1.4 ratio (237.4 ppm), and a similar trend was shown by nitrogen. Phosphorus content was the highest at the lowest nitrogen–potassium ratio, whereas potassium content decreased with increasing N:K ratio (from 5816.1 ppm at 0.6 to 3002.73 ppm at 1.4).

Nitrate reduction is a stage in nitrate assimilation affecting the latter process [36]; nitrate reductase is the related enzyme that is activated in response to light conditions, and a strong correlation exists between the activity of nitrate reductase and the concentration of nitrate in plants [37]. In addition, the expression of nitrate absorption and reduction systems is induced by both endogenous nitrogen and plant development [38]. According to Chen et al. [39], the growth of cabbage plants is strongly influenced by the amount of nitrogen they receive. When the nitrogen supply exceeded the optimal threshold, the plants experienced a decrease in growth and an increase in nitrate accumulation leading to toxicity symptoms caused by high levels of NO_3^- concentration, which in turn caused a biomass decrease. However, over the mentioned threshold, there was a slower rate of rise in nitrate concentration, maybe due to a reduction in nitrate absorption as a consequence of a decrease in plant growth, which also resulted in yield loss. The efficiency of the net nitrate absorption rate significantly improves with the increase of plant relative growth rate, as observed by Ter Steege et al. [40], which highlights the critical role of plant growth in nitrate uptake. While most studies have demonstrated that nitrate reductase is an enzyme induced by its substrate [41], a few investigations have shown that even a small amount of nitrate is enough to trigger nitrate reductase induction [42]. Additionally, the expression of nitrate reductase may be regulated by either the flow of nitrogen or the nitrogen status of plants [43]. In the trials conducted by Chen et al. [39], it was shown that nitrate had a beneficial impact on nitrate reductase activity only when the nitrate supplies were low. However, when the nitrogen inputs were at their optimal level, the nitrate reductase activity either achieved a plateau value or even dropped.

Recent research elucidated the intricate dynamics between nitrogen, phosphorus, and potassium applications and their effects on cauliflower growth, yield, maturity, and nutrient content. In fact, Sony et al. [44] highlighted the significant N-P interaction on growth parameters and yield, suggesting that tailored N-P management can strongly impact nitrate, N, P, and K accumulation in plants whereas, in contrast, in Everaarts' [45] investigation, the increased N application rates did not augment the nitrogen content.

Metwaly [46] studied the nutrient dynamics by examining the effects of P and K on cauliflower, demonstrating that the relationship between these nutrients significantly affected vegetative growth, leaf chemistry, and head yield and quality, thereby emphasizing the importance of optimizing P and K levels along with N management to enhance cauliflower production and nutrient profile optimization.

3.3. *Quality Parameters*

The later transplant elicited a higher concentration of ascorbic acid (4.6 mg 100 g^{-1} f.w.) and total phenols (78.4 mg equivalent GAE 100 g^{-1} f.w.) compared to the earlier one (3.7 mg 100 g^{-1} f.w. for ascorbic acid and 64.2 mg equivalent GAE 100 g^{-1} f.w. for total phenols; Table 3). The antioxidant activity did not show a significant difference between the two transplanting times.

Table 3. Effect of transplanting time and N:K ratio on ascorbic acid, antioxidant activity, and total polyphenols of cauliflower curds.

Experimental Treatment	Ascorbic Acid (mg 100 g^{-1} f.w.)	Antioxidant Activity (μmol Trolox Equivalent 100 g^{-1} f.w.)	Total Phenols (mg Equivalent GAE 100 g^{-1} f.w.)
Transplanting			
9 September	3.74 b	368.06	64.17 b
7 October	4.55 a	362.97	78.43 a
		n.s.	
N:K			
0.6	3.37 c	330.44 b	55.57 c
0.8	3.95 b	368.69 ab	67.25 b
1.0	3.92 b	346.49 b	69.28 b
1.2	4.88 a	387.42 ab	79.78 ab
1.4	4.63 a	394.50 a	84.66 a

n.s. not significant; within each column, mean values followed by different letters are significantly different at $p \leq 0.05$ according to Tukey's test.

The ascorbic acid content was the lowest at the 0.6 ratio (3.4 mg 100 g^{-1} f.w.) and increased progressively with a rising nitrogen–potassium ratio, peaking at 1.2 with 4.9 mg 100 g^{-1} f.w. The antioxidant activity and total polyphenols also showed an increasing trend from the lowest to the highest nitrogen–potassium ratio. Both the highest antioxidant activity and total polyphenol content were recorded at 1.4 (394.5 μmol Trolox equivalent 100 g^{-1} f.w. and 84.7 mg equivalent GAE 100 g^{-1} f.w.).

The mentioned findings reflect the favourable conditions associated with delayed transplanting and an increased nitrogen–potassium ratio for enhancing the phytochemical content of plants, augmenting their nutritional value and health benefits.

Some studies highlighted the significant impact of transplanting time and the applied nitrogen–potassium (N:K) ratio on the vitamin C content, antioxidant activity, and polyphenol levels in cauliflower [47]. Contrary to the trend recorded in our study, Lisiewska and Kmiecik [47] found that the increasing nitrogen fertilization decreased the vitamin C content in cauliflower. Furthermore, Giri [48] reported that a lower dose of nitrogen, when combined with organic manures, improved the post-harvest quality of cauliflower, including higher vitamin C content, thus valorising sustainable fertilization strategies. Šlosár et al. [49] observed that nitrogen and sulphur fertilization is a key factor to enhance the accumulation of health-promoting compounds, positively affecting the content of vitamin C and β-carotene in cauliflower. Moreover, a previous study carried out on coloured potato cultivars [50], reported that nitrogen was more effective than potassium in increasing anthocyanin content, and the modulation of the ratio between these two elements was an effective way to increase the expression of polyphenolic compounds.

3.4. Nitrogen and Potassium Use Efficiency

In Figure 2, the trend of nitrogen and potassium use efficiency related to the average yield of cauliflower heads is illustrated. The nitrogen use efficiency showed a negative trend from the lowest to the highest nitrogen dose applied. Differently, the potassium use efficiency displayed a slightly positive trend from 0.8 to 1.2 (N:K), then decreasing from 1.2 to 1.4. The decreasing trend of nitrogen use efficiency (NUE) with increasing N:K ratios, along with a stable potassium use efficiency (KUE), might be due to a complex interplay of factors. Plants require a balanced nutrition which can be altered by an excess of nitrogen supplied, especially if potassium, which is crucial for stress tolerance and enzyme activation, is not increased proportionally. The mentioned imbalance could lead to excessive consumption of nitrogen, causing plant absorption to be higher than needed which does not turn out into an increased head yield, thus leading to inefficiency. The relatively stable efficiency of potassium suggests that this element was remarkably used by the plants, indicating its sufficiency across the ratios. Furthermore, nutrient competition at the root uptake sites may limit the availability and absorption of some elements, which are also influenced by the soil properties, such as pH and microbial activity.

Nitrogen use efficiency (NUE) is an essential parameter to measure the efficiency of plant utilization of available nitrogen (N) for growth and yield, encompassing the processes of N uptake, assimilation, and use within a plant system. Studies of different plant species and environmental conditions reveal that NUE is a complex trait influenced by both genetic and environmental factors, showing variability under different levels of nitrogen supply [51]. In *Helianthus annuus*, a trade-off between nitrogen productivity (NP) and mean retention time of nitrogen (MRT) suggests that plants optimize their nitrogen use strategies based on available N levels, either by maximizing nitrogen conservation or utilization [51]. At the molecular level, as observed in *Brassica napus* under low nitrogen (LN) conditions, NUE is enhanced through the regulation of root architecture and nitrogen transporter activity, specifically through the upregulation of nitrate transporters like BnNRT1.5 and downregulation of BnNRT1.8, facilitating efficient nitrate uptake and translocation. These adjustments are part of a plant's adaptive response to varying N availability, aiming to maintain growth and yield with minimal nitrogen input, reflecting an intricate balance between the demand for nitrogen and its availability in the environment [52].

According to previous works [46,53], the combined effects of nitrogen and potassium enhance the growth and yield of cauliflower by increasing the uptake of nutrients, such as nitrogen for protein synthesis, and potassium for regulating water use through stomatal conductance, and enhancing resistance to stress. Once the optimal point is reached, any additional increases in N and K concentrations do not result in further growth or yield advantages, maybe because of the limitation of the plant's capacity to uptake nutrients, inefficiencies in water usage and stomatal regulation, and imbalances in nutrient distribution. The mentioned factors reflect a delicate balance between nutrient, water, and physiological requirements for maximizing plant yield.

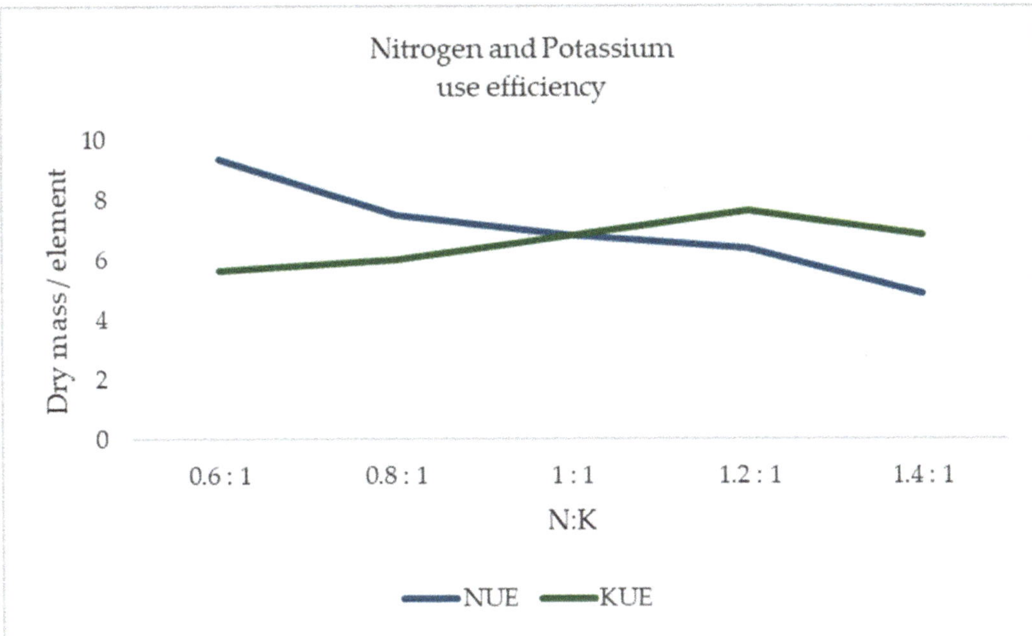

Figure 2. Nitrogen use efficiency (NUE) and potassium use efficiency (KUE) in relation to N:K.

4. Conclusions

From this research it can be inferred that the choice of both the most appropriate transplanting time and nitrogen–potassium ratio is crucial for the cauliflower landrace Gigante di Napoli, under the perspective to sustain 'niche' crops that are useful both for the economy closely linked to an agricultural area and for its biodiversity. The earlier transplanting time (9 September) proved to be the most effective for head yield, while the later one (7 October) for the quality characteristics. A nitrogen–potassium ratio above 1.2 was not effective to enhance either yield or curd quality. Nitrogen and potassium use efficiency showed different trends, with the first one decreasing with the augmentation of nitrogen supplied, and the second one increasing from 0.8 to 1.2 N:K ratio. From this study, a significant effect of the nitrogen–potassium ratio arose on the yield and quality of the cauliflower landrace Gigante di Napoli, whose importance closely relates to biodiversity and land preservation. The latter aim should be pursued, targeting the compromise between efficient plant nutrition and territory safeguarding, and choosing the best fitting transplanting time both to enhance production and face climate changes.

Author Contributions: Conceptualization, A.V.T. and G.C.; methodology, A.V.T., A.S. (Antonio Salluzzo), A.S. (Agnieszka Sekara), O.C.M. and L.d.P.; software, A.V.T., E.C. and P.L.; validation, R.P. and G.C.; formal analysis, A.V.T., E.C., A.S. (Antonio Salluzzo) and L.d.P.; investigation, A.V.T., E.C., L.d.P., P.L. and A.C.; data curation, A.V.T., E.C. and A.S. (Antonio Salluzzo); writing—original draft preparation, A.V.T.; writing—review and editing, A.S. (Agnieszka Sekara), R.P., O.C.M. and G.C.; visualization, A.V.T.; supervision, G.C.; project administration, A.V.T., L.V. and G.C.; funding acquisition, L.V. and G.C. All authors have read and agreed to the published version of the manuscript.

Funding: This research received no external funding.

Data Availability Statement: Data is contained within the article.

Conflicts of Interest: Author Lorenzo Vecchietti was employed by the company Hydro Fert S.r.l. The remaining authors declare that the research was conducted in the absence of any commercial or financial relationships that could be construed as a potential conflict of interest.

References

1. Salgotra, R.K.; Chauhan, B.S. Genetic diversity, conservation, and utilization of plant genetic resources. *Genes* **2023**, *14*, 174. [CrossRef] [PubMed]
2. Dudley, N.; Alexander, S. Agriculture and biodiversity: A review. *Biodiversity* **2017**, *18*, 45–49. [CrossRef]
3. Frison, E.; Cherfas, J.; Hodgkin, T. Agricultural biodiversity is essential for a sustainable improvement in food and nutrition security. *Sustainability* **2011**, *3*, 238–253. [CrossRef]
4. Falco, S. On the value of agricultural biodiversity. *Annu. Rev. Resour. Econ.* **2012**, *4*, 207–223. [CrossRef]
5. Gabriel, D.; Sait, S.M.; Kunin, W.E.; Benton, T.G. Food production vs. biodiversity: Comparing organic and conventional agriculture. *J. Appl. Ecol.* **2013**, *50*, 355–364. [CrossRef]
6. Everaarts, A. Nitrogen balance during the growth of cauliflower. *Sci. Hortic.* **2000**, *83*, 173–186. [CrossRef]
7. Kodithuwakku, D.; Kirthisinghe, J.P. The effect of different rates of nitrogen fertilizer application on the growth, yield, and postharvest life of cauliflower. *Trop. Agric. Res.* **2010**, *21*, 110–114. [CrossRef]
8. Xie, Y.; Kristensen, H. Overwintering grass-clover as intercrop and moderately reduced nitrogen fertilization maintain yield and reduce the risk of nitrate leaching in an organic cauliflower (*Brassica oleracea* L. var. botrytis) agroecosystem. *Sci. Hortic.* **2016**, *206*, 71–79. [CrossRef]
9. Boogaard, R.; Thorup-Kristensen, K. Effects of nitrogen fertilization on growth and soil nitrogen depletion in cauliflower. *Acta Agric. Scand. Sect. B—Soil Plant Sci.* **1997**, *47*, 149–155.
10. Meng, H.; Xinsheng, H.; Jing, W.; Jiang, D.-F. Study of the effect of nitrogen, phosphorus, and potassium fertilizer combined application on the yield of peanut intercrop cauliflower. *J. Qingdao Agric. Univ.* **2010**, *27*, 285–288.
11. Kalisz, A.; Gil, J.; Kunicki, E.; Sękara, A.; Sałata, A.; Caruso, G. Different temperature regimes influenced the quality of broccoli seedlings, which caused a change in the chemical composition of mature heads. *Agronomy* **2021**, *11*, 1806. [CrossRef]
12. Castoldi, R.; Charlo, H.C.O.; Vargas, P.F.; Braz, L.T. Growth, nutrients accumulation and crop productivity of cauliflower. *Hortic. Bras.* **2009**, *27*, 438–446. [CrossRef]
13. Figas, A.; Jagosz, B.; Rolbiecki, S.; Rolbiecki, R.; Ptach, W. Effect of the predicted climate changes on the water needs of cauliflower cultivated in Central Poland. In Proceedings of the 18th International Scientific Conference "Eng. Rural Dev.", Jelgava, Latvia, 22–24 May 2019.
14. Wang, R.; Wang, Y.; Zhang, Z.; Pan, H.; Lan, L.; Huang, R.; Peng, Y. Effects of Exponential N Application on Soil Exchangeable Base Cations and the Growth and Nutrient Contents of Clonal Chinese Fir Seedlings. *Plants* **2023**, *12*, 851. [CrossRef]
15. Xu, X.; Du, X.; Wang, F.; Sha, J.; Chen, Q.; Tian, G.; Zhu, Z. Effects of potassium levels on plant growth, accumulation and distribution of carbon, and nitrate metabolism in apple dwarf rootstock seedlings. *Front. Plant Sci.* **2020**, *11*, 904. [CrossRef] [PubMed]
16. Li, Z.; Xia, S.; Zhang, R.; Zhang, R.; Chen, F.; Liu, Y. N2O emissions and product ratios of nitrification and denitrification are altered by K fertilizer in acidic agricultural soils. *Environ. Pollut.* **2020**, *265 Pt B*, 115065. [CrossRef]
17. Wang, M.; Zheng, Q.; Shen, Q.; Guo, S. The critical role of potassium in plant stress response. *Int. J. Mol. Sci.* **2013**, *14*, 7370–7390. [CrossRef] [PubMed]
18. Ali, S.; Hafeez, A.; Ma, X.; Tung, S.A.; Chattha, M.S.; Shah, A.N.; Luo, D.; Ahmad, S.; Liu, J. Equal potassium-nitrogen ratio regulated the nitrogen metabolism and yield of high-density late-planted cotton (*Gossypium hirsutum* L.) in Yangtze River valley of China. *Ind. Crops Prod.* **2019**, *129*, 231–241. [CrossRef]
19. Lester, G.; Jifon, J.; Makus, D. Impact of potassium nutrition on postharvest fruit quality: Melon (*Cucumis melo* L) case study. *Plant Soil* **2010**, *335*, 117–131. [CrossRef]
20. Peel, M.C.; Finlayson, B.L.; McMahon, T.A. Updated world map of the Köppen-Geiger climate classification. *Hydrol. Earth Syst. Sci.* **2007**, *11*, 1633–1644. [CrossRef]
21. Di Mola, I.; Ottaiano, L.; Cozzolino, E.; El-Nakhel, C.; Rippa, M.; Mormile, P.; Mori, M. Assessment of yield and nitrate content of wall rocket grown under diffuse-light-or clear-plastic films and subjected to different nitrogen fertilization levels and biostimulant application. *Horticulturae* **2022**, *8*, 138. [CrossRef]
22. Jagota, S.K.; Dani, H.M. A new colorimetric technique for the estimation of vitamin C using Folin phenol reagent. *Anal. Biochem.* **1982**, *127*, 178–182. [CrossRef]
23. Kaur, C.; Kapoor, H.C. Antioxidant activity and total phenolic content of some Asian vegetables. *Int. J. Food Sci. Technol.* **2002**, *37*, 153–161. [CrossRef]
24. Singleton, V.L.; Orthofer, R.; Lamuela-Raventós, R.M. Analysis of total phenols and other oxidation substrates and antioxidants by means of folin-ciocalteu reagent. In *Methods in Enzymology*; Academic Press: Cambridge, MA, USA, 1999; Volume 299, pp. 152–178.

25. Devi, P.; Sharma, A.; Saini, A.; Sharma, S. Effect of different spacing and transplanting dates on the growth and yield of cauliflower (*Brassica oleracea* var. botrytis L.). *Int. J. Adv. Res.* 2022. [CrossRef]
26. Kartika, L.N.F.; Lakitan, B. Growth responses and yield of cauliflower (*Brassica oleracea* var. botrytis L.) to the delayed transplanting and drought stress. *E3S Web Conf.* **2021**, *306*, 01007. [CrossRef]
27. Sultana, J.; Mannan, M.A.; Ahmed, S.; Khan, K.U. Planting date affects growth and yield performance of cauliflower (*Brassica oleracea* var. botrytis). *South Asian J. Agric.* 2019. [CrossRef]
28. Kalisz, A.; Sękara, A.; Cebula, S.; Grabowska, A.; Kunicki, E. Impact of low-temperature transplant treatment on yield and quality of cauliflower curds in late spring production. *Sci. Hortic.* **2014**, *176*, 134–142. [CrossRef]
29. Rahman, M.; Imran, M.; Ikrum, M.; Rahman, M.; Rabbani, M. Effects of Planting Date and Growth Hormone on Growth and Yield of Cauliflower. *J. Environ. Sci. Nat. Resour.* **2017**, *9*, 143–150. [CrossRef]
30. Wadan, D.; Khan, M.; Hassan, I.; Jan, M. Effect of different levels of nitrogen and potassium on yield and yield components of cauliflower. *Int. J. Life Sci.* **2008**, *2*, 579–583.
31. Yildirim, E. Effects of plant growth-promoting rhizobacteria (PGPR) and different fertilizer combinations on yield and quality properties in cauliflower (*Brassica oleracea* L. var. botrytis). *Akad. Ziraat Derg.* **2022**, *11*, 35–46. [CrossRef]
32. Devienne-Barret, F.; Richard-Molard, C.; Chelle, M.; Maury, O.; Ney, B. Ara-rhizotron: An effective culture system to study simultaneously root and shoot development of Arabidopsis. *Plant Soil* **2006**, *280*, 253–266. [CrossRef]
33. Marschner, P. *Marschner's Mineral Nutrition of Higher Plants*; Academic Press: Cambridge, MA, USA, 2012. [CrossRef]
34. Katuwal, D.R.; Bhandari, N.; Pokhrel, A. Response of different levels of nitrogen on growth and yield of cauliflower (*Brassica oleracea* var. botrytis) varieties. *Asian J. Agric. Hortic. Res.* **2023**, *10*, 105–114. [CrossRef]
35. Cutcliffe, J.A.; Munro, D.C. Effects of nitrogen, phosphorus and potassium on yield and maturity of cauliflower. *Can. J. Plant Sci.* **1976**, *56*, 127–131. [CrossRef]
36. Ali, A. Nitrate assimilation pathway in higher plants: Critical role in nitrogen signalling and utilization. *Plant Sci. Today* **2020**, *7*, 182–192. [CrossRef]
37. Signore, A.; Bell, L.; Santamaria, P.; Wagstaff, C.; Van Labeke, M.C. Red light is effective in reducing nitrate concentration in rocket by increasing nitrate reductase activity, and contributes to increased total glucosinolates content. *Front. Plant Sci.* **2020**, *11*, 532456. [CrossRef]
38. Zhang, S.; Zhang, Y.; Wang, Y.; Hao, Y.; Su, W.; Sun, G.; Song, S. Nitrogen absorption pattern detection and expression analysis of nitrate transporters in flowering Chinese cabbage. *Horticulturae* **2022**, *8*, 188. [CrossRef]
39. Chen, B.M.; Wang, Z.H.; Li, S.X.; Wang, G.X.; Song, H.X.; Wang, X.N. Effects of nitrate supply on plant growth, nitrate accumulation, metabolic nitrate concentration and nitrate reductase activity in three leafy vegetables. *Plant Sci.* **2004**, *167*, 635–643. [CrossRef]
40. Ter Steege, M.W.; Stulen, I.; Wiersema, P.K.; Posthumus, F.; Vaalburg, W. Efficiency of nitrate uptake in spinach: Impact of external nitrate concentration and relative growth rate on nitrate influx and efflux. *Plant Soil* **1999**, *208*, 125–134. [CrossRef]
41. Kaya, C. Nitrate reductase is required for salicylic acid induced water stress tolerance of pepper by upraising the AsA-GSH pathway and glyoxalase system. *Physiol. Plant* **2021**, *172*, 351–370. [CrossRef]
42. Matt, P.; Geiger, M.; Walch-Liu, P.; Engels, C.; Krapp, A.; Stitt, M. The immediate cause of the diurnal changes of nitrogen metabolism in leaves of nitrate-replete tobacco: A major imbalance between the rate of nitrate reduction and the rates of nitrate uptake and ammonium metabolism during the first part of the light period. *Plant Cell Environ.* **2001**, *24*, 177–190.
43. Fu, Y.F.; Zhang, Z.W.; Yuan, S. Putative connections between nitrate reductase S-nitrosylation and NO synthesis under pathogen attacks and abiotic stresses. *Front. Plant Sci.* **2018**, *9*, 474. [CrossRef]
44. Sony, S.; Islam, F.; Akter, M.; Rahman, M.; Rahman, M. Interaction effect of nitrogen and phosphorus on curd yield and seed production of cauliflower. *J. Exp. Agric. Int.* **2020**, *42*, 216–225. [CrossRef]
45. Everaarts, A.P.; Booi, R. The effect of nitrogen application on nitrogen utilization by white cabbage (*Brassica oleracea* var. capitata) and on nitrogen in the soil at harvest. *J. Hortic. Sci. Biotechnol.* **2000**, *75*, 705–712. [CrossRef]
46. Metwaly, E. Influence of phosphorus and potassium on growth and yield of cauliflower. *J. Plant Prod.* **2017**, *8*, 329–334. [CrossRef]
47. Lisiewska, Z.; Kmiecik, W. Effects of level of nitrogen fertilizer, processing conditions and period of storage of frozen broccoli and cauliflower on vitamin C retention. *Food Chem.* **1996**, *57*, 267–270. [CrossRef]
48. Giri, H. Effect of different doses and sources of nitrogen on postharvest quality of cauliflower. *Asian J. Agric. Hortic. Res.* **2021**, *8*, 19–26. [CrossRef]
49. Šlosár, M.; Uher, A.; Čekey, N. The effect of nitrogen and sulphur nutrition on the yield and content of antioxidants in broccoli. *Acta Hortic. Regiotectuare* **2013**, *16*, 14–17. [CrossRef]
50. Michalska, A.; Wojdyło, A.; Bogucka, B. The influence of nitrogen and potassium fertilisation on the content of polyphenolic compounds and antioxidant capacity of coloured potato. *J. Food Compos. Anal.* **2016**, *47*, 69–75. [CrossRef]
51. Yuan, Z.; Li, L.; Huang, J.; Han, X.; Wan, S. Effect of nitrogen supply on the nitrogen use efficiency of an annual herb, *Helianthus annuus* L. *J. Integr. Plant Biol.* **2005**, *47*, 539–548. [CrossRef]

52. Wu, Z.; Luo, J.; Han, Y.; Hua, Y.; Guan, C.; Zhang, Z. Low nitrogen enhances nitrogen use efficiency by triggering NO_3^- uptake and its long-distance translocation. *J. Agric. Food Chem.* **2019**, *67*, 6736–6747. [CrossRef]
53. Singh, J.P.; Singh, S.; Singh, V. Soil potassium fractions and response of cauliflower and onion to potassium. *J. Indian Soc. Soil Sci.* **2010**, *58*, 384–387.

horticulturae

MDPI

Article

Plant-Growth-Promoting Rhizobacteria and Arbuscular Mycorrhizal Fungi May Improve Soil Fertility and the Growth, Nutrient Uptake, and Physiological Performance of Batavia Lettuce (*Lactuca sativa* L. var. *longifolia*) Plants

Theocharis Chatzistathis [1,*], Konstantinos Zoukidis [2], Christos Vasilikiotis [2], Antonios Apostolidis [2], Anastasia E. Giannakoula [3], Areti Bountla [1] and Apostolos Chatziathanasiadis [4]

1 Hellenic Agricultural Organization (ELGO) DIMITRA, Institute of Soil and Water Resources, 57001 Thessaloniki, Greece; a.bountla@elgo.gr
2 Department of Agro-Environmental Systems Management, Perrotis College, American Farm School, 57001 Thessaloniki, Greece; kzouki@afs.edu.gr (K.Z.); cvasil@afs.edu.gr (C.V.); 1960roughsea@gmail.com (A.A.)
3 Department of Agriculture, International University of Greece, 54700 Thessaloniki, Greece; agianna@cp.teithe.gr
4 Comerco Agrotechnology, Panepistimiou 44, 10679 Athens, Greece; comerco@otenet.gr
* Correspondence: chchatzista@gmail.com or t.chatzistathis@elgo.gr

Abstract: A greenhouse experiment was realized to investigate the effects of plant-growth-promoting rhizobacteria (PGPR) and arbuscular mycorrhizal fungi (AMF) on soil quality and the growth, nutrient uptake, and physiological performance of Batavia lettuce (*Lactuca sativa* L. var. *longifolia*). For this purpose, six fertilization treatments were applied: (i) inorganic fertilization (I.F.—control, i.e., usual/conventional fertilization), (ii) I.F. + PGPR, (iii) I.F. + PGPR + AMF, (iv) manure (MAN), (v) MAN + PGPR, and (vi) MAN + PGPR + AMF. Soil fertility was influenced by the treatments, and soil respiration (CO_2 flux) was significantly increased after applications of PGPR and AMF. Following MAN fertilization in particular, an approximately sixfold higher soil respiration value was recorded following the MAN + PGPR + AMF treatment compared to the control (I.F.). Root, leaf, and total biomass dry weights were significantly affected by the PGPR and AMF applications, mainly following the I.F. treatment. In contrast, K, Mg, and Fe uptake was significantly influenced by PGPR and AMF application following the MAN treatment. The SPAD value, performance index (PI), photosynthetic rate, and stomatal conductance were significantly higher in the I.F. + PGPR + AMF -treated plants compared to the control plants. Overall, these data prove the beneficial roles of PGPR and AMF in soil quality and fertility and the nutrient uptake and physiological performance of lettuce plants. However, further clarification is needed in the near future to test the interaction effects between PGPR, AMF, and the type of fertilizer used (organic or inorganic).

Keywords: soil quality; soil respiration; plant nutrition; growth promoters; inorganic fertilization; organic fertilization; CO_2 assimilation; vegetable crops

1. Introduction

Lactuca sativa L. is one of the most economically important leafy vegetable crops; it belongs, to the Asteraceae family and contains polyphenols, flavonoids, vitamins A, C, and E, carotenoids, Ca, Fe, and other nutrients. It is traditionally consumed fresh [1–4]. It is widely cultivated in temperate areas, and more than 21 million tons are produced annually [3]. Lettuce plants have a quick and short growing period and require high fertilization rates to satisfy their urgent nutritional needs and support their growth rate.

In recent years, there has been an increasing interest in the benefits of sustainable agriculture and a demand for high-quality, healthier fruits and vegetables. Sustainable agriculture not only preserves environmental resources, producing high-quality products and protecting human health [5], but also addresses major environmental and agronomic challenges [6], such as decreases in high inorganic fertilization rates and the enhancement of sustainable crop management [7,8]. In this direction, the use of biofertilizers or microbial fertilizers (e.g., the use of *Rhizobium leguminosarum*, isolated from *Phaseolus vulgaris* nodules, or the production of rock biofertilizer that is mixed with organic matter, inoculated with diazotrophic bacteria and the fungus *Cunninghamella elegans*), proved to be a promising tool for improving nutrient uptake (mainly N and P) and plant growth for some vegetable species, such as lettuce and carrot crops [9,10]. Similarly, the inoculation of lettuce plants with N_2-fixing plant-growth-promoting rhizobacteria led to a reduction in crop N fertilizer requirements [11], while the use of native phosphate-solubilizing rhizobacteria enhanced lettuce and tomato crop productivity and sustainability [12].

In the context of sustainable agriculture, plant-growth-promoting rhizobacteria (PGPR), biostimulants, and arbuscular mycorrhizal fungi (AMF) were used in an *L. sativa* crop and were found to play a crucial role in (i) boosting plant biomass and yield, as well as improving nutritional quality and impact on human health [13–16], (ii) increasing nutrient (especially N, P, Mg) [15,16] and water uptake (by also ensuring plant adaptation to drought conditions) [17], (iii) increasing plant tolerance to severe salt stress [18–21], and limiting the detrimental effects of salinity on soil structural stability [22]. The colonization of greenhouse lettuce plants by AMF may induce the accumulation of secondary metabolites, vitamins, and minerals in leaves; thus, AMF could allow the intake of minerals and compounds with antioxidant properties to be enhanced, without increasing the consumption of lettuce by humans [14]. In another study, the inoculation of lettuce plants with AMF was found to increase their yields (by an average of 186%), but resulted in no substantial change in lettuce nutritional quality; in addition, the AMF inoculants did not alter natural soil communities [22]. In contrast, other biostimulants (based on *Trichoderma*) were found to modulate rhizosphere microbial populations and improve the nutritional quality of leafy vegetables [16].

Apart from AMF and PGPR, the organic production of lettuce has relied on the application of organic materials, such as manures, humus-rich composts, and seaweed extracts (in foliar applications), in order to boost plant growth and yield, increase nutrient uptake and mycorrhizal colonization, and improve the antioxidant responses of *L. sativa* [23–25]. In addition, urban waste compost application improved soil properties and increased lettuce production [26]. In another comparative study, organic amendments applied to lettuce and orchardgrass were shown to supply adequate levels of N while resulting in good productivity [27]. Other studies focused on the mineral nutrition of lettuce plants grown under hydroponic conditions in a recirculating nutrient solution [28].

Despite the significant effects of manure and PGPR application and AMF inoculation on the stimulation of the growth, nutrient uptake, and nutritional quality of *L. sativa* plants, to the best of our knowledge, no attention has been systematically paid to their combined application until now, and almost no comparative studies have been conducted on the inorganic and organic fertilization of Batavia lettuce. In addition, the influence of AMF inoculation and PGPR application on lettuce productivity and physiological performance, in relation to soil health and fertility, has not been studied.

Thus, the aims of our research were as follows: (i) to investigate the effects of PGPR and AMF on soil health and Batavia lettuce productivity, nutrient uptake, and plant physiology under inorganic and cow manure regimes and (ii) to perform a comparative study on the effects of inorganic and organic fertilization on soil respiration and lettuce growth, nutrient uptake, photosystem II activity (F_v/F_m ratio; performance index), photosynthetic and transpiration rates, and intrinsic water use efficiency (WUEi). Batavia (also known as French) lettuce was chosen for study since it is one of the most economically important leafy green genotypes for salads with good tolerance to heat and contains significant quantities of vitamins A and C, antioxidants, and K, Ca, and Fe.

2. Materials and Methods

2.1. Plant Material and Conditions of Experimental Greenhouse

The plant material consisted of Batavia lettuce plants (*L. sativa* L. var. *longifolia*), which were grown inside a plastic-covered experimental greenhouse at Perrotis College of the American Farm School (AFS) in Thessaloniki (latitude, 40.341134°; longitude, 22.592796°; altitude, 65 m above sea level), Macedonia, Northern Greece. The plants were obtained as seedlings and randomized based on the initial number and surface of leaves. They were divided into six groups coinciding with the six fertilization treatments described below and were transplanted into 0.60 × 3.30 m plots containing 65 plants. The lettuce plants were grown inside the experimental greenhouse under natural light conditions for 65 days (from the middle of October to the middle of December 2019). During the experiment, the lettuce plants were irrigated with an automated drip irrigation system (for one hour to reach a soil moisture level of approximately 70% of its water-holding capacity) twice per week, providing them with high-quality tap water (0.72 mS/cm). The minimum and maximum temperatures during the experiment were 7 °C (in early December) and 24 °C (in early November), respectively, while the average temperature for the whole experimental period was approximately 17 °C. Finally, the relative humidity inside the greenhouse ranged from 65% to 77%.

2.2. Treatments and Methodology of PGPR and AMF Application

Six treatments were applied: (i) inorganic fertilization (I.F.—control, i.e., usual/conventional fertilization), (ii) I.F. + PGPR, (iii) I.F. + PGPR + AMF, (iv) manure (MAN), (v) MAN + PGPR, and (vi) MAN + PGPR + AMF. Each treatment consisted of 65 plants with an area of 2 m^2. In the I.F. treatments, a 25-10-0 (N-P-K) fertilizer (300 kg/ha N and 120 kg/ha P) was combined with a 0-0-30 (N-P-K) fertilizer (275 kg/ha K) in each of the treatments receiving inorganic fertilizer; the fertilizers were mixed with soil and incorporated at a 20 cm depth before the initiation of the experiment. For the organic fertilizer applications, well-digested and composted cow manure (MAN) (2 ton/ha), was incorporated at a 20 cm depth before the initiation of the experiment. The nutrient composition of the cow manure (MAN), as well as the initial and final fertility of the greenhouse soil (GS) are described in Table 1.

In order to study the influence of PGPR and AMF on the nutrition and physiological performance of the Batavia lettuce plants, the following commercial products were used: (a) Micoseeds Plus, 1 kg (powder of mycorrhiza fungi, containing the mycelium and seeds of the AMF *Glomus* sp. enriched with beneficial microorganisms, *Trichoderma* sp., *Bacillus* spp., *Streptomyces* sp., and *Pseudomonas* sp.), provided by Comerco Agrotechnology (Athens, Greece); (b) Clonotri, 0.25 mL (a commercial liquid product provided by Comerco Agrotechnology containing rhizospheric bacteria, PGPR, and other beneficial microorganisms;; (c) Strepse, 0.25 mL (provided by Comerco Agrotechnology; a liquid product, containing rhizospheric bacteria, PGPR, and other beneficial microorganisms); and (d) Nutryaction, 1 L (an activator of microorganisms and a biostimulant of plant growth; provided by Comerco Agrotechnology).

The PGPR containing the commercial products Clonotri, Strepse, and Nutryaction were prepared as follows: 50 mL of Clonotri, 50 mL of Strepse, and 150 mL of Nutryaction were mixed with a limited volume of water (2–3 L) at room temperature (20–25 °C). The three products were stirred very well and allowed to stand for 2 h in order for the microorganisms to be activated. The ratio of Clonotri, Strepse, and Nutryaction in the final volume solution was 2:2:6 (i.e., 2 mL/L Clonotri, 2 mL/L Strepse, and 6 mL/L Nutryaction). The rhizosphere of each lettuce seedling was soaked in this final solution for a few seconds before the establishment of the experiment. The same solution was prepared and added by fertigation 7 days after the initiation of the experiment. To investigate the additional effects of AMF, the lettuce seedlings were first soaked in the PGPR solution, and a small quantity (1–2 g) of granular Micoseeds Plus was added on the rhizosphere of each seedling prior to their transplantation.

Table 1. The properties of cow manure and greenhouse soil (initial fertility) before the beginning of the experiment (i.e., before manure, PGPR, and AMF applications), as well as at the end of the experimentation (final fertility, i.e., after applying the treatment).

	pH	Organic Matter	$CaCO_3$	NO_3-N	P	K	Ca	Mg	Fe	Mn	Zn	Cu	B
		%						mg kg^{-1}					
Manure (MAN)	7.01	50.06	8.49	126.81	323	10,234	2907	2808	51.27	56.06	93.05	11.65	28.80
Greenhouse soil (GS)—initial fertility	7.26	2.25	11.80	76.09	79	408	2468	662	7.39	7.15	8.23	5.88	0.92
Final soil fertility													
Inorganic fertilization (IF)	(7.52 ± 0.62) a	(1.86 ± 0.36) a	(10.80 ± 1.99) a	(50.00 ± 8.96) a	(79 ± 13.45) a	(298 ± 43.27) a	(2111 ± 212.26) a	(397 ± 44.04) b	(1.04 ± 0.18) b	(2.97 ± 0.55) b	(1.58 ± 0.26) ab	(1.74 ± 0.29) a	(0.70 ± 0.08) bc
IF + PGPR	(7.63 ± 1.01) a	(1.60 ± 0.44) a	(11.80 ± 2.06) a	(49.00 ± 8.12) a	(70 ± 12.27) ab	(180 ± 32.22) b	(2205 ± 192.20) a	(422 a ± 50.56) ab	(0.78 ± 0.16) bc	(2.64 ± 0.48) bc	(1.96 ± 0.35) ab	(0.75 ± 0.12) b	(0.62 ± 0.10) c
IF + PGPR + AMF	(7.54 ± 0.88) a	(1.73 ± 0.28) a	(11.30 ± 2.20) a	(55.00 ± 12.03) a	(84 ± 10.91) a	(188 ± 44.05) b	(2196 ± 122.28) a	(423 ± 28.12) ab	(0.54 ± 0.12) c	(1.99 ± 0.31) c	(1.49 ± 0.30) b	(0.73 ± 0.09) b	(0.85 ± 0.11) bc
Manure (MAN)	(7.66 ± 0.90) a	(2.04 ± 0.46) a	(9.80 ± 1.97) a	(12.00 ± 2.70) c	(53 ± 8.86) b	(197 ± 40.12) b	(2159 ± 301.17) a	(427 ± 46.06) ab	(0.31 ± 0.07)d	(2.33 ± 0.63) bc	(1.90 ± 0.32) ab	(0.76 ± 0.17) b	(0.99 ± 0.21) ab
MAN + PGPR	(7.76 ± 1.10) a	(2.00 ± 0.38) a	(9.30 ± 1.68) a	(21.00 ± 2.97) b	(52 ± 10.05) b	(207 ± 47.25) ab	(2104 ± 290.15) a	(485 ± 30.25) a	(0.21 ± 0.05)d	(0.68 ± 0.11)d	(0.20 ± 0.03) c	(0.25 ± 0.05) c	(1.17 ± 0.23) a
MAN + PGPR + AMF	(7.60 ± 0.96) a	(1.91 ± 0.40) a	(12.00 ± 2.34) a	(16.00 ± 3.27) bc	(66 ± 11.35) ab	(174 ± 39.11) b	(2117 ± 167.13) a	(416 ± 37.11) ab	(1.74 ± 0.34) a	(4.39 ± 0.77) a	(2.20 ± 0.28) a	(1.45 ± 0.39) a	(0.98 ± 0.20) ab

The values of the treatments (final fertility) are the means of 3 replicates; different letters in the same column indicate statistically significant differences among the 6 treatments according to Duncan's multiple range test at $p \leq 0.05$.

2.3. Soil Sampling, Manure Nutrient Composition, and Lab Analyses

A composite soil sample from the upper 20 cm of the soil surface of the experimental area was collected and analyzed, and its chemical properties are shown in Table 1. A composite sample from well-digested and composted cow manure, produced at the American Farm School, was analyzed. Its nutrient composition is presented in Table 1. Soil samples were dried at room temperature, stones were removed, and the samples were sieved to pass through a 10-mesh screen; the screened samples were sent to the Soil and Water Resources Institute's lab (ELGO DIMITRA) for chemical analyses of the following parameters: pH, organic matter, % $CaCO_3$ content, NO_3-N, available P, exchangeable cations (Ca, Mg, and K) and micronutrients (Fe, Mn, Zn, Cu, and B). The parameters were determined as follows: pH was determined in a 1:1 soil—distilled water paste [29]; particle size was analyzed according to the Bouyoucos method [30]; organic matter was determined using potassium dichromate [31]; the NO_3-N concentration was determined according to the VCl_3/Griess method [32]; and available P was determined according to the Olsen method [33]. The exchangeable cations were extracted using CH_3COONH_4 [34]; Fe, Mn, Zn, and Cu were extracted using a DTPA solution at a pH of 7.3 [35]; and the B concentration was determined according to the method described by Wolf [36]. Finally, the % $CaCO_3$ content was determined according to the acid neutralization method [37]. Potassium, Ca, Mg, Fe, Mn, Zn, and Cu concentrations were measured by ICP spectrometry (OPTIMA 2100 DV optical emission spectrometer, Perkin Elmer, Waltham, MA, USA) [38].

2.4. Methodology of Soil Respiration Rate Measurement

Soil respiration (Rs) was measured in situ in each of the six treatments using a LI-COR LI-6800 soil CO_2 flux chamber (LI-COR, Lincoln, NE, USA). A PVC collar (with a 19.5 cm inner diameter and a height of 11.5 cm) was inserted into the soil to a depth of 5 cm in each location at least 24 h prior to the measurements, and it was kept in place for the duration of the experiment. Each measurement was obtained in triplicate and comprised a 90 s measurement, including a 25 s dead-band measurement and a 45 s post-purge measurement. For each location soil temperature and soil moisture were simultaneously measured using a Li-COR Stevens probe (LICOR, Lincoln, NE, USA).

2.5. Plant Growth Measurements

At the end of the experiment, all plants were harvested and separated into root systems and leaves to determine the total plant, root, and leaf biomass values and leaf/root ratio. After weighting all the plant tissues separately and determining their fresh weight (F.W.), they were washed twice (with high-quality tap water followed by distilled water) and dried at 75 °C for 72 h to determine their dry weight (D.W.) values. By adding the leaf and root dry weights, the total plant D.W., i.e., the total plant biomass, was calculated.

2.6. Nutrient Uptake and Tissue Nutrient Concentrations

The dried vegetal tissues were ground to a fine powder and passed through a 30-mesh screen. A 0.5 g. portion of the fine powder of each sample was dry-ashed in a muffle furnace at 515 °C for 5 h. The ash was dissolved in 3 mL of 6 N HCl and diluted with double-distilled water up to 50 mL. Phosphorus, K, Ca, Mg, Fe, Mn, Zn, and Cu concentrations were determined by ICP spectrometry (OPTIMA 2100 DV optical emission spectrometer, Perkin Elmer, Waltham, MA, USA) [38], while those of N and B were determined by the Kjeldahl and azomethine-H methods, respectively [39,40]. Macronutrient concentrations were expressed in % D.W., while those of micronutrients were expressed in mg kg^{-1}.

2.7. Chlorophyll Fluorescence, Gas Exchange Measurements, and Intrinsic Water Use Efficiency (WUEi)

At the end of the experiment, the following parameters were determined using a PAM-2000 fluorometer (HeinzWalz GmbH, Effeltrich, Germany): F_v/F_m, the maximum quantum yield of primary photochemistry; F_0, the minimum fluorescence; F_m, the maximum fluorescence; and $F_v = F_m - F_0$, the variable fluorescence. In addition, the performance index (PI) was determined, reflecting the functionality of both photosystems I and II and providing quantitative information on the current state of plant performance under stress conditions [41]. The above-mentioned parameters were determined on the youngest mature, fully expanded plant leaves. Before performing the measurements with the fluorometer, the leaves were preconditioned under dark conditions for 20 min [42].

For the gas exchange measurements, i.e., photosynthetic and transpiration rates, stomatal conductance, and the intercellular CO_2 concentration, an LC PRO portable gas-exchange-measuring system (ADC Bioscientific Ltd., Hoddesdon, UK) was used. All the measurements were carried out during the time period from 10:00 to 12:00 a.m. (under natural light intensity) on the youngest mature, fully expanded leaves. Finally, the intrinsic water use efficiency (WUEi, showing the water efficiency at the plant level) was determined as the net photosynthetic rate per stomatal conductance.

2.8. Statistical Analysis

The experimental design consisted of a 6×1 completely randomized factorial with six treatments and one plant species. In each of the six treatments, 10 plant replicates were randomly selected for analyses (thus, the total number of experimental plants was 60). For the determination of final soil fertility, three sample replicates were included per treatment. The data were statistically analyzed via a one-way ANOVA using the SPSS statistical program (version 28, IBM, Armonk, NY, USA); for the comparison of mean values among the treatments, Duncan's multiple range test, with $p \leq 0.05$, was used.

3. Results

3.1. Initial and Final Soil Fertility

The properties and nutrient contents of the cow manure used, as well as the initial and final soil fertility, are shown in Table 1.

As indicated by the results in Table 1, the MAN was a good source of nutrients (particularly NO_3-N, P, K, and micronutrients), contributing highly to the enrichment of the GS with organic C. At the end of the experiment (final fertility), high amounts of nutrients were removed (compared to the initial soil fertility), as indicated in Table 1. Significantly higher NO_3-N concentrations were determined in the IF treatments compared to the organic (MAN, MAN + PGPR, and MAN + PGPR + AMF) ones. Similarly, for the Olsen-determined P, significantly higher concentrations were recorded following the IF and IF + PGPR + AMF treatments compared to the MAN and MAN + PGPR treatments. The highest exchangeable K concentration was found in soil receiving the IF treatment (298 mg kg^{-1}), and it was significantly higher than those determined following the other two IF treatments, MAN and MAN + PGPR + AMF (Table 1). Regarding micronutrients, the highest Fe concentration (1.74 mg kg^{-1}) was recorded for the MAN + PGPR + AMF treatment, followed by the IF treatment (1.04 mg kg^{-1}). These concentrations were significantly higher than those found in soil following the MAN, MAN + PGPR, and IF + PGPR + AMF treatments (Table 1). Similar to Fe, the highest Mn and Zn concentrations were determined in the MAN + PGPR + AMF treated soil (4.39 mg kg^{-1} and 2.20 mg kg^{-1}, respectively). Finally, the B concentration was significantly higher in the MAN + PGPR-treated soil (1.17 mg kg^{-1}) compared to the inorganic treatments (IF, IF + PGPR, and IF + PGPR + AMF) (Table 1).

3.2. Soil Respiration Data

Figure 1 presents soil respiration data among the six treatments. As shown for the organic fertilization (MAN) treatments, significantly higher CO_2 emissions were recorded compared to the relevant (with or without PGPR and AMF application) inorganic treatments. For both fertilization regimes (either inorganic or organic), PGPR and AMF significantly and synergistically boosted CO_2 emissions; the highest CO_2 emission was determined for the MAN + PGPR + AMF treatment (Figure 1).

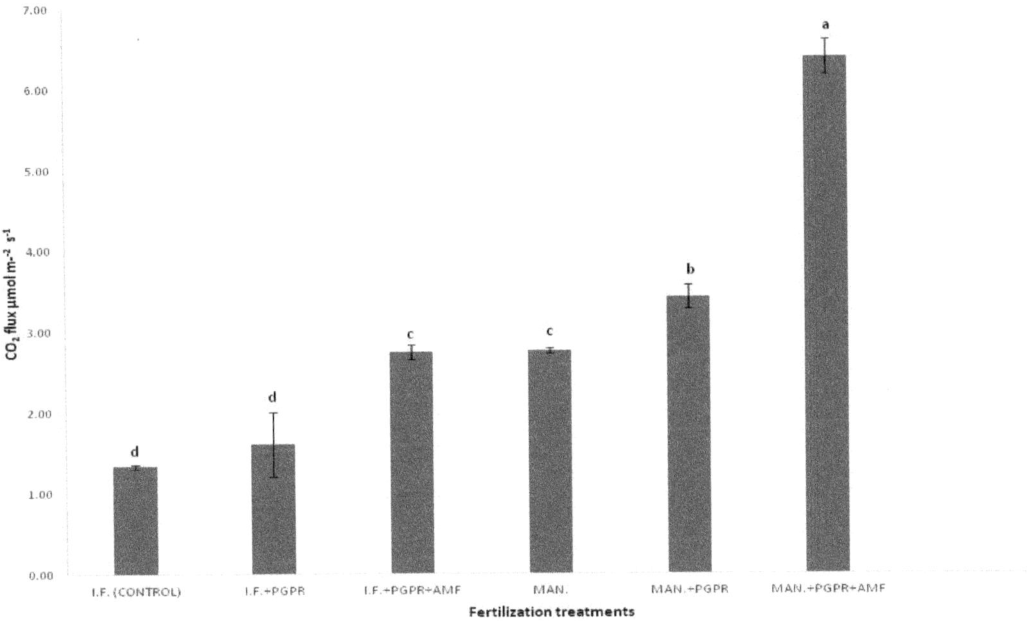

Figure 1. Soil respiration (μmol $m^{-2}s^{-1}$ CO_2 flux) under different fertilization treatments. Different letters on bars (means ± standard deviations) indicate statistically significant differences among treatments.

3.3. *Plant Growth*

From Table 2, it can be concluded that the addition of PGPR and AMF to inorganic fertilizer (I.F.) positively influenced the dry weights of the roots, leaves, and total biomass. In particular, between IF and the IF + PGPR + AMF treatment, significant differences in the three dry weights were recorded. In the MAN treatments, only PGPR application (i.e., MAN + PGPR) positively influenced the dry weights of the roots, leaves, and total biomass, while AMF addition (i.e., MAN + PGPR + AMF) had a negative effect (Table 2).

3.4. *Plant Tissue Nutrient Concentrations*

Leaf and root nutrient concentrations among treatments are shown in Tables 3 and 4. Nitrogen and P concentrations were not significantly influenced among the treatments, whereas significant differences in K concentration were found in the organic fertilization regime between the MAN and MAN + PGPR + AMF treatments (Table 3). The lowest leaf Mg concentration was observed in the MAN treatment (0.39% D.W.); significant differences in Mg uptake were only found between the (i) MAN and (ii) MAN + PGPR, MAN + PGPR + AMF treatments (Table 3).

In both the I.F. and organic (MAN) fertilization regimes, PGPR and AMF application significantly increased leaf Fe concentrations (Table 3). Despite the beneficial role of PGRP and AMF in Fe uptake, a similar influence was not found for the other micronutrients. In the IF regime, for example, foliar Mn, Zn, Cu, and B concentrations were significantly decreased after PGPR and/or AMF application, while in the MAN regime, the addition of PGPR and AMF did not significantly affect the uptake of these micronutrients (Table 3).

Table 2. Fresh and dry weights of different plant tissues, total biomass, and leaf/root ratio values of *L. sativa* L. (var. *longifolia*) plants among 6 treatments.

	Root		Leaves		Leaf/Root		Total Biomass	
	F.W.	D.W.	F.W.	D.W.	F.W.	D.W.	F.W.	D.W.
Treatment		g.						g.
Inorganic fertilization (IF)	(11.14 ± 2.36) ab	(0.71 ± 0.06) b	(248.20 ± 13.80) b	(9.49 ± 1.39) b	(22.61 ± 3.89) a	(13.41 ± 2.23) ab	(259.30 ± 30.34) b	(10.20 ± 1.40) b
IF + PGPR	(11.95 ± 2.53) a	(0.88 ± 0.10) a	(275.00 ± 42.30) ab	(11.75 ± 1.33) ab	(22.47 ± 2.12) a	(12.52 ± 1.81) b	(276.10 ± 45.91) ab	(12.67 ± 1.51) ab
IF + PGPR + AMF	(12.76 ± 2.07) a	(0.99 ± 0.14) a	(325.10 ± 35.47) a	(13.97 ± 1.40) a	(25.89 ± 3.94) a	(14.39 ± 2.06) a	(337.83 ± 34.35) a	(14.96 ± 1.43) a
Manure (MAN)	(9.01 ± 1.03) b	(0.84 a ± 0.05) b	(271.00 ± 36.78) ab	(12.13 ± 1.37) ab	(28.05 ± 3.69) a	(13.34 ± 1.18) ab	(281.09 ± 36.70) ab	(13.06 ± 1.46) ab
MAN + PGPR	(11.17 ± 1.59) ab	(0.96 ± 0.21) a	(323.50 ± 45.14) a	(12.66 ± 1.32) a	(28.03 ± 2.98) a	(13.52 ± 2.21) ab	(335.10 ± 38.64) a	(13.62 ± 1.45) a
MAN + PGPR + AMF	(10.01 ± 1.18) ab	(0.68 ± 0.09) b	(253.30 ± 28.85) b	(9.81 ± 1.23) b	(25.66 ± 3.31) a	(15.95 ± 2.65) a	(263.34 ± 28.47) b	(10.44 ± 1.24) b

The values are the means of 10 plants; different letters in the same column indicate statistically significant differences among the 6 treatments, according to Duncan's multiple range test, at $p \leq 0.05$.

Table 3. Leaf nutrient concentrations of *L. sativa* L. (var. longifolia) plants among 6 treatments.

	N	P	K	Ca	Mg	Fe	Mn	Zn	Cu	B
Treatment			**% D.W.**					**mg kg^{-1}**		
Inorganic fertilization (IF)	(3.91 ± 0.11) a	(0.47 ± 0.03) a	(8.11 ± 0.20) ab	(1.20 ± 0.10) a	(0.51 ± 0.06) ab	(85 ± 0.72) b	(93 ± 7.35) a	(49 ± 12.76) a	(11 ± 1.20) ab	(32 ± 3.06) a
IF + PGPR	(4.05 ± 0.16) a	(0.51 ± 0.03) a	(7.50 ± 0.75) ab	(0.96 ± 0.05) ab	(0.52 ± 0.06) ab	(146 ± 22.66) a	(56 ± 5.53) bc	(36 ± 2.79) b	(10 ± 2.19) b	(20 ± 1.13) b
IF + PGPR + AMF	(4.08 ± 0.06) a	(0.48 ± 0.03) a	(8.52 ± 1.12) a	(1.04 ± 0.10) ab	(0.51 ± 0.02) ab	(141 ± 26.48) a	(44 ± 2.28) c	(41 ± 5.54) ab	(7 ± 1.59) c	(28 ± 6.25) ab
Manure (MAN)	(3.82 ± 0.12) a	(0.41 ± 0.09) a	(6.38 ± 1.32) b	(0.79 ± 0.17) b	(0.39 ± 0.08) b	(73 ± 11.06) b	(61 ± 2.40) b	(46 ± 6.83) a	(13 ± 2.17) a	(26 ± 8.53) ab
MAN + PGPR	(3.85 ± 0.07) a	(0.51 ± 0.02) a	(7.52 ± 0.67) ab	(1.13 ± 0.14) ab	(0.57 ± 0.06) a	(152 ± 44.24) a	(67 ± 3.37) b	(47 ± 9.06) a	(12 ± 2.65) ab	(26 ± 0.37) ab

Table 3. *Cont.*

	N	P	K	Ca	Mg	Fe	Mn	Zn	Cu	B
Treatment			% D.W.					mg kg^{-1}		
MAN + PGPR + AMF	(4.25 ± 0.21) a	(0.50 ± 0.05) a	(8.23 ± 0.90) a	(1.11 ± 0.09) ab	(0.55 ± 0.03) a	(186 ± 47.44) a	(63 ± 7.09) b	(40 ± 3.00) ab	(13 ± 2.86) a	(26 ± 4.57) ab

Different letters in the same column indicate statistically significant differences among the 6 treatments, according to Duncan's multiple range test, at $p \leq 0.05$.

Table 4. Root nutrient concentrations of *L. sativa* L. (var. longifolia) plants among 6 treatments.

	N	P	K	Ca	Mg	Fe	Mn	Zn	Cu	B
Treatment			% D.W.					mg kg^{-1}		
Inorganic fertilization (IF)	(1.98 ± 0.15) a	(0.45 ± 0.11) a	(6.38 ± 1.65) a	(0.89 ± 0.09) a	(0.42 ± 0.08) a	(835 ± 211) b	(66 a ± 12.02) b	(46 ± 1.94) ab	(21 ± 4.56) b	(29 ± 3.43) ab
IF + PGPR	(2.02 ± 0.27) a	(0.34 ± 0.04) a	(4.57 ± 0.27) b	(0.75 ± 0.07) a	(0.40 ± 0.02) a	(561 ± 162) c	(54 ± 7.87) bc	(36 ± 3.65) b	(19 ± 2.74) b	(19 ± 2.30) c
IF + PGPR + AMF	(2.18 ± 0.20) a	(0.37 ± 0.05) a	(4.79 ± 0.58) ab	(0.96 ± 0.19) a	(0.46 ± 0.10) a	(944 ± 247) ab	(75 ± 18.57) a	(45 ± 6.94) ab	(15 ± 3.29) b	(30 ± 6.96) ab
Manure (MAN)	(1.92 ± 0.05) a	(0.30 ± 0.04) a	(4.00 ± 0.56) b	(0.82 ± 0.13) a	(0.36 ± 0.05) a	(479 ± 163.96) c	(47 ± 6.52) c	(45 ± 10.34) ab	(34 ± 5.09) a	(24 ± 3.72) bc
MAN + PGPR	(1.99 ± 0.11) a	(0.36 ± 0.06) a	(4.13 ± 0.60) b	(0.89 ± 0.07) a	(0.38 ± 0.01) a	(886 ± 261) b	(71 ± 6.96) a	(41 ± 7.23) ab	(19 ± 3.66) b	(26 ± 7.47) b
MAN + PGPR + AMF	(1.91 ± 0.12) a	(0.37 ± 0.10) a	(4.92 ± 0.88) ab	(1.08 ± 0.24) a	(0.46 ± 0.07) a	(1232 ± 322.18) a	(80 ± 15.57) a	(52 ± 10.42) a	(18 ± 2.12) b	(35 ± 9.54) a

Different letters in the same column indicate statistically significant differences among the 6 treatments, according to Duncan's multiple range test, at $p \leq 0.05$.

3.5. Physiological Performance (SPAD, Performance Index-PI, Photosynthetic Rate, Intercellular CO_2, and Water Use Efficiency—WUEi) of Lettuce Plants

As shown in Figure 2, significantly higher values of SPAD were determined for the combined application of PGPR and AMF in the I.F. regime compared to the control (inorganic fertilization only). Similarly, the highest value for the performance index (PI) was recorded for the I.F. + PGPR + AMF treatment, and it was significantly higher than the values determined for the single I.F. and MAN applications (Figure 3B). The photosynthetic rate was significantly higher in the MAN + PGPR treatment group compared to the single MAN and I.F. applications as well as the I.F. + PGPR treatment (Figure 4A). Stomatal opening was significantly higher following the combined application of PGPR and AMF in the I.F. regime compared to the single applications of I.F. and MAN as well as the I.F. + PGPR and MAN + PGPR treatments (Figure 4B). Water use efficiency (WUEi) was significantly higher following the single MAN application compared to the MAN + PGPR + AMF treatment (Figure 4E).

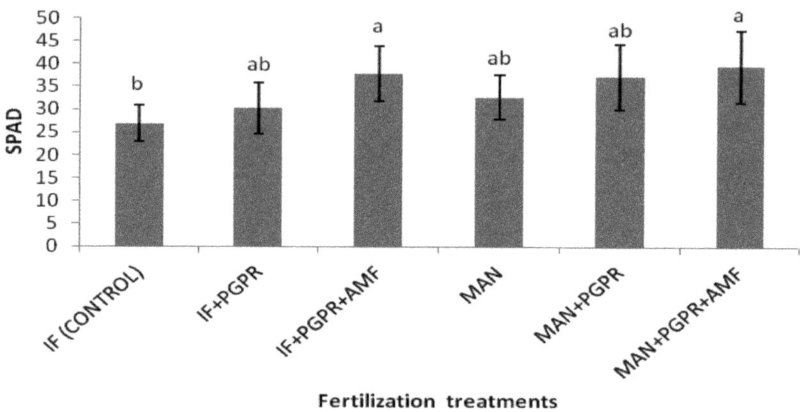

Figure 2. SPAD values of Batavia lettuce plants under different fertilization treatments. Different letters on bars (means ± standard deviations) indicate statistically significant differences among treatments.

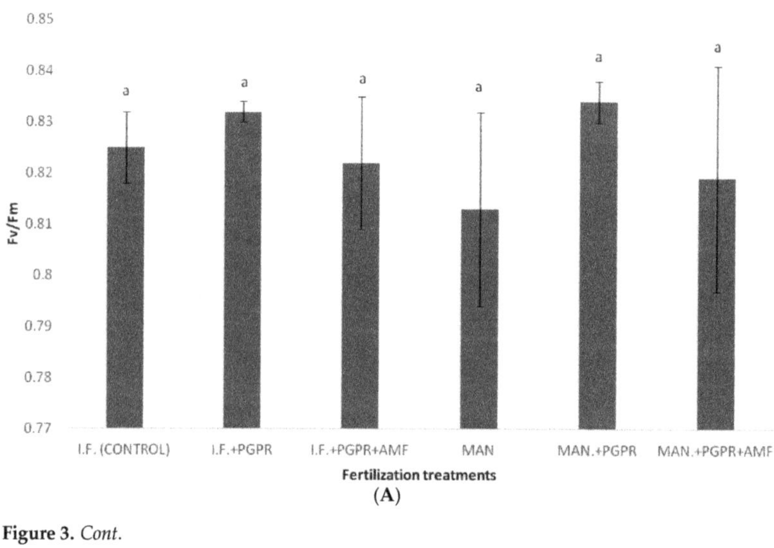

(A)

Figure 3. *Cont.*

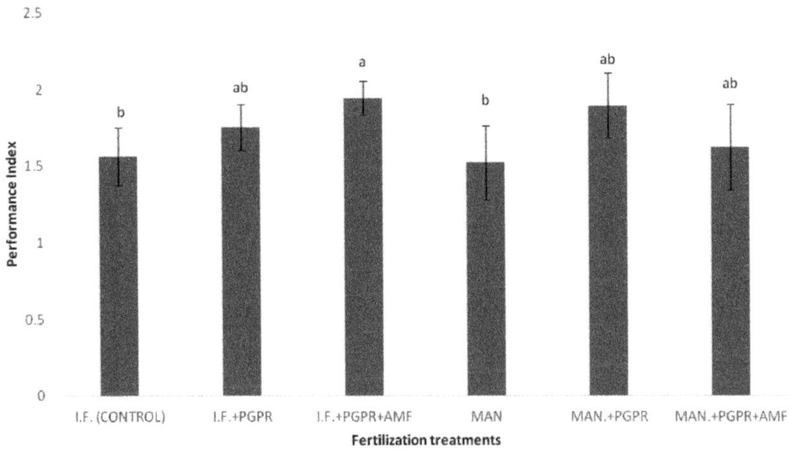

(B)

Figure 3. Photosystem II (PSII) activity of Batavia lettuce plants, as indicated by parameters F_v/F_m (**A**) and performance index (PI) (**B**), following different fertilization treatments. Different letters on bars (means ± standard deviations) indicate statistically significant differences among treatments.

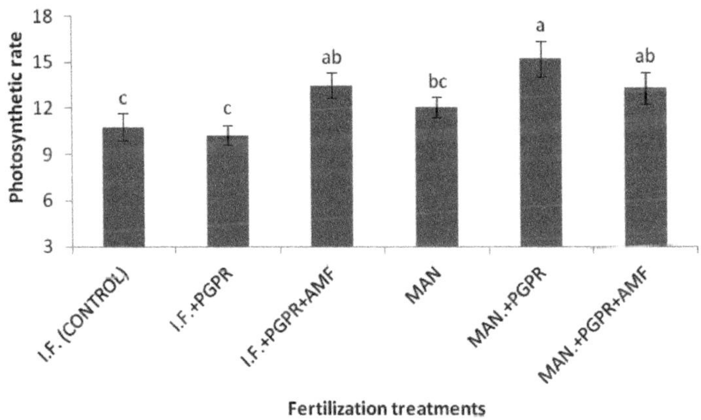

(A)

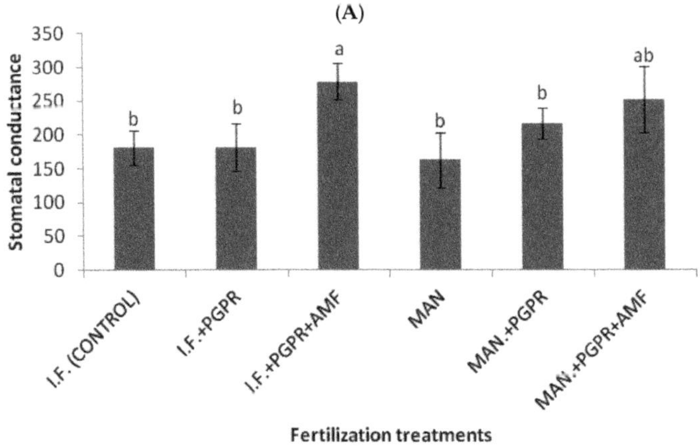

(B)

Figure 4. *Cont.*

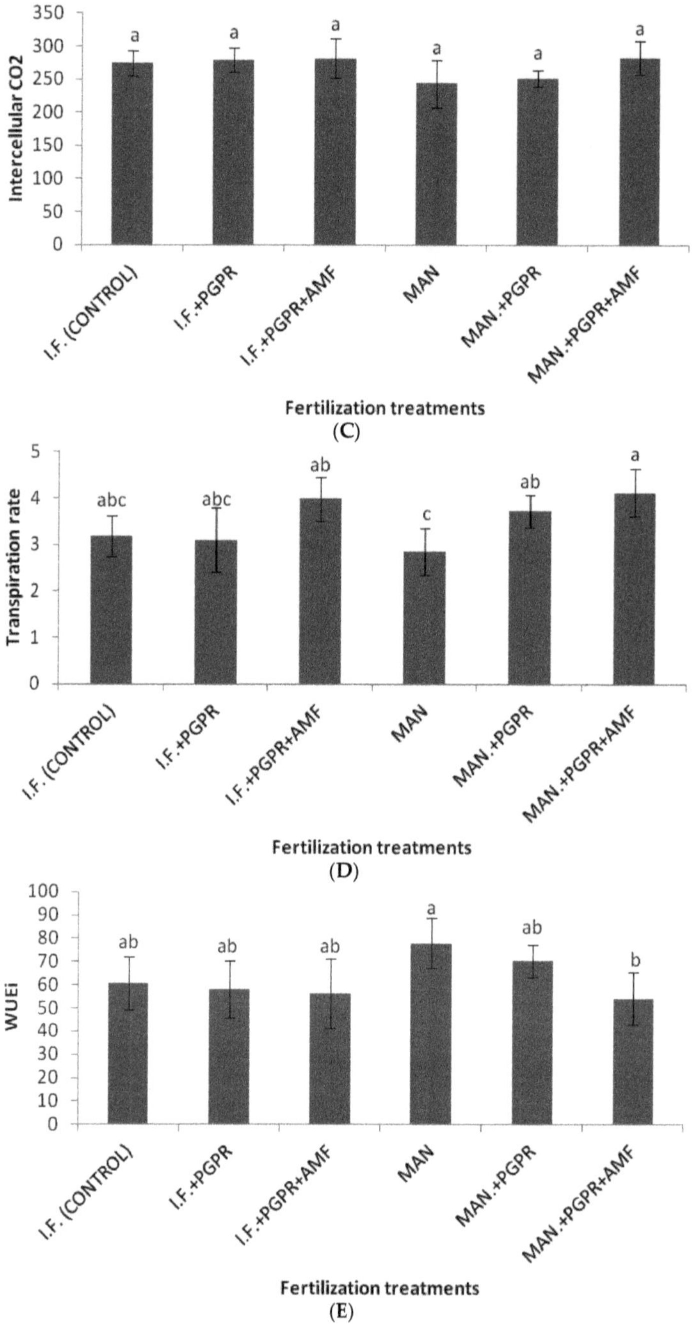

Figure 4. Photosynthetic rate (μmol CO_2 $m^{-2}s^{-1}$) (**A**), stomatal conductance (mmol $m^{-2}s^{-1}$) (**B**), intercellular CO_2 concentration (μmol CO_2 mol air^{-1}) (**C**), transpiration rate (mmol H_2O $m^{-2}s^{-1}$) (**D**), and intrinsic water use efficiency (WUEi) (μmol CO_2 mol H_2O^{-1}) (**E**) of Batavia lettuce plants following different fertilization treatments. Different letters on bars (means ± standard deviations) indicate statistically significant differences among treatments.

4. Discussion

As shown in Table 1, it is clear that cow manure (MAN) was a good organic source of nutrients (particularly NO_3-N, P, K, and micronutrients), contributing highly to the enrichment of the greenhouse soil (GS) with organic C. In terms of final soil fertility, significantly higher NO_3-N concentrations were determined for the IF treatments compared to the organic ones (MAN, MAN + PGPR, and MAN + PGPR + AMF) (Table 1). This could be ascribed to the slow N mineralization rate of organic matter and agrees with other published data [43]. Similar conclusions can be drawn for the Olsen-determined P (lower concentrations were determined in the organic treatments compared to the IF ones) (Table 1). The differences in the Olsen-determined P concentrations between the inorganic and organic (MAN) regimes are probably due to the direct application of the 25-10-0 fertilizer in the IF treatment before the experimentation combined with the slow organic P mineralization rate in the MAN treatments, something which is also in agreement with other published data [44].

PGPR and AMF applications boosted soil respiration (CO_2 flux); in the organic fertilization (MAN) regime in particular, differences among the three treatments were significant, and the highest CO_2 flux was recorded for the simultaneous application (co-application) of PGPR and AMF (the MAN + PGPR + AMF treatment) (Figure 1). This is in agreement with other studies, which concluded that PGPR application boosted soil respiration [45,46]. Both the above-mentioned studies concluded that the combined application (co-application) of PGPR and organic fertilizers had a positive impact on soil quality. The boost in soil respiration after the combined application of biochar and PGPR was found to be dependent on soil type (a 100% increase in soil respiration in acidic soils and an approximately 50% increase in calcareous soils) [45]. In our study, approximately three- and sixfold higher soil respiration values were determined after the combined application of PGPR and AMF in the IF and MAN fertilization regimes, respectively, compared to the control (Figure 1), which is quite impressive. In contrast to the above data, it was found that PGPR application did not significantly influence soil respiration in Ginger (*Zingiber officinale* Rosc.) plantations compared to the application of a simple NPK fertilizer [47]. The differences in the effect of PGPR on soil respiration data in the above studies could possibly be ascribed to the different genera/species of PGPR used for experimentation.

PGPR and AMF additions in the inorganic fertilization (IF) regime positively influenced root, leaf, and total biomass dry weights. In particular, between the IF and IF + PGPR + AMF treatments, significant differences in the three dry weights were recorded (Table 2). Although other studies concluded that the simultaneous application of PGPR and organic fertilizers positively influenced plant growth and biomass [45,46], there seems to be a lack of published data on the co-application of inorganic fertilizers and PGPR as well as on the co-application of IF with PGPR and AMF. From this point of view and to the best of our knowledge, our study is the first one showing that the combined application of inorganic fertilizer with PGPR and AMF can have a significant impact on the leaf d.w. and total biomass of lettuce plants (Table 2). This was not observed in the organic fertilization (MAN) regime, in which the combined application effect of MAN, PGPR, and AMF was not significant compared to the MAN application only (Table 2). In contrast, significantly lower root, leaf, and total biomass dry weights were recorded for the MAN + PGPR + AMF treatment compared to the MAN + PGPR treatment (Table 2), indicating that the single application of PGPR with MAN was more beneficial than the one with PGPR and AMF combined. This result is quite interesting and needs to be further investigated in the near future. Other studies have shown that treatments involving biochar applications (at doses of 3 t ha^{-1}) resulted in a 70% decrease in AMF infection [45]. It seems possible that a similar mechanism could be responsible in our case, i.e., MAN fertilization could have impeded root system infection by AMF. Of course, this hypothesis needs to be further investigated in the near future.

Significant differences in K uptake were found in the organic fertilization regime between the MAN and MAN + PGPR + AMF treatments (Table 3). The increase in K uptake after the simultaneous application of PGPR and AMF led to an increase in stomatal opening (Figure 4B), photosynthetic (Figure 4A) and transpiration (Figure 4D) rates. Potassium nutrition affects stomatal opening and photosynthetic and transpiration rates, as shown by other researchers [48,49]. In addition, significant differences between the (i) MAN treatment and the (ii) MAN + PGPR and MAN + PGPR + AMF treatments were found in Mg uptake (Table 3). Magnesium is part of the chlorophyll molecule; thus, the differences in Mg concentrations among the three organic (MAN) treatments could have led to differences in chlorophyll content, which could possibly explain the differences in the photosynthetic rate (Figure 4A). This hypothesis is also confirmed by the significantly higher SPAD values determined following the co-application of the PGPR and AMF treatments in the IF regime (Figure 2). Thus, the previous data may explain the beneficial effect of PGPR and AMF co-application on K and Mg uptake (in the organic (MAN) fertilization regime), as well as on SPAD (Figure 2), the photosynthetic rate (Figure 4A), and stomatal opening (Figure 4B). Of course, further research is needed to test this hypothesis regarding the influence of PGPR and AMF on Mg uptake by Batavia lettuce plants, as well as on their chlorophyll content and photosynthetic rates. In other studies with *L. sativa* [15,16], applications of AMF and *Trichoderma* sp. biostimulants showed increased N, P, and Mg uptake, which is in partial disagreement with our data. Maybe the differences between our results and those of the other researchers [15,16] could be ascribed to differences in the growth experimental conditions among the three studies (greenhouse soil experimentation with organic or inorganic fertilization and PGPR and AMF applications, a soilless system with sub-optimal P, and greenhouse experimentation with sub-optimal, optimal, and supra-optimal levels of N, respectively).

The significant increase in Fe uptake after PGPR and AMF addition could influence SPAD values (Figure 2) since Fe plays a crucial role in chlorophyll synthesis without being part of its molecule. The enhanced SPAD values following the PGPR and AMF treatments could influence the photosynthetic rate of lettuce plants (Figure 4A); however, further investigation is needed into the roles of PGPR and AMF in Fe uptake and its subsequent effect on chlorophyll content and photosynthesis. The co-application of Azolla compost, Azolla biochar, and PGPR (*P. fluorescens*) has been found to increase most nutrient concentrations in rosemary plant tissues (shoot and root nutrient concentrations), while the highest shoot and root nutrient concentrations were determined in plants following the Azolla biochar + PGPR treatment [46], which is in partial agreement with our results, especially those concerning Fe (Tables 3 and 4).

Although non-significant differences were found among the treatments in the F_v/F_m ratio, the performance index (PI) was a more reliable index, showing quicker reactions of PSII to various stresses, as also shown in previous studies [50]. In our study, the co-application of PGPR and AMF in the IF regime positively influenced the PI, causing significant a difference in its values between the IF and IF + PGPR + AMF treatments (Figure 3B). In contrast, in the MAN fertilization regime, the addition of PGPR and AMF did not significantly increase the PI (Figure 3B). Thus, it seems that PGPR and AMF co-application did not work as efficiently in the MAN fertilization regime as it worked in the IF regime, something that also confirms our plant growth data (Table 2). Similar to the PI, the photosynthetic rate showed a significant increase after PGPR and AMF co-application only in the IF regime and not in the MAN fertilization regime; in that case, only PGPR addition positively influenced the photosynthetic rate of the plants (Figure 4A). The conclusions for stomatal conductance, in which the influence of PGPR and AMF co-application was significant only in the IF and not in the MAN fertilization regime, were similar (although the addition of both PGPR and AMF increased stomatal opening, this tendency was non-significant) (Figure 4B). Thus, further research is needed to investigate and clarify the effects of PGPR and AMF on PI, the photosynthetic rate, and stomatal conductance in lettuce and other species. The intrinsic water use efficiency (WUEi), which shows how efficiently water is used at the plant level, showed a significant decline in plants in the MAN fertilization regime after simultaneous PGPR + AMF application (Figure 4E). Other studies have shown that PGPR (AG-54-*Pseudomonas* sp. and AG-70-*Bacillus* sp., applied separately or combined) improved water stress tolerance in wheat plants [51], while the WUEi was influenced by both AMF inoculation and by N and K fertilization [52,53].

5. Conclusions and Future Perspectives

The application of PGPR and AMF positively influenced soil quality and fertility as well as plant growth, nutrient uptake, SPAD, performance index, photosynthetic rate, and stomatal opening in Batavia lettuce plants under both inorganic (IF) and organic (MAN) fertilization conditions. In some cases, the results were more promising in the IF treatment; however, this tendency needs further investigation in the near future in order to clarify the effect(s) of the type of fertilization (organic or inorganic) on the effects of PGPR and AMF. Thus, their roles as soil improvers and plant growth promoters should be further examined in the frame of sustainable agriculture with the aim of decreasing high conventional fertilization rates.

Author Contributions: Conceptualization, T.C. and A.C.; methodology, T.C., K.Z., C.V., and A.C.; software: T.C.; validation, T.C. and A.C.; formal analysis, T.C.; investigation, T.C., K.Z., C.V., A.A., A.B., and A.E.G.; resources, T.C., K.Z., and C.V.; data curation, T.C., K.Z., C.V., A.E.G. and A.B.; writing—original draft preparation, T.C.; writing—review and editing, T.C., A.C., and C.V.; visualization, T.C. and K.Z.; supervision, T.C.; project administration, T.C.; funding acquisition, T.C. All authors have read and agreed to the published version of the manuscript.

Funding: This research was partially funded by small-scale projects of the Institute of Soil and Water Resources (ELGO DIMITRA).

Data Availability Statement: The data presented in this study are available upon request from the corresponding author. The data are not publicly available due to privacy reasons.

Acknowledgments: The authors are thankful to the enterprise "COMERCO AGROTECHNOLOGY" for providing them with the commercial products "MICOSEEDS PLUS", "CLONOTRI", "STREPSE", and "NUTRYACTION" for the needs of this research.

Conflicts of Interest: The research was conducted in the absence of any commercial or financial relationships that could be construed as a potential conflict of interest.

References

1. Mou, B. Nutritional quality of lettuce. *Curr. Nutr. Food Sci.* **2012**, *8*, 177–187. [CrossRef]
2. Ahmed, Z.F.; Alnuaimi, A.K.; Askri, A.; Tzortzakis, N. Evaluation of Lettuce (*Lactuca sativa* L.) production under hydroponic system: Nutrient solution derived from fish waste vs. inorganic nutrient solution. *Horticulturae* **2021**, *7*, 292. [CrossRef]
3. Shatilov, M.V.; Razin, A.F.; Ivanova, M.I. Analysis of the world lettuce market. *IOP Conf. Ser. Earth Environ. Sci.* **2019**, *395*, 012053. [CrossRef]
4. Ayuso-Calles, M.; Garcia-Estevez, I.; Jimenez-Gomez, A.; Flores-Felix, J.D.; Escribano-Bailón, M.T.; Rivas, R. Rhizobium laguerreae Improves Productivity and Phenolic Compound Content of Lettuce (*Lactuca sativa* L.) under Saline Stress Conditions. *Foods* **2020**, *9*, 1166. [CrossRef] [PubMed]
5. Maitra, S.; Hossain, A.; Brestic, M.; Skalicky, M.; Ondrisik, P.; Gitari, H.; Brahmachari, K.; Shankar, T.; Bhadra, P.; Palai, J.B.; et al. Intercropping—A low input agricultural strategy for food and environmental security. *Agronomy* **2021**, *11*, 343. [CrossRef]
6. Ahmad, M.A.; Gupta, L.M.; Gupta, M. Effect of integrated nutrient management on growth and yield of Aloe barbadensis. *Indian J. Agric. Sci.* **2016**, *86*, 91–95. [CrossRef]
7. Chatzistathis, T.; Tzanakakis, V.; Giannakoula, A.; Psoma, P. Inorganic and organic amendments affect soil fertility, nutrition, photosystem II activity and fruit weight and may enhance the sustainability of *Solanum lycopersicon* L. (cv. 'Mountain Fresh') crop. *Sustainability* **2020**, *12*, 9028. [CrossRef]
8. Chatzistathis, T.; Tsaniklidis, G.; Papaioannou, A.; Giannakoula, A.; Koukounaras, A. Comparative approach on the effects of soil amendments and a Controlled release fertilizer application on the growth, nutrient uptake, physiological performance and fruit quality of pepper (*Capsicum annuum* L.) plants. *Agronomy* **2022**, *12*, 1935. [CrossRef]
9. Flores-Felix, J.D.; Menendez, E.; Rivera, L.P.; Marcos-Garcia, M.; Martinez-Hidalgo, P.; Mateos, P.F.; Martinez-Molina, E.; De la Encarnacion Velazquez, M.; Garcia-Fraile, P.; Rivas, R. Use of *Rhizobium leguminosarum* as a potential biofertilizer for *Lactuca sativa* and *Daucus carota* crops. *J. Plant Nutr. Soil Sci.* **2013**, *176*, 876–882. [CrossRef]

10. Stamford, N.P.; Felix, F.; Oliveira, W.; Silva, E.; Carolina, S.; Arnaud, T.; Dolores Freitas, A. Interactive effectiveness of microbial fertilizer enriched in N on lettuce growth and on characteristics of an Ultisol of the rainforest region. *Sci. Hortic.* **2019**, *247*, 242–246. [CrossRef]
11. Kaymak, H.C.; Aksoy, A.; Kotan, R. Inoculation with N_2 fixing plant growth promoting rhizobacteria to reduce nitrogen fertilizer requirement of lettuce. *Acta Sci. Pol. Hortorum Cultus* **2020**, *19*, 23–35. [CrossRef]
12. Maldonado, S.; Rodriguez, A.; Avila, B.; Morales, P.; Gonzalez, M.P.; Araya Angel, J.P.A.; Olalde, V.; Bravo, J.; Jana, C.; Sierra, C.; et al. Enhanced crop productivity and sustainability by using native phosphate solubilizing rhizobacteria in the agriculture of arid zones. *Front. Sustain. Food Syst.* **2020**, *4*, 607355. [CrossRef]
13. Baslam, M.; Garmendia, I.; Goicoechea, N. The arbuscular mycorrhizal symbiosis can overcome reductions in yield and nutritional quality in greenhouse-lettuces cultivated at inappropriate growing seasons. *Sci. Hortic.* **2013**, *164*, 145–154. [CrossRef]
14. Baslam, M.; Garmendia, I.; Goicoechea, N. Enhanced accumulation of vitamins, nutraceuticals and minerals in lettuces associated with arbuscular mycorrhizal fungi (AMF): A question of interest for both vegetables and humans. *Agriculture* **2013**, *3*, 188–209. [CrossRef]
15. Cela, F.; Avio, L.; Giordani, T.; Vangelisti, A.; Cavallini, A.; Turrini, A.; Sbrana, C.; Pardossi, A.; Incrocci, L. Arbuscular mycorrhizal fungi increase nutritional quality of soilless grown lettuce, while overcoming low phosphorus supply. *Foods* **2022**, *11*, 3612. [CrossRef]
16. Fiorentino, N.; Ventorino, V.; Woo, S.L.; Pepe, O.; De Rosa, A.; Glola, L.; Romano, I.; Lombardi, N.; Napolitano, M.; Colla, G.; et al. *Trichoderma*-based biostimulants modulate rhizosphere microbial populations and improve N uptake efficiency, yield, and nutritional quality of leafy vegetables. *Front. Plant Sci.* **2018**, *9*, 743. [CrossRef] [PubMed]
17. Azcon, R.; Gomez, M.; Tobar, R. Physiological and nutritional responses by *Lactuca sativa* L. to nitrogen sources and mycorrhizal fungi under drought conditions. *Biol. Fertil. Soils* **1996**, *22*, 156–161. [CrossRef]
18. Kohler, J.; Hernandez, J.A.; Caravaca, F.; Roldan, A. Induction of antioxidant enzymes is involved in the greater effectiveness of a PGPR versus AM fungi with respect to increasing the tolerance of lettuce to severe salt stress. *Environ. Exp. Bot.* **2009**, *65*, 245–252. [CrossRef]
19. Kohler, J.; Caravaca, F.; Roldan, A. An AM fungus and a PGPR intensify the adverse effects of salinity on the stability of rhizosphere soil aggregates of *Lactuca sativa*. *Soil Biol. Biochem.* **2010**, *42*, 429–434. [CrossRef]
20. Miceli, A.; Moncada, A.; Vetrano, F. Use of microbial biostimulants to increase the salinity tolerance of vegetable transplants. *Agronomy* **2021**, *11*, 1143. [CrossRef]
21. Moncada, A.; Vetrano, F.; Miceli, A. Alleviation of salt stress by plant growth-promoting bacteria in hydroponic leaf lettuce. *Agronomy* **2020**, *10*, 1523. [CrossRef]
22. Epelde, L.; Urra, J.; Anza, M.; Gamboa, J.; Garbisu, C. Inoculation of arbuscular mycorrhizal fungi increases lettuce yield without altering natural soil communities. *Arch. Agron. Soil Sci.* **2022**, *68*, 413–430. [CrossRef]
23. Brito, L.M.; Pinto, R.; Mourao, I.; Coutinho, J. Organic lettuce, rye/vetch, and Swiss chard growth and nutrient uptake response to lime and horse manure compost. *Org. Agric.* **2012**, *2*, 163–171. [CrossRef]
24. Rasouli, F.; Amini, T.; Asadi, M.; Hassanpouraghdam, M.B.; Aazami, M.A.; Ercisli, S.; Skrovankova, S.; Mlcek, J. Growth and antioxidant responses of lettuce (*Lactuca sativa* L.) to arbuscular mycorrhiza inoculation and seaweed extract foliar application. *Agronomy* **2022**, *12*, 401. [CrossRef]
25. Solaiman, Z.M.; Yang, H.; Archdeacon, D.; Tippett, O.; Tibi, M.; Whiteley, A.S. Humus-rich compost increases lettuce growth, nutrient uptake, mycorrhizal colonization and soil fertility. *Pedosphere* **2019**, *29*, 170–179. [CrossRef]
26. De Nobile, F.B.; Calero Hurtado, A.; De Melo Prado, R.; De Souza, H.A.; Anunciacao, M.G.; Palaretti, L.F.; Sizuki Nociti Dezem, L.A. A novel technology for processing urban waste compost as a fast-releasing nitrogen source to improve soil properties and broccoli and lettuce production. *Waste Biomass Valorization* **2021**, *12*, 6191–6203. [CrossRef]
27. Hammermeister, A.M.; Astatkie, T.; Jeliazkova, E.A.; Warman, P.R.; Martin, R.C. Nutrient supply from organic amendments applied to unvegetated soil, lettuce and orchardgrass. *Can. J. Soil Sci.* **2006**, *86*, 21–33. [CrossRef]
28. Neocleous, D.; Savvas, D. The effects of phosphorus supply limitation on photosynthesis, biomass production, nutritional quality and mineral nutrition in lettuce grown in a recirculating nutrient solution. *Sci. Hortic.* **2019**, *252*, 379–387. [CrossRef]
29. McLean, E. Soil pH and lime requirement. In *Methods of Soil Analysis, Part 2: Chemical and Microbiological Properties*; Page, A.L., Miller, R.H., Keeney, D.R., Eds.; Agronomy Monograph; ASA; SSSA: Madison, WI, USA, 1982; pp. 199–224.
30. Gee, G.; Bauder, J. Particle-size analysis. In *Methods of Soil Analysis, Part 1: Physical and Mineralogical Methods*; Klute, A., Ed.; ASA; SSSA: Madison, WI, USA, 1986; pp. 383–411.
31. Nelson, D.W.; Sommers, L.E. Total carbon, organic carbon and organic matter. In *Methods of Soil Analysis, Part 2: Chemical and Microbiological Properties*; Page, A.L., Miller, R.H., Keeney, D.R., Eds.; Agronomy Monograph; ASA; SSSA: Madison, WI, USA, 1982; pp. 539–547.
32. Hood-Nowotny, R.; Umana, N.H.-N.; Inselbacher, E.; Oswald-Lachouani, P.; Wanek, W. Alternative Methods for Measuring Inorganic, Organic, and Total Dissolved Nitrogen. *Soil Sci. Soc. Am. J.* **2010**, *74*, 1018–1027. [CrossRef]
33. Olsen, S.; Sommers, L. Phosphorus. In *Methods of Soil Analysis, Part 2: Chemical and Microbiological Properties*; Page, A.L., Miller, R.H., Keeney, D.R., Eds.; Agronomy Monograph; ASA; SSSA: Madison, WI, USA, 1982; pp. 403–430.
34. Thomas, G.W. Exchangeable cations methods of soil analysis. In *Methods of Soil Analysis, Part 2: Chemical and Microbiological Properties*; Page, A.L., Miller, R.H., Keeney, D.R., Eds.; Agronomy Monograph; ASA; SSSA: Madison, WI, USA, 1982; pp. 159–166.

35. Lindsay, W.L.; Norvell, W.A. Development of a DTPA soil test for zinc, iron, manganese, and copper. *Soil Sci. Soc. Am. J.* **1978**, *42*, 421–428. [CrossRef]
36. Wolf, B. The determination of boron in soil extracts, plant materials, composts, manures, water and nutrient solutions. *Com. Soil Sci. Plant Anal.* **1971**, *2*, 363–374. [CrossRef]
37. Van Reeuwijk, L.P. *Procedures for Soil Analysis*, 6th ed.; International Soil Reference and Information Centre: Wageningen, The Netherlands, 2002.
38. Hansen, T.H.; De Bang, T.C.; Laursen, K.H.; Pedas, P.; Husted, S.; Schjoerring, J.K. Multielement plant tissue analysis using ICP spectrometry. In *Plant Mineral Nutrients. Methods in Molecular Biology (Methods and Protocols)*; Maathuis, F., Ed.; Humana Press: Totowa, NJ, USA, 2013; Volume 953.
39. Chapman, H.D.; Pratt, P.F. *Methods of Analysis for Soils, Plants and Waters*; Division of Agricultural Sciences, University of California: Riverside, CA, USA, 1961; p. 309.
40. Gaines, T.P.; Mitchell, G.A. Boron determination in plant tissue by the azomethine-H method. *Commun. Soil Sci. Plant Anal.* **1979**, *10*, 1099–1108. [CrossRef]
41. Strasser, R.J.; Tsimilli-Michael, M.; Srivastava, A. Analysis of the chlorophyll a fluorescence transient. In *Chlorophyll a Fluorescence: A Signature of Photosynthesis*; Papageorgiou, G., Govindjee, G., Eds.; Springer: Dordrecht, The Netherlands, 2004; pp. 321–362.
42. Schreiber, U.; Schliwa, U.; Bilger, W. Continuous recording of photochemical and non-photochemical chlorophyll fluorescence quenching with a new type of modulation fluorometer. *Photosynth. Res.* **1986**, *10*, 51–62. [CrossRef] [PubMed]
43. Chatzistathis, T.; Papadakis, I.E.; Papaioannou, A.; Chatzissavvidis, C.; Giannakoula, A. Comparative study effects between manure application and a Controlled-release fertilizer on the growth, nutrient uptake, photosystem II activity and photosynthetic rate of *Olea europaea* L. (cv. 'Koroneiki'). *Sci. Hortic.* **2020**, *264*, 109176. [CrossRef]
44. Chatzistathis, T.; Monokrousos, N.; Psoma, P.; Tziachris, P.; Metaxa, I.; Strikos, G.; Papadopoulos, F.H.; Papadopoulos, F.H. How fully productive olive trees (*Olea europaea* L., cv. 'Chondrolia Chalkidikis') manage to over-satisfy their P nutritional needs under low Olsen P availability in soils? *Sci. Hortic.* **2020**, *265*, 109251. [CrossRef]
45. Rekasi, M.; Szili-Kovacs, T.; Takacs, T.; Bernhadrt, B.; Puspan, I.; Kovacs, R.; Kutasi, J.; Draskovits, E.; Molnar, S.; Molnar, M.; et al. Improving the fertility of sandy soils in the temperate region by combined biochar and microbial inoculants treatments. *Arch. Agron. Soil Sci.* **2019**, *65*, 44–57. [CrossRef]
46. Sadegh-Kasmaei, L.; Yasrebi, J.; Zarei, M.; Ronaghi, A.; Ghasemi, R.; Saharkhiz, M.J.; Ahmadabadi, Z.; Schnug, E. Influence of plant growth promoting rhizobacteria, compost and biochar of Azolla on Rosemary (*Rosmarinus officinalis* L.) growth and some soil quality indicators in a calcareous soil. *Commun. Soil Sci. Plant Anal.* **2019**, *50*, 119–131. [CrossRef]
47. Dinesh, R.; Anandaraj, M.; Kumar, A.; Srinivasan, V.; Bini, Y.K.; Subila, K.P.; Aravind, R.; Hamza, S. Effects of plant growth-promoting rhizobacteria and NPK fertilizers on biochemical and microbial properties of soils under ginger (Zingiber officinale) cultivation. *Agric. Res.* **2013**, *2*, 346–353. [CrossRef]
48. Arquero, O.; Barranco, D.; Benlloch, M. Potassium starvation increases stomatal conductance in olive trees. *HortScience* **2006**, *41*, 433–436. [CrossRef]
49. Therios, I. Olives. In *Crop Production Science in Horticulture*; CAB: International, UK, 2009; pp. 156–163.
50. Chatzistathis, T.; Papadakis, I.E.; Papaioannou, A.; Dichala, O.; Giannakoula, A.; Kostas, S.; Tziachris, P. Genotypic tolerance of two *Punica granatum* L. cultivars (Wonderful and Acco) to serpentine stress. *Sci. Hortic.* **2019**, *247*, 344–355. [CrossRef]
51. Mutumba, F.A.; Zagal, E.; Gerding, M.; Castillo-Rosales, D.; Paulino, L.; Schoebitz, M. Plant growth promoting rhizobacteria for improved water stress tolerance in wheat genotypes. *J. Soil Sci. Plant Nutr.* **2018**, *18*, 1080–1096. [CrossRef]
52. Zhang, B.B.; Zhang, H.; Jing, Q.; Wu, Y.X.; Xiao, S.Z.; Wang, M.M. Effect of mycorrhiza fungi inoculation and N fertilization on physiological characteristics, growth and nitrogen and phosphorus uptake of wheat under two distinct water regimes. *Agric. Res. Arid Areas* **2019**, *37*, 214–220.
53. Zhu, L.D.; Shao, X.H.; Zhang, Y.C.; Zhang, H.; Hou, M.M. Effects of K fertilizer application on photosynthesis and seedling growth of sweet potato under drought stress. *J. Food Agric. Environ.* **2012**, *10*, 487–491.

horticulturae

MDPI

Article

Evaluation of Nitrogen and Water Management Strategies to Optimize Yield in Open Field Cucumber (*Cucumis sativus* L.) Production

Adewale Suraj Bello [1], Samsul Huda [2], Zhong-Hua Chen [2,3], Muhammad Fasih Khalid [1,4], Mohammed Alsafran [5] and Talaat Ahmed [1,*]

[1] Environmental Science Center, Qatar University, Doha 2713, Qatar; a.suraj@qu.edu.qa (A.S.B.); fasihkh65@gmail.com (M.F.K.)
[2] School of Science, Western Sydney University, Richmond, NSW 2753, Australia; s.huda@westernsydney.edu.au (S.H.); z.chen@westernsydney.edu.au (Z.-H.C.)
[3] Hawkesbury Institute for the Environment, Western Sydney University, Penrith, NSW 2751, Australia
[4] Southwest Florida Research and Education Center, Horticultural Sciences Department, Institute of Food and Agricultural Science, University of Florida, Immokalee, FL 34142, USA
[5] Agricultural Research Station, Qatar University, Doha 2713, Qatar; m.alsafran@qu.edu.qa
* Correspondence: t.alfattah@qu.edu.qa

Abstract: Countries in arid climates, such as Qatar, require efficient water-saving strategies and nitrogen treatment for vegetable production. Vegetable importation constituted approximately USD 352 million of Qatar's 2019 GDP; hence, enhancing local production is essential. This study investigated the effect of varying nitrogen and water levels on cucumber (*Cucumis sativus* L.) fruit yield. Various water management strategies were also evaluated. A split plot design was employed with two water levels (W1: 50% deficit irrigation, W2: 100% full irrigation) and three nitrogen levels (N1: 50 kg N ha^{-1}, N2: 70 kg N ha^{-1}, N3: 100 kg N ha^{-1}) to examine cucumber yield and physiological response. Our findings revealed that using minimal drip irrigation and reducing nitrogen levels significantly enhanced the growth, SPAD index, fruit characteristics, and yield components of cucumber. Drip irrigation had a greater influence on cucumber production than nitrogen levels. Shoot height increased by 4% from W2N1 (T1) to W1N3 (T6) and 4.93% from W2N2 (T2) to W1N2 (T5). Fruit length and width increased by 10.63% and 13.41% from T2 and T1 to T5, respectively. The highest total yield occurred at T5, followed by T6, T2, W2N3 (T3), W1N1 (T4), and T1 at 34.5, 29.1, 27.6, 25.8, 25.2, and 20.4 t/ha, respectively. The optimal combination comprised 50% deficit irrigation (W1) and 70 kg N ha^{-1} (N2) nitrogen. These results suggest the importance of optimizing drip irrigation for achieving maximum cucumber fruit yield in arid climates.

Keywords: *Cucumis sativus* L.; field trial; nitrogen fertilizer; drip irrigation; Qatar

1. Introduction

In most parts of the world, particularly arid and semi-arid regions, population increase and climatic change have posed serious threats to vegetable production. Indeed, droughts may become more frequent, severe, and prolonged due to climate change [1,2]. Most countries in this region, such as Qatar and Kuwait, are characterized by harsh weather conditions with low rainfall and a higher chance of drought occurrence, poor soil composition coupled with high temperatures, which invariably hinder agriculture production [3]. Thus, water availability is a major constraint to vegetable production in this region. Meanwhile, it has been projected that agricultural production will continue to increase in arid regions due to a rise in population and demand for more arable land, ultimately leading to an increase in demand for irrigation water resources [4,5]. Also, climate change has made the future uncertain as the predicted water balance between precipitation and evapotranspiration is anticipated to be unfavorable, requiring a large amount of water for irrigation [6]. Indeed, the predicted occurrence of drought in Qatar is high as its annual rainfall is often lower than the prevailing annual evapotranspiration [7]; hence, water supply remains the major constraint to crop production.

The availability of water resources is essential for the sustainable existence of humans, while the persistent increase in the global population coupled with industrialization has resulted in a noticeable increase in its consumption [8,9]. The situation is even more dire in arid and semi-arid regions. Thus, water is a scarce resource in Qatar, and its uses must be rationed among industry, municipal, and agriculture sectors [10]. Agricultural water usage accounts for 85.40% of the total water usage or consumption in Qatar, followed by municipal (11.84%) and industrial (2.76%) use [10].

Similarly, soil fertility is crucial to sustainable vegetable production and requires adequate maintenance and improvement [11,12]. Fertilization impacts soil nutrients and fertility, thus increasing crop yield and product quality. However, excessive usage of chemical fertilizers often results in low efficiency; imbalanced fertilization can result in micronutrient deficiencies [13]. In an attempt to increase agricultural output, excessive fertilizer application has led to soil degradation, groundwater contamination, and other ecological and environmental issues, while overdrawing groundwater is consequential to the groundwater table, increasing the salinity of the soil and exposing the soil to other stresses [14]. Moreover, excessive use of nitrogen (N) fertilizers can pose serious health and environmental hazards, as over 70% of N fertilizer applied in food production is lost to the environment [12,15]. Although N is an essential nutrient for crop productivity, achieving optimal application levels can be challenging due to the dynamic availability of N to plants during growth [16].

Owing to its unique taste, nutritional benefits, and high yield, cucumber is an important horticultural crop produced and consumed globally. Cucumber provides vitamins, minerals, and nutrients essential for human growth and development [14,17]. Vegetable production is a top priority outlined in the Qatar National Food Security Strategy 2018–2023. Although Qatar currently ranks 140th worldwide, it strives for self-sufficiency in vegetable production. Indeed, a significant increase in vegetable production, from 65,117 tons in 2017 to 101,881 tons in 2021, has been reported comprising four major vegetables: cucumber (19.48%), tomato (15.35%), bell pepper (6.02%), and eggplant (4.64%) [18]. Cucumber can be eaten raw and is often incorporated in fruit juice preparation. It offers certain health benefits as it possesses a significant amount of water and is extremely low in caloric content. More specifically, cucumber exhibits potential benefits in managing diabetes, reducing lipids, and acting as an antioxidant. Cucumber also aids in detoxification by eliminating stored waste products and harmful chemical toxins, while its fresh juice is beneficial for skin nourishment [19]. Cucumber is extensively cultivated in Qatar due to its short growth cycle and flavorful taste coupled with huge economic value. However, efforts to increase cucumber production may lead to poor fertilizer application and excessive water irrigation, resulting in water and nitrogen waste, increased soil salinity, nitrogen leaching, pollution, and consequently, reduced cucumber output and quality [20]. The impact of N availability on vegetable yield and quality is eminent. Thus, its application should be well-tailored and managed in cucumber (*Cucumis sativus* L.) to enhance yield and quality while minimizing its negative impacts [14,21]. In the open field, cultivated cucurbits, such as cucumber, melon, pumpkin, and squash, exhibit variation in optimal nitrogen rates depending on the cucurbit location and species, ranging from 80 to 200 kg N ha^{-1} [16,22–24]. However, in Qatar, the recommended rate of N fertilizer for general cucurbit cultivation has not been properly documented. As such, it is unclear whether the current use is under, over, or within the reported range for application as basal or surface treatment. Consequently, it is unclear whether the N rate use is suitable for open-field cucumber production in Qatar.

Water and nitrogen remain the most essential limiting factors that influence the accumulation of crop biomass, invariably resulting in crop yield variation [25–29]. Accordingly, the current adopted a sustainable irrigation technique to conserve water and an effective nitrogen management strategy to sustain crop cultivation/production in arid and semi-arid regions. We found that the rational utilization of nitrogen and irrigation water can substantially mitigate the suppressive impact of harsh weather conditions on crops under adequate agronomic practices, increasing the crop yield. We propose that our results will optimize nitrogen and irrigation application to obtain crops with better yields while protecting the environment from pollution.

2. Materials and Methods

2.1. Site Location

The experiment was conducted in field conditions at the Agriculture Research Station of Qatar University, situated 60 km north of Qatar off Shamal road in Rawdat Al-Faras within the Al Shamal Municipality, between 2 November 2022 and 10 February 2023. The farm is located at 25°48′ N latitude and 51°20′ E longitude. The study area can be characterized as a subtropical arid desert climate, characterized by minimal annual precipitation of 58.15 mm, temperature fluctuations ranging from 25.60 °C to 33.49 °C, humidity levels varying between 32% and 72%, average wind speed of 6 knots, mean sea level (MSL) atmospheric pressure of 6 hPa, and an average of 7.8 daily sunshine hours per year [18,30]. However, most rainfall occurred during November and December up to January 2023. The water source for irrigation is underground brackish water; thus efficient use of available water is crucial for preserving land and water resources. Anions and cations in water samples were quantified using the Metrohm 850 Ion Chromatography (IC) system in the Central Laboratory Unit, Qatar University, following the methodology outlined by Wilson et al. (2011), while the pH and electrical conductivity were determined using YSI Exo-2 Sonde (YSI Inc., Yellow Springs, OH, USA) [31].

2.2. Soil Properties, Land Preparation, and Seeding Transplanting

At the experimental site, the available soil is a combination of sand, loam, and clay. Before transplanting seedlings, soil samples were extracted from the experimental field at a depth ranging from 0 to 30 cm using an auger. These soil samples were subjected to a comprehensive chemical analysis at the sediment lab of the Environmental Science Center, Qatar University, to assess various soil properties, including organic matter content, pH, electrical conductivity (EC), as well as phosphorus (P), calcium (Ca), copper (Cu), magnesium (Mg), sodium (Na), zinc (Zn), boron (B), potassium (K), and manganese (Mn) concentrations. The analysis procedure involved utilizing a heating block and acid digestion method to determine the total metal content in the soil. During this procedure, 0.25 g of fine soil samples were mixed with 9 mL of HNO_3 acid and heated at 95 °C for 30 min. Subsequently, 3 mL of HF acid was added, and the mixture was heated for an additional 30 min at the same temperature. The temperature was then increased to 135 °C for 60 min to reduce the volume, followed by a slight increase to 155 °C until the sample was nearly dried. Subsequently, 3 mL of HNO_3 acid and 20 mL of deionized water were added, and the mixture was heated until it became clear. After cooling, the solution was quantitatively transferred into a 100 mL volumetric flask and diluted with deionized water to the desired volume. The ion concentrations were determined using a Perkin Elmer ICPOES 5300 DV instrument (Perkin-Elmer Optima 5300 DV, PerkinElmer, Waltham, MA, USA). Additionally, the soil's organic matter content was evaluated by estimating the organic carbon content using the Walkley and Black method. Although the soil does not have serious issues with salinity or poor drainage, the plot of land was selected based on prior support for vegetable production. The pre-planting operation, including clearing of the field, followed by plowing and furrowing, was performed to achieve good soil structure.

After land preparation and a 3-day acclimatization period, the seedlings were transplanted and subjected to the same treatment to ensure uniformity before the experimental treatments.

2.3. Experimental Design

The experiment was carried out under main and sub-plot arrangements. Irrigation was applied to the main plots while nitrogen (N) was applied to the sub-plots, with three replicates in a split-plot design. Cucumber cultivation/production was evaluated by the drip irrigation practices applied under different rates: full (100%) and deficit (50%). Three nitrogen levels were studied, representing varying percentages of the recommended 100 kg N ha^{-1} for cucumber drip irrigation. Urea was the nitrogen source, with concentrations of 100%, 70%, and 50% of the recommended rate (Table 1). In Qatar, the conventional method is extensively accepted by most farmers for scheduling irrigation and is based on the duration required to irrigate a plot. The cucumber hybrid seedling was transplanted on 2 November 2022, and the experiment ran until January/February 2023, with a 70 m × 50 cm row-to-row plant-to-plant spacing arrangement. Each experimental plot comprised four rows, 3 m long and 2.10 m wide, with five cucumber seedlings planted in each row, resulting in a total of twenty cucumber seedlings per plot. Phosphorus (70 kg P_2O_5 ha^{-1}) and potassium (60 kg $Ca(H_2PO_4)_2 \cdot H_2O$ ha^{-1}) were applied to all treatments. Drip irrigation was applied every other day each week until the required quantity of water was reached. The components of the drip system included polyethylene laterals with 12 mm diameters fixed with in-line drippers at a distance of 50 cm. Nitrogen was applied at seedling, flowering, fruit enlarging, and fruit maturity stages in equal doses with the drip irrigation system starting 5 days after transplantation (Table 2). All agronomic activities, including weeding and plant protection measures, were performed similarly to local practices. The fruits were harvested between 17 December 2022 and 10 January 2023 when they reached their full size and were green.

Table 1. Details of experimental treatments with irrigation and nitrogen fertilizer.

Treatment	Treatment Designation
Irrigation method (independent)	
Deficit irrigation–50%	W1
Full irrigation–100%	W2
50 kg N ha^{-1} (50% of recommended N1)	N1
70 kg N ha^{-1} (70% of recommended N2)	N2
100 kg N ha^{-1} (100% of recommended N3)	N3
Interaction	
Full irrigation, 100% nitrogen (W2N3)	T_1
Full irrigation, 70% nitrogen (W2N2)	T_2
Full irrigation, 50% nitrogen (W2N1)	T_3
Deficit irrigation, 100% nitrogen (W1N3)	T_4
Deficit irrigation, 70% nitrogen (W1N2)	T_5
Deficit irrigation, 50% nitrogen (W1N1)	T_6

Table 2. Application of nitrogen fertilizer at different growth stages of cucumber. N1: 50% of recommended N, N2: 70% of recommended N, N3: 100% of recommended N.

N Fertilizer Application	Seedling	Flowering	Fruit Enlarging	Fruit Maturity
N1 50 kg N ha^{-1} (50%)	5	15	15	15
N2 70 kg N ha^{-1} (70%)	7	21	21	21
N3 100 kg N ha^{-1} (100%)	10	30	30	30

2.4. Growth, Fruit, and Yield Parameters

The shoot height of the cucumber plant was determined on the 17th day after transplantation by measuring from the assumed base (above soil level) to the highest point where the leaves were completely opened using a ruler. The average was recorded at the beginning and end of the experiment. A few days after transplantation, the leaves per plant were manually counted. The leaf length and width were measured with a ruler, and the average length and width were recorded. Four weeks after transplantation, the stem width was manually measured with a digital vernier caliper (Clarke CM145 Digital Vernier Caliper) and used to calculate the plant diameter. When the leaves were green and healthy during the second week after transplantation, the chlorophyll content was measured using a common diagnostic tool, Soil Plant Analysis Development (SPAD) chlorophyll meter (SPAD–2: Konica Minolta Optics, Osaka, Japan), and expressed as the SPAD Index [32]. The fruit length and width were manually determined with a digital vernier caliper at each harvest. The summation of the harvested fruit weight was determined at each harvest and the end of the cultivation period with a laboratory scale and recorded in grams per fruit. The number of fruits collected per experimental unit was manually counted and recorded at each harvest, and the sum was calculated for the total number of fruits per experimental unit for all treatments. Thus, the fruit parameters, viz. length, width, number/plant, average weight, and fresh yield were recorded at each harvest.

2.5. Statistical Analysis

The effects of irrigation and different levels of N application coupled with their interaction were determined via analysis of variance (ANOVA) using Minitab 19 (Minitab Inc., State College, PA, USA). Mean differences were compared using Tukey's post hoc test at a 5% significance level ($p \leq 0.05$).

3. Results

The impact of different levels of irrigation and nitrogen treatment on cucumber vegetative growth parameters, yield constituents, and total yield are shown in Figure 1A–I. and Table 3. The results presented in Table 4 indicate statistically significant differences in the measured parameters.

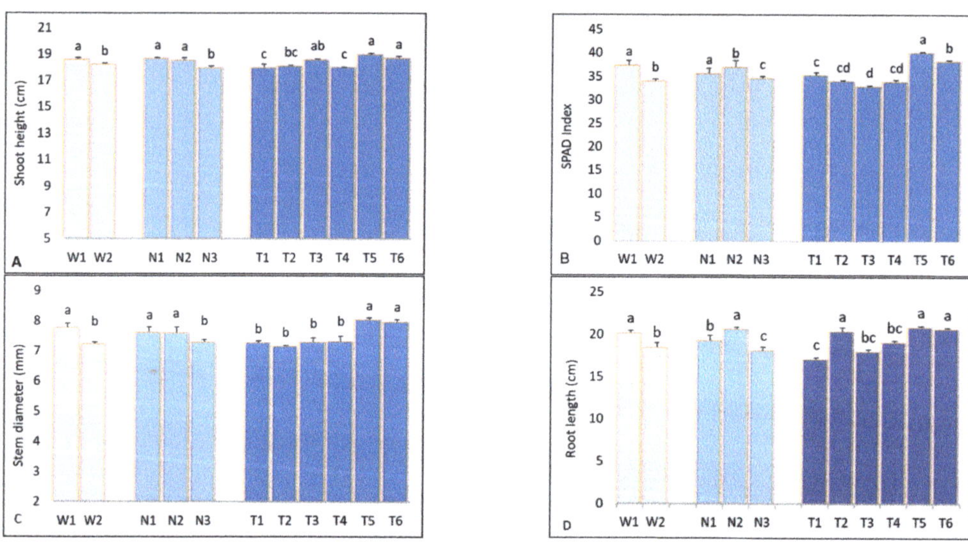

Figure 1. *Cont.*

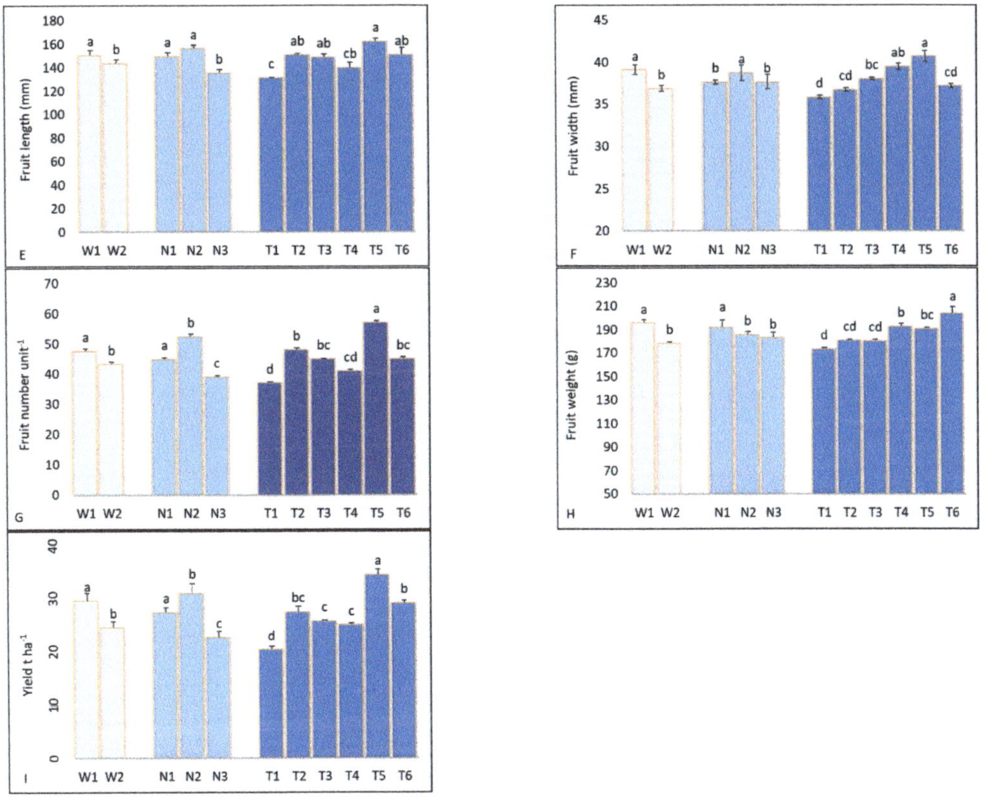

Figure 1. Shoot height (**A**), SPAD index (**B**), stem diameter (**C**), root length (**D**), fruit length (**E**), fruit width (**F**), number of fruit unit^{-1} (**G**), fruit weight (**H**), and yield (ton ha^{-1}) (**I**) of cucumber-treated plants treated with 50, 70, and 100% nitrogen deficit (N1, N2, and N3) and 50 and 100% full drip irrigation (W1 and W2). The interaction study is represented by T1 (Full irrigation, 100% nitrogen), T2 (Full irrigation, 70% nitrogen), T3 (Full irrigation, 50% nitrogen), T4 (Deficit irrigation, 100% nitrogen), T5 (Deficit irrigation, 70% nitrogen), and T6 (Deficit irrigation, 50% nitrogen) respectively with three replicates per treatment. The different letters indicate significance according to the Tukey's test ($p < 0.05$).

Table 3. Effect of drip irrigation and N treatment on the vegetative growth parameters and yield of cucumber. Analysis of variance (ANOVA) p values are shown ($p < 0.05$, significance; $p > 0.05$, no significance). Different lowercase letters in the same column represent significant differences between N treatments or drip irrigation methods or interactions ($p < 0.05$); the same letters represent non-significant differences between N treatments or drip irrigation methods or interactions ($p > 0.05$) with three replicates per treatment.

Treatments	Shoot Height (cm)	SPAD Index	Stem Width (mm)	Root Length (cm)	Fruit Length (mm)	Fruit Width (mm)	Number of Fruit/Unit	Fruit Weight (g)	Yield (ton/ha)
Drip irrigation levels									
W50	18.58 ± 0.15 a	37.42 ± 0.92 a	7.78 ± 0.13 a	20.2 ± 0.29 a	150.98 ± 3.91 a	39.09 ± 0.56 a	48 ± 0.84 a	196.01 ± 2.68 a	29.58 ± 1.41 a
W100	18.23 ± 0.12 b	34.07 ± 0.39 b	7.25 ± 0.05 b	18.5 ± 0.52 b	143.77 ± 3.16 b	36.86 ± 0.33 b	43 ± 0.58 b	178.36 ± 1.23 b	24.58 ± 1.13 b
Nitrogen treatments									
N50	18.67 ± 0.07 a	35.60 ± 1.20 b	7.63 ± 0.17 a	19.32 ± 0.60 b	149.87 ± 2.84 a	37.58 ± 0.22 b	45 ± 0.26 a	192.46 ± 5.79 a	27.47 ± 0.78 a
N70	18.54 ± 0.20 a	37.02 ± 1.35 a	7.61 ± 0.20 a	20.63 ± 0.24 a	156.56 ± 2.77 a	38.70 ± 0.93 a	53 ± 0.76 b	185.91 ± 2.26 b	31.05 ± 1.68 b
N100	18 ± 0.11 b	34.61 ± 0.44 c	7.31 ± 0.07 b	18.10 ± 0.46 c	135.7 ± 2.7 b	37.64 ± 0.83 b	39 ± 0.37 c	183.18 ± 4.45 b	22.73 ± 1.08 c
Interaction									
T1	17.99 ± 0.24 c	35.28 ± 0.63 c	7.29 ± 0.04 b	17.1 ± 0.17 c	131.45 ± 0.36 c	35.84 ± 0.19 d	37 ± 0.33 d	173.60 ± 0.75 a	20.39 ± 0.55 d
T2	18.10 ± 0.05 bc	34.00 ± 0.11 cd	7.16 ± 0.02 b	20.4 ± 0.47 a	150.85 ± 0.83 ab	36.73 ± 0.22 cd	48 ± 0.58 b	180.89 ± 0.27 c	27.56 ± 0.96 bc
T3	18.58 ± 0.04 ab	32.93 ± 0.12 d	7.29 ± 0.15 b	18 ± 0.27 bc	149.00 ± 2.13 ab	38.00 ± 0.13 bc	45 ± 0.00 bc	180.59 ± 0.63 cd	25.80 ± 0.09 c
T4	18.00 ± 0.02 c	33.94 ± 0.35 cd	7.32 ± 0.16 b	19.1 ± 0.17 bc	139.95 ± 4.28 cb	39.44 ± 0.39 ab	41 ± 0.33 cd	192.76 ± 2.60 b	25.07 ± 0.28 c
T5	18.99 ± 0.09 a	40.04 ± 0.11 a	8.05 ± 0.05 a	20.87 ± 0.15 a	162.27 ± 2.27 a	40.67 ± 0.65 a	57 ± 0.58 a	190.92 ± 0.62 bc	34.54 ± 0.97 a
T6	18.75 ± 0.12 a	38.27 ± 0.26 b	7.97 ± 0.07 a	20.63 ± 0.15 a	150.73 ± 5.93 ab	37.16 ± 0.23 cd	45 ± 0.58 bc	204.34 ± 5.12 a	29.14 ± 0.45 b

Table 4. Analysis of variance showing statistically significant effects of various treatments on cucumber shoot height, SPAD index, stem width, root length, fruit length, fruit width, number of fruit/plant, fruit weight, and yield. Three replicates per treatment. * Indicates highly significant differences at $p < 0.05$ level, ns = not significant.

Source of Variation	Adjusted Mean Square								
	Shoot Height (cm)	SPAD Index	Stem Width (mm)	Root Length (cm)	Fruit Length (mm)	Fruit Width (mm)	Number of Fruit/Plant	Fruit Weight (g)	Yield (ton/ha)
N treatment	0.76 *	8.83 *	0.20 *	9.64 *	680.36 *	2.36 *	30.50 *	136.54 *	104.50 *
Drip irrigation levels	0.56 *	50.41 *	1.28 *	13.01 *	234.43 *	22.49 *	9.39 *	1400.79 *	112.51 *
N × W levels	0.33 *	24.90 *	0.29 *	1.87 *	37.07 ^{ns}	10.68 *	3.39 *	73.21 *	5.09 *
SE	0.04	0.31	0.03	0.2	31.97	0.37	0.61	17.17	1.24
SD	0.17	1.33	0.12	0.84	135.55	1.55	2.59	72.8	5.25

3.1. Soil of the Cultivated Plot and Irrigation Water

Soil samples were analyzed via ion chromatography, YSI Meter, and MOOPAM. A substantial amount of anions and cations was detected, with moderate and suitable pH levels, and total organic matter for proper cucumber growth (Table 5). Additionally, the electrical conductivity (EC) and pH of the irrigation water source were quantified. The ion composition of the irrigation water source comprised sodium, magnesium, potassium, calcium, chlorine, and sulfate (Table 5).

Table 5. Certain soil and water characteristics in the open field experiment with three replicates per treatment.

Chemical Constituents	Analytical Method	Soil Sample (depth 0–30 cm)(mg kg^{-1} DW)	Water Sample (mg L^{-1})
Sodium (Na)	Ion chromatography (IC)andInductively coupled plasma–optical Emission spectrometry (ICP–OES)	10,294 ± 97.4	274.5 ± 8.7
Magnesium (Mg)		32,031 ± 134.7	98.5 ± 2.2
Potassium (K)		13,445 ± 40.5	24.5 ± 1.7
Calcium (Ca)		62,330 ± 126.6	297.7 ± 4.7
Boron (B)		3849 ± 31.7	nd
Copper (Cu)		30 ± 0.1	nd
Iron (Fe)		26,378 ± 129.4	nd
Manganese (Mn)		585 ± 2.8	nd
Zinc (Zn)		76 ± 0.5	nd
Phosphorus (P)		1223 ± 11.4	nd
Chlorine (Cl)		nd	433.5 ± 10.6
Sulfate (SO_4)		nd	422.5 ± 11.8
EC (mS/cm)	YSI Meter	2.25 ± 1.4	3.01 ± 0.5
pH		7.82 ± 0.2	7.07 ± 0.4
TOM (%)	MOOPAM	0.5 ± 0.2	nd

EC, electrical conductivity; pH, potential of hydrogen; TOM (%), total organic matter; nd, not determined with three replicates.

3.2. Effect of Irrigation and Nitrogen on Cucumber Growth Parameters and SPAD Index

A range of cucumber growth parameters, yield constituents, and fruit yield was observed based on the amount of water (W) used in drip irrigation at various nitrogen levels (Table 3). Plant shoot height significantly increased with decreasing N levels up to N1. In drip irrigation, the tallest plant shoot was found at N2 and N1 in W1 (T5 and T6) while in W2 the tallest was under N1 treatment (T3).

The SPAD index in plant leaves revealed that the chlorophyll content significantly increased with decreasing fertilizer application; the SPAD index was highest under N2 and N1 in W1 (T5 and T6). The highest SI was achieved with N3 treatment in W2. The stem diameter followed the same trend, with the largest diameter recorded under N2 and N1 in W1 (T5 and T6), while the largest in the W2 treatment was found under N1 (T3). Also, the root length was significantly increased with decreasing fertilizer application; the root length was longest with N1 and N2 in W1 (T5 and T6). Meanwhile, in the W2 plots, the longest roots were achieved under the N3 treatment. Similarly, the stem diameter was the largest under N2 and N1 in W1 (T5 and T6) and under N1 in W2 (T3).

3.3. Effect of Irrigation and Nitrogen on Fruit Yield

Figure 1E–H showed a significant difference in the fruit yield constituents, including the length, width, number of fruits per plant, and weight of the cucumbers based on the N treatment and W level. In W1, the longest fruit was observed with N2 and N1 treatments and the shortest with N3 treatments. In the W2 treatment, the longest fruit was recorded at N2. followed by N1, and the shortest was obtained following N3 treatment (Figure 1E). The fruit width showed a similar pattern, with the largest obtained under N3 in W1 and N1 in W2 and the smallest found under N1 in W1 and N3 in W2 treatments (Figure 1F). The fruit weight followed the same trend; the heaviest fruit was obtained under N1, followed by N2 in W1, and N2 and N1 in W2, while the lightest fruit was found under N3 in W1 and W2 treatments.

The fruit yield was noticeably larger in W1 compared to W2 under N1 and N2 treatments. The highest fruit yield was obtained under N2, followed by N1 (T5 and T6) in W1, and the lowest was achieved under N3. W2 produced the maximum yield at N2 (T2) and the lowest yield under N3 (T1) (Figure 1I). However, the yields produced under W1 and N2 (34.5421 t ha^{-1}), N1 (29.137 t ha^{-1}), and N3 (25.0734 t ha^{-1}) were higher than those under W2 and N2 (27.5621 t ha^{-1}), N1 (25.7986 t ha^{-1}), and N3 (20.3913 t ha^{-1}). The most fruitful yield was obtained under treatment with W1 and N2, which was 22.48% larger than W2 and N2, suggesting that W1 and N2 (T5) was the ideal combination. Hence, drip irrigation (W1) produces a 22.48% greater fruit yield with a 50% water deficit and 30 kg of nitrogen per ha compared to the highest yield achieved with W2 at the same N level. These findings suggest that optimizing water quantity applied via drip under different N levels in the field is crucial for maximizing drip irrigation benefits.

4. Discussion

As unfavorable climate change continues to make water-saving techniques increasingly important in arid and semi-arid areas where water availability and supply are limited, optimizing the conventional or full irrigation system leads to enhanced crop production and water use efficiency [33–38]. At the end of the current study, the tallest shoot was observed in the treatment subjected to W1 drip irrigation (deficit water irrigation system), while the W2 treatment showcased the shortest plant (Table 3). These findings agree with those of Harmanto et al. (2005) [39] and Rahil and Qanadillo (2015) [20], who reported that at 70% irrigation deficit, the plant height was highest for tomato and cucumber cultivated to optimize irrigation and N consumption (Abdalhi et al., 2016) [40]. In contrast, Hashem et al. (2011) [41] contended that for cucumber plants, a 100% irrigation level succeeded by 80% and 120% irrigation levels across two growing seasons, leading to increased vegetative parameters. Additionally, Ngouajio et al. (2007) mentioned that augmented vegetative parameters under the 100% irrigation level were attributed to the appropriate irrigation volume, particularly during early growth stages, fostering a deeper and more expansive root system [42]. Similarly, the SPAD index value for the leaf chlorophyll content increased to 37.42 with water deficit irrigation in W1 (Table 3), indicating that excessive or deficit irrigation can achieve high or low chlorophyll content. Interestingly, our finding was supported by Amer et al. (2009) [43], who found that under deficit irrigation, the chlorophyll content statistically increases. Similarly, the findings agreed with the work of Abdalhi et al. (2016) [40], who stated that leaf chlorophyll content decreased significantly as the water irrigation deficit increased; the maximum leaf chlorophyll was obtained when the irrigation deficit was maintained [40]. Deficient irrigation as a water-saving strategy has exhibited a positive impact on yield compared with full or traditional irrigation. It has been argued successfully in different studies conducted across the world. In our study, the increased yield at W1 compared with W2 is in line with that reported by Dogan et al. (2008) who achieved the highest cv. Ananas fruit yields at 75% irrigation (43.8 Mg ha^{-1}) compared to 100% (38.9 Mg ha^{-1}) [44]. In another study conducted on cotton cultivation, the adoption of deficit irrigation resulted in a reduction of water consumption by 31–34%, leading to a 13–24% increase in cottonseed yield [45–47]; our findings supported this. However, our findings contradict others, who reported that the quantity of irrigation leads to a linear increase in onion fruit yields [41,48,49]. Hence, our findings support the notion that reduced irrigation can lead to higher crop yields.

Different studies have widely reported that applying fertilizer to various crops significantly positively affects healthy plant growth and yield [50–52]. Our findings showed that fertilization application and two-drip irrigation practices resulted in higher yields with a minimal water supply and N required (Table 3). However, under the open-field planting of cucumber, the yield exhibited a significant increase with a slight increase in N level up to N2, below the recommended N nutrient requirement for cucumber cultivation. Thus, the yield increased from N1 to N2 (13.03%) (i.e., 50 to 70 Kg ha^{-1}) at a stage where the yield was assumed to be constant before decreasing (11.43%) at N3 (1000 Kg ha^{-1}). These results suggest that the optimum fertilizer treatment is likely 70 Kg ha^{-1}. The maximal yields were evidenced by increased fruit quantity and enhanced individual fruit weight. However, no significant difference in fruit weight was observed between N1 and N3 plots, indicating that, within the confines of our experimental parameters, N1 fertilization influenced these yield constituents. Conversely, excessive quantities of N in N3 plots impact fruit weight. Our findings aligned with those of Cabello et al. (2009), who found a substantial increment in yield (18%) upon transitioning from 30 to 85 kg N ha^{-1}. Subsequently, the yield demonstrated constancy before a marginal decline (13%) at the 139 kg ha^{-1} level [53]. Contrarily, it was reported that under arid and semi-arid climatic conditions, cucumber plants responded to as much as 240 kg N ha^{-1} [54].

Considering the interaction between water and N, at N70 in W1 (T5), considered an optimum combination, the highest cucumber fruit yield was obtained compared to the other treatments, leading to a 50% reduction in water use and 30 and 50 Kg N ha^{-1} (Table 3). This study's findings agree with those of Rahil et al. (2015), who discovered that crop yields increased when irrigation was reduced by as much as 70% [20]. Alternatively, no significance was recorded in yield between T2, T3, and T4. In summary, the highest fruit yield was achieved under the 'T5 treatment, which was 40.87, 20.00, 25.32, 27.41, and 15.66% higher than T1, T2, T3, T4, and T6, respectively. Hence, the cucumber fruit yield was higher under deficit irrigation treatment, which could result from the crop's potential to adapt to areas with lower water availability or reduced water supply. In arid and semi-arid regions, these methods can be utilized as a sustainable strategy to maximize water use, allowing farmers to achieve favorable crop yields while conserving water resources. To achieve maximum productivity and long-term viability, irrigation strategies must consider crop-specific requirements and local environmental conditions, as demonstrated by this study.

5. Conclusions

Our findings show that cucumber growth performance is sensitive to fertigation and irrigation levels. The application of deficit drip water irrigation at 70% N level is considered the sustainable and optimal standard to obtain higher fruit yield in an arid area like Qatar, characterized by harsh climatic conditions. Moreover, improving the application of N fertilizer and reducing water usage in field cucumber production produces fresh vegetables and reduces environmental pollution from leaching and contamination by excessive N fertilizer use. In our forthcoming research, we will emphasize the utilization of omics techniques to delve deeper into the intricate dynamics underlying nutrient management and cucumber cultivation. This approach will expand our knowledge of these complex, multifaceted dynamic processes and enhance cucumber production.

Author Contributions: Conceptualization, S.H., Z.-H.C., M.A. and T.A.; methodology, A.S.B., M.F.K. and T.A.; software, A.S.B. and T.A.; formal analysis, A.S.B. and T.A.; investigation, A.S.B.; resources M.A. and T.A.; writing—original draft preparation, B.AS.; writing—review and editing, A.S.B., S.H., Z.-H.C., M.A. and T.A.; supervision, T.A.; funding acquisition, S.H., Z.-H.C., M.A. and T.A. All authors have read and agreed to the published version of the manuscript.

Funding: This publication was made possible by MME grant MME01-0826-190018 from the Qatar National Research Fund (a member of Qatar Foundation). The findings achieved herein are solely the responsibility of the authors.

Data Availability Statement: Access to the data used in this study is subject to certain restrictions. The data were obtained under rules and are not publicly available. However, interested individuals may request access to the data by contacting the corresponding author (T.A.) and making a reasonable request, subject to approval.

Acknowledgments: The authors are grateful to the Director of the Agricultural Research Station for providing the necessary support. Also, we are grateful to AGRICO for supplying us with the inputs (seedlings) used. The soil samples of the field analysis using Inductively coupled plasma-optical emission spectrometry (ICP-OES) were facilitated by the Environmental Science Center. Also, water samples analysis by the Ion Chromatography analysis was facilitated by the Central Laboratories unit, at Qatar University.

Conflicts of Interest: The authors declare no conflict of interest.

References

1. Jones, E.; van Vliet, M.T. Drought impacts on river salinity in the southern US: Implications for water scarcity. *Sci. Total Environ.* **2018**, *644*, 844–853. [CrossRef] [PubMed]
2. Parkash, V.; Singh, S.; Deb, S.K.; Ritchie, G.L.; Wallace, R.W. Effect of deficit irrigation on physiology, plant growth, and fruit yield of cucumber cultivars. *Plant Stress* **2021**, *1*, 100004. [CrossRef]
3. Ben Hassen, T.; El Bilali, H. Sustainable Agriculture and Food Security in Qatar: International Threats and Local Constraints. In *Sustainable Agriculture and Food Security*; Springer: Berlin/Heidelberg, Germany, 2022; pp. 425–442.
4. Sauer, T.; Havlík, P.; Schneider, U.A.; Schmid, E.; Kindermann, G.; Obersteiner, M. Agriculture and resource availability in a changing world: The role of irrigation. *Water Resour. Res.* **2010**, *46*, 1–12. [CrossRef]
5. Amare, D. Review on Effect of Different Irrigation Method on Water Use Efficiency, Yield Productivity and Nitrogen Application. *J. Biol. Agric. Health* **2020**, *10*, 1–9.
6. Bisbis, M.B.; Gruda, N.; Blanke, M. Potential impacts of climate change on vegetable production and product quality–A review. *J. Clean. Prod.* **2018**, *170*, 1602–1620. [CrossRef]
7. Issaka, A.I.; Paek, J.; Abdella, K.; Pollanen, M.; Huda, A.; Kaitibie, S.; Goktepe, I.; Haq, M.; Moustafa, A.T. Analysis and calibration of empirical relationships for estimating evapotranspiration in Qatar: Case study. *J. Irrig. Drain. Eng.* **2017**, *143*, 05016013. [CrossRef]
8. Van Vuuren, D.P.; Bijl, D.L.; Bogaart, P.; Stehfest, E.; Biemans, H.; Dekker, S.C.; Doelman, J.C.; Gernaat, D.E.; Harmsen, M. Integrated scenarios to support analysis of the food–energy–water nexus. *Nat. Sustain.* **2019**, *2*, 1132–1141. [CrossRef]
9. Bello, A.S.; Zouari, N.; Da'ana, D.A.; Hahladakis, J.N.; Al-Ghouti, M.A. An overview of brine management: Emerging desalination technologies, life cycle assessment, and metal recovery methodologies. *J. Environ. Manag.* **2021**, *288*, 112358. [CrossRef]
10. Ahmad, A.Y.; Al-Ghouti, M.A. Approaches to achieve sustainable use and management of groundwater resources in Qatar: A review. *Groundw. Sustain. Dev.* **2020**, *11*, 100367. [CrossRef]
11. Çolpan, E.; Zengin, M.; Özbahçe, A. The effects of potassium on the yield and fruit quality components of stick tomato. *Hortic. Environ. Biotechnol.* **2013**, *54*, 20–28. [CrossRef]
12. Souri, M.K.; Yarahmadi, B. Effect of Amino Chelates Foliar Application on Growth and Development of Marigold (*Calendula officinalis*) plant. *Iran. J. Plant Prod. Technol.* **2015**, *15*, 109–119.
13. Souri, M.k.; Sooraki, F.Y.; Moghadamyar, M. Growth and quality of cucumber, tomato, and green bean under foliar and soil applications of an aminochelate fertilizer. *Hortic. Environ. Biotechnol.* **2017**, *58*, 530–536. [CrossRef]
14. Li, J.; Yang, X.; Zhang, M.; Li, D.; Jiang, Y.; Yao, W.; Zhang, Z. Yield, Quality, and Water and Fertilizer Partial Productivity of Cucumber as Influenced by the Interaction of Water, Nitrogen, and Magnesium. *Agronomy* **2023**, *13*, 772. [CrossRef]
15. Souri, M.K. Aminochelate fertilizers: The new approach to the old problem; a review. *Open Agric.* **2016**, *1*, 118–123. [CrossRef]
16. Van Eerd, L.L.; O'Reilly, K.A. Yield, nitrogen dynamics, and fertilizer use efficiency in machine-harvested cucumber. *HortScience* **2009**, *44*, 1712–1718. [CrossRef]
17. Zhao, H.; Zhai, X.; Guo, L.; Yang, Y.; Li, J.; Ren, C.; Wang, K.; Liu, X.; Zhan, R.; Wang, K. Comparing protected cucumber and field cucumber production systems in China based on emergy analysis. *J. Clean. Prod.* **2019**, *236*, 117648. [CrossRef]
18. PSA. Planning and Statistic Authority. PSA Statistical Subjects. 2021. Available online: https://www.psa.gov.qa/en/statistics1/StatisticsSite/Pages/default.aspx (accessed on 20 September 2023).
19. Mukherjee, P.K.; Nema, N.K.; Maity, N.; Sarkar, B.K. Phytochemical and therapeutic potential of cucumber. *Fitoterapia* **2013**, *84*, 227–236. [CrossRef]
20. Rahil, M.; Qanadillo, A. Effects of different irrigation regimes on yield and water use efficiency of cucumber crop. *Agric. Water Manag.* **2015**, *148*, 10–15. [CrossRef]
21. Nakano, Y.; Noda, S.; Miyake, Y.; Kogoshi, M.; Sato, F.; Iijima, W. Applicability of Variable-Rate Nitrogen Top Dressing Based on Measurement of the Within-Field Variability of Soil Nutrients for Cabbage Production. *Horticulturae* **2023**, *9*, 506. [CrossRef]
22. Jamil Mohammad, M. Squash yield, nutrient content and soil fertility parameters in response to methods of fertilizer application and rates of nitrogen fertigation. *Nutr. Cycl. Agroecosystems* **2004**, *68*, 99–108. [CrossRef]

23. Esmel, C.E.; Santos, B.M.; Gilreath, J.P. Effects of nitrogen rates on summer squash growth and yield. *HortScience* **2006**, *41*, 506C-506. [CrossRef]
24. Ferrante, A.; Spinardi, A.; Maggiore, T.; Testoni, A.; Gallina, P.M. Effect of nitrogen fertilisation levels on melon fruit quality at the harvest time and during storage. *J. Sci. Food Agric.* **2008**, *88*, 707–713. [CrossRef]
25. Hamzei, J.; Soltani, J. Deficit irrigation of rapeseed for water-saving: Effects on biomass accumulation, light interception and radiation use efficiency under different N rates. *Agric. Ecosyst. Environ.* **2012**, *155*, 153–160. [CrossRef]
26. Yang, G.; Tang, H.; Tong, J.; Nie, Y.; Zhang, X. Effect of fertilization frequency on cotton yield and biomass accumulation. *Field Crops Res.* **2012**, *125*, 161–166. [CrossRef]
27. Fondio, L.; N'Gbesso, M.; Coulibaly, N.D. Effect of mineral fertilization on African eggplant (*Solanum* spp.) Productivity in Côte d'Ivoire. *J. Agric. Sci. Technol.* **2016**, *6*, 188–195. [CrossRef]
28. Du, Y.-D.; Cao, H.-X.; Liu, S.-Q.; Gu, X.-B.; Cao, Y.-X. Response of yield, quality, water and nitrogen use efficiency of tomato to different levels of water and nitrogen under drip irrigation in Northwestern China. *J. Integr. Agric.* **2017**, *16*, 1153–1161. [CrossRef]
29. Mwinuka, P.R.; Mbilinyi, B.P.; Mbungu, W.B.; Mourice, S.K.; Mahoo, H.F.; Schmitter, P. Optimizing water and nitrogen application for neglected horticultural species in tropical sub-humid climate areas: A case of African eggplant (*Solanum aethiopicum* L.). *Sci. Hortic.* **2021**, *276*, 109756. [CrossRef]
30. Shomar, B.; Darwish, M.; Rowell, C. What does integrated water resources management from local to global perspective mean? Qatar as a case study, the very rich country with no water. *Water Resour. Manag.* **2014**, *28*, 2781–2791. [CrossRef]
31. Wilson, M.B.; Zhang, C.; Gandhi, J. Analysis of inorganic nitrogen and related anions in high salinity water using ion chromatography with tandem UV and conductivity detectors. *J. Chromatogr. Sci.* **2011**, *49*, 596–602. [CrossRef]
32. Bello, A.S.; Saadaoui, I.; Ahmed, T.; Hamdi, H.; Cherif, M.; Dalgamouni, T.; Al Ghazal, G.; Ben-Hamadou, R. Enhancement in Bell Pepper (*Capsicum annuum* L.) Plants with Application of *Roholtiella* sp. (Nostocales) under Soilless Cultivation. *Agronomy* **2021**, *11*, 1624. [CrossRef]
33. Intrigliolo, D.S.; Castel, J.R. Response of grapevine cv.'Tempranillo'to timing and amount of irrigation: Water relations, vine growth, yield and berry and wine composition. *Irrig. Sci.* **2010**, *28*, 113–125. [CrossRef]
34. Fernández, V.; Brown, P.H. From plant surface to plant metabolism: The uncertain fate of foliar-applied nutrients. *Front. Plant Sci.* **2013**, *4*, 289. [CrossRef] [PubMed]
35. Dong, Y.; Zhao, Z.; Zhang, X.; Liu, X.; Li, Q. Improvement of alternate partial root-zone drip irrigation on photosynthesis and water use efficiency of cucumbers. *J. Plant Nutr. Fertil.* **2016**, *22*, 269–276.
36. Huang, P.F.; Yin, G.H.; Gu, J.; Liu, Z.X.; Guo, J.L. Effects of alternate subsurface drip irrigation on yield and water use efficiency of spring maize. *Ying Yong Sheng Tai Xue Bao= J. Appl. Ecol.* **2016**, *27*, 2507–2512.
37. Liu, X.; Liu, B.; Cui, Q.; Zhang, W.; Li, M.; Ai, X.; Bi, H.; Li, Q. Effects of alternative drip irrigation and nitrogen fertilization on growth, photosynthesis, yield, and water-nitrogen use efficiency of cucumbers in solar greenhouse. *Plant Physiol. J.* **2016**, *52*, 905–916.
38. Luo, H.; Li, F. Tomato yield, quality and water use efficiency under different drip fertigation strategies. *Sci. Hortic.* **2018**, *235*, 181–188. [CrossRef]
39. Salokhe, V.; Babel, M.; Tantau, H. Water requirement of drip irrigated tomatoes grown in greenhouse in tropical environment. *Agric. Water Manag.* **2005**, *71*, 225–242.
40. Abdalhi, M.A.; Cheng, J.; Feng, S.; Yi, G. Performance of drip irrigation and nitrogen fertilizer in irrigation water saving and nitrogen use efficiency for waxy maize (*Zea mays* L.) and cucumber (*Cucumis sativus* L.) under solar greenhouse. *Grassl. Sci.* **2016**, *62*, 174–187. [CrossRef]
41. Hashem, F.; Medany, M.; El-Moniem, E.M.A.; Abdallah, M.M.F. Influence of green-house cover on potential evapotranspiration and cucumber water requirements. *Arab Univ. J. Agric. Sci.* **2011**, *19*, 205–215.
42. Ngouajio, M.; Wang, G.; Goldy, R. Withholding of drip irrigation between transplanting and flowering increases the yield of field-grown tomato under plastic mulch. *Agric. Water Manag.* **2007**, *87*, 285–291. [CrossRef]
43. Amer, K.H.; Midan, S.A.; Hatfield, J.L. Effect of deficit irrigation and fertilization on cucumber. *Agron. J.* **2009**, *101*, 1556–1564. [CrossRef]
44. Dogan, E.; Kirnak, H.; Berekatoglu, K.; Bilgel, L.; Surucu, A. Water stress imposed on muskmelon (*Cucumis melo* L.) with subsurface and surface drip irrigation systems under semi-arid climatic conditions. *Irrig. Sci.* **2008**, *26*, 131–138. [CrossRef]
45. Du, T.; Kang, S.; Zhang, J.; Li, F.; Hu, X. Yield and physiological responses of cotton to partial root-zone irrigation in the oasis field of northwest China. *Agric. Water Manag.* **2006**, *84*, 41–52. [CrossRef]
46. Du, T.; Kang, S.; Zhang, J.; Li, F. Water use and yield responses of cotton to alternate partial root-zone drip irrigation in the arid area of north-west China. *Irrig. Sci.* **2008**, *26*, 147–159. [CrossRef]
47. Chai, Q.; Gan, Y.; Turner, N.C.; Zhang, R.-Z.; Yang, C.; Niu, Y.; Siddique, K.H. Water-saving innovations in Chinese agriculture. *Adv. Agron.* **2014**, *126*, 149–201.
48. Zhang, H.-X.; Chi, D.-C.; Qun, W.; Jun, F.; Fang, X.-Y. Yield and quality response of cucumber to irrigation and nitrogen fertilization under subsurface drip irrigation in solar greenhouse. *Agric. Sci. China* **2011**, *10*, 921–930. [CrossRef]
49. Sahin, U.; Kuslu, Y.; Kiziloglu, F. Response of cucumbers to different irrigation regimes applied through drip-irrigation system. *JAPS J. Anim. Plant Sci.* **2015**, *25*, 198–205.
50. Souri, M.K.; Hatamian, M. Aminochelates in plant nutrition: A review. *J. Plant Nutr.* **2019**, *42*, 67–78. [CrossRef]

51. Dimkpa, C.O.; Fugice, J.; Singh, U.; Lewis, T.D. Development of fertilizers for enhanced nitrogen use efficiency–Trends and perspectives. *Sci. Total Environ.* **2020**, *731*, 139113. [CrossRef]
52. Padilla, F.M.; Farneselli, M.; Gianquinto, G.; Tei, F.; Thompson, R.B. Monitoring nitrogen status of vegetable crops and soils for optimal nitrogen management. *Agric. Water Manag.* **2020**, *241*, 106356. [CrossRef]
53. Cabello, M.J.; Castellanos, M.; Romojaro, F.; Martinez-Madrid, C.; Ribas, F. Yield and quality of melon grown under different irrigation and nitrogen rates. *Agric. Water Manag.* **2009**, *96*, 866–874. [CrossRef]
54. Cui, B.-J.; Niu, W.-Q.; Du, Y.-D.; Zhang, Q. Response of yield and nitrogen use efficiency to aerated irrigation and N application rate in greenhouse cucumber. *Sci. Hortic.* **2020**, *265*, 109220. [CrossRef]

horticulturae

MDPI

Article

Impact of Soil Biodisinfection Techniques in Horticultural Crops on Profitability within the Framework of the Circular Economy

Francisco José Castillo-Díaz [1,2], Luis J. Belmonte-Ureña [2,*], Ana Batlles-delaFuente [2] and Francisco Camacho-Ferre [1]

[1] Department of Agronomy, Sustainable Protected Agriculture Research Network, University of Almería La Cañada de San Urbano, 04120 Almería, Spain; fcastillo@ual.es (F.J.C.-D.); fcamacho@ual.es (F.C.-F.)
[2] Department of Economy and Business, Sustainable Protected Agriculture Research Network, University of Almería La Cañada de San Urbano, 04120 Almería, Spain; anabatlles@ual.es
* Correspondence: lbelmont@ual.es

Abstract: Agricultural systems must identify fertilization strategies in line with the principles of sustainable development and the circular economy to achieve environmentally friendly food production and to meet the fertilizer-reduction targets set by some territories, such as the European Union (EU). The objective of this work was to identify the multifunctionality and cost-effectiveness in the use of the biodisinfection technique in greenhouse horticultural crops. For this purpose, a cost comparison of the different cultivation strategies was conducted, based on a systematic review of the existing literature. The results of this work suggest that the biodisinfection of soil containing agricultural biomass serves as a fertilization strategy that allows a reduction in a large part of inorganic fertilizers in protected horticultural crops of intensive nature, in addition to mitigating the expression of soil pathogens. Reducing the use of fertilizers and phytosanitary products, management fees, and important means of production, such as water, allows a reduction of up 6.1% in production costs. The biodisinfection technique is a methodology that can promote circularity in farms while favoring the use of waste, such as agricultural biomass, which can damage the environment when not properly treated.

Keywords: agriculture; stakeholder engagement; sustainable development; waste management

1. Introduction

Agriculture has generated serious environmental impacts on ecosystems, mainly since the middle of the 20th century with the advent of the Green Revolution [1]. The primary negative impacts include erosion, loss of genetic diversity, loss of quality of water bodies (above and below ground), emission of greenhouse gases, and ozone layer degradation, among others [2–7].

Environmental damage has been caused by the intensification of production processes, which have expanded dependence on inputs such as fertilizers, phytosanitary products, and energy. This has led to increased consumption of natural resources and polluted ecosystems. The objective of this intensification was to expand the carrying capacity of the planet to supply food and fiber to a steadily increasing population. We are currently in a similar situation where productivity needs to be increased again as the world population is expected to reach 9.7 billion in 2050 [1,8].

However, despite the similarity of the situations discussed above, the production principle has changed. Today, there is a desire to produce food in a way that increases field productivity while reducing dependence on agri-food inputs and respecting ecosystem and socioeconomic balances [9]. This approach derives from the international agreements reached by the United Nations (UN) member states through the 2030 Agenda signed in 2015. This document brings together the main economic, social, and environmental issues on a global scale through the 17 Sustainable Development Goals (SDGs) and their 169 targets. The sustainability of agricultural production is reflected in several SDGs, but mainly in the second and twelfth [10].

Faced with this situation, several countries have reformulated their structural policies. The EU has positioned itself as a vanguard territory and has published several ambitious strategies and action plans [11]. These include the European Green Pact, which seeks environmental neutrality by 2050 while aiming to make its net carbon emissions disappear [12], and the implementation of an economic system based on the circular economy that seeks to reduce the consumption of inputs, reuse by-products generated in production processes, and recycle the waste that is generated [13,14]. Spain is one of the most advanced territories regarding the implementation of the circular economy [15].

Despite this radical change, the EU has divided it into phases. The first phase ends in 2030. In this period, agriculture is and will be affected by the Farm-to-Table and Biodiversity Strategies, with a 20% reduction in fertilizer usage and a 50% reduction in pesticides projected, as well as an expansion of organic farming to 25% of agricultural fields, and the maintenance of biodiversity [16,17]. These objectives imply that farmers must get rid of inputs that have been in place for more than six decades. Farmers must, therefore, be offered tools to cope with the proposed paradigm shift, particularly regarding the reduction in soil fumigants and fertilizers, in order not to reduce the profitability and productivity of agricultural systems and not to compromise the sovereignty and food security of the European territory [18,19]. The European Parliament has recently approved the Nature Restoration Law, which aims to restore 20% of degraded ecosystems by 2030 and practically 100% by 2050. This law will establish binding, mandatory, and common objectives among EU Member States, which may affect the use of agrochemicals. The main professional agricultural organizations at the European level (Copa-Cogeca and Europêche) have indicated that it is necessary to identify effective and feasible alternatives to avoid decreased productivity of agricultural fields before establishing such drastic measures. One of the measures proposed in the new law is the obligation to leave 10% of agricultural land as biodiversity reservoirs. For this reason, it is of great interest to identify and communicate the benefits of production techniques that help achieve this objective without reducing crop yields [20–22].

Greenhouse production models are one of the agricultural systems that may be affected by this change [18], although these systems are one of the alternatives that will safeguard food sovereignty and security in the Eurozone and other parts of the world. About 210,000 hectares of greenhouses have been identified in the EU, with a concentration of 76,000 hectares in Spain. In the Spanish territory, the main production area is the southeastern portion of the peninsula, where the province of Almeria concentrates 32,827 hectares [23–25]. Almeria's protected production system specializes in the horticultural production of eight products (tomato, sweet pepper, watermelon, cucumber, eggplant, melon, zucchini, and bean) and accounts for 6.5% of the EU's horticultural production. The Almeria horticultural system is the sixth-largest vegetable producer in Europe, generating more merchandise than 22 EU Member States [18,26–28]. The productive importance of the Almeria Model leads it to generate an estimated income of EUR 2200 million per year, making protected agriculture and its auxiliary industry Almeria's socioeconomic engine [29,30].

In addition, the production techniques used in protected agriculture in Almeria make it an agri-food system with a smaller energy footprint than other greenhouse systems in the EU, which is possible thanks to the benevolent climate and the use of greenhouses without climate control [30,31]. Agriculture in Almeria is based in the framework of integrated production and agro-ecological techniques, such as biological control [32].

However, the Almeria Model has some negative externalities. For example, the management and subsequent treatment of agricultural biomass were some of the main problems that afflicted the protected Almeria model at the end of the 20th century because of the indiscriminate dumping of this material in natural areas. Nowadays, although the situation has improved significantly, some significant problems regarding the management of the residual agricultural biomass have been identified [33]. To begin with, there is high production volume (i.e., 1.8 million tons per year), which is seasonal given that about 80% of the organic material is generated in only three months (February, May, and June). This issue prevents adequate sizing of the management plants. In addition, farms are located far from the production centers, which generates increased costs for the transport of the material. Finally, vegetal debris is usually mixed with other products, mainly plastic waste (i.e., raffia and rings for grooving) [34], complicating its management even further. Thus, some of the alternatives studied to solve the problem do not offer an adequate solution to this issue (i.e., bioenergy or animal feed) [35,36]. The EU has defined agricultural biomass as wasted material [37], that being the main agricultural waste generated in protected agri-food systems and surpassing plastics [38].

Because of the need for sustainable production with low dependence on fertilizers to promote circularity, several previous studies have tried to develop a fertilization strategy adapted to horticultural crops in the southeast of the Iberian Peninsula through the use of agricultural biomass on the farm, which can be extrapolated to other agri-food systems. This fertilization plan consists of nourishing crops with nutrients obtained from the agricultural biomass generated in the same shed using the biodisinfection technique. The communication of the technical and economic benefits of these techniques has focused mainly on the long-cycle tomato crop, and they have not been identified in other horticultural crops [18,39].

In this context, the objective of this research was to identify the fertilization capacity of the biodisinfection technique in soils both affected and unaffected by telluric diseases to be used as an alternative in the southeastern portion of the Iberian Peninsula to the reduction in fertilizers proposed by the EU. This study also aimed to verify the effects of the biodisinfection technique on the decrease in production costs in the cultivation of tomato, sweet pepper, and watermelons.

To this end, this article is subdivided into the following sections. First, the methodology is described. Next, the theoretical framework is indicated, characterizing the productive benefits of soil biodisinfection in horticultural crops. Then, a cost analysis of the use of the biodesinfection technique utilizing the agricultural biomass generated in the same greenhouse is presented. Finally, the conclusions are presented.

2. Materials and Methods

2.1. General Technical Characteristics of Greenhouse Agriculture in Almeria

This study identified the fertilization capacity of the biodesinfection technique in protected agriculture in Almeria. This production system is characterized by intensive agriculture under a greenhouse that combines the execution of crops per season in a long cycle (i.e., a 320-day-after-transplanting (DAT) crop) or in two short cycles (i.e., two crops of 160 DAT each) [18,40]. This protected cultivation system specializes in producing eight vegetable species: tomato, sweet pepper, watermelon, zucchini, cucumber, eggplant, melon, and beans. Several sustainable techniques, such as integrated pest and disease control, biological control, and grafting, are implemented in the production process [30]. In addition, solarization is used by 95.2% of farmers, either as the principal method or in combination with chemical fumigants [41].

Before describing the biodisinfection techniques and their effect on crop productivity, it is necessary to define the concept of "solarization" because of its close relationship with the aforementioned techniques [18]. This method is widely used for soil disinfection. For example, in greenhouse agriculture in Almeria, it is used [41,42].

The production follows international standards such as GLOBALGAP [29,32,43], although organic production represents almost 10% of the greenhouse area [44].

In addition, due to its benevolent climatic conditions, production is carried out in greenhouses that do not require climatic correction, as opposed to other European greenhouse systems, which require more energy [31]. Crops are grown in "raspa y amagado"-type modules. The predominant productive structures in the long cycles have heights of 6.0, 5.0, and 4.7 m at the ridge, under the canal, and in the trellis, respectively, while the short cycles have heights of 4.5, 3.5, and 3.0 at the ridge, under the canal, and in the trellis, respectively. These structures have lateral windows around the perimeter and zenithal windows on the ridge. The irrigation system consists of high-frequency irrigation equipment with emitters having a nominal flow rate of 3 $L \cdot h^{-1}$ and an irrigation controller consisting of five tanks where the inorganic or ecological fertilizers added through the fertigation equipment are deposited. Pest and disease control follows the Integrated Production regulations, except in farms that produce under the Organic Production regulations. The climate control in the sheds has a passive character thanks to the benevolent climate of Almeria, so it requires less energy than other European greenhouse production systems [30,32,40,45].

2.2. Systematic Literature Review

First, a systematic review of the scientific literature was performed. This procedure allows for synthesizing information on a specific topic [46,47]. This step aimed to describe biodisinfection techniques and their benefits on crop yield, both in disease and disease-free soils, and in crops with and without the addition of inorganic fertilizers. The literature review was performed using the "snowball" or "chain of references method" [48].

In this process, we used scientific literature published in English and submitted to a peer-review process before publication. The database used was Scopus [49], The terms introduced were biodisinfection, biosolarization, biofumigation, solarization, or their synonyms. The bibliographic review focused on publications that analyzed their effect on horticultural crops. In addition, publications derived from field trials performed at the "Catedrático Eduardo Fernández" Experimental Farm of the UAL-ANECOOP Foundation were reviewed. These works addressed the evaluation of a fertilization strategy based exclusively on the use of agricultural biomass with the application of the biodisinfection technique [18,39,50,51]. The literature review was carried out from January to March 2023.

2.3. Cost Analysis

Furthermore, an economic impact analysis was performed to identify the influence and economic benefits of using agricultural biomass with biodisinfection in the horticultural system of Almeria, the most important agricultural production area in the southeast of the peninsula. For this purpose, an analysis of total and partial costs was used. The study parameters selected were partial and total fixed and variable costs, as recommended in previous research [18,52–63].

The crops chosen were tomato in a long cycle, serving as a reference value [18]; the alternative of tomato and watermelon crops, both in a short cycle; and the alternative of sweet pepper and watermelon crops, also in a short cycle (Table 1). These three crop alternatives were evaluated via two production methods, conventional production (C), described by Camacho-Ferre [40], and via the production method using soil biodisinfection with the use of agricultural biomass (A) (100% reduction in fertilizer, 100% reductions in the use of chemical soil fumigants, 37.2% decrease in water consumption, elimination of external management costs of agricultural biomass). These criteria were used because they were recommended in previous research [18]. Table S1 shows the specific production criteria for each plant species.

Table 1. Species and crops evaluated.

Type	Crop	
Long cycle	Tomato	
	Autumn–Winter	Spring–Summer
Short cycle	Tomato	Watermelon
	Sweet pepper	Watermelon

Cost Structure

We used the same cost structure as the Experimental Farm "Catedrático Eduardo Fernández" of the UAL-ANECOOP Foundation. The production costs for the cultivation of the plant species evaluated in this work were obtained from those reported by Camacho-Ferre and Toresano [64] for the economic valuation carried out for Agroseguros, S.A. This entity specializes in managing insurance on behalf of the insurance companies that are part of the coinsurance pool [65]. The procedure for determining production costs has been described in previous research. To calculate the costs, the means of production and demand for specific inputs for the production model are used [18], as indicated in Section 2.1. All production costs used in this research were obtained under the same production criteria. Therefore, they are comparable to each other. The values were updated to current prices through the ECOICOP Index [66]. The value of the total cost of cultivation was calculated by the sum of fixed and variable costs using the following mathematical expression:

$$TC = VC + FC, \quad (1)$$

- TC: total costs (EUR/ha·year);
- VC: variable costs (EUR/ha·year);
- FC: fixed costs (EUR/ha·year).

In addition, the rate of variation was calculated:

$$RC = \frac{FV - IV}{IV}, \quad (2)$$

- RC: variation rate (%);
- IV: initial value (EUR/ha·year);
- FV: final value (EUR/ha·year).

3. Theoretical Framework

3.1. Biodesinfection Techniques

3.1.1. Soil Solarization

The solarization technique was originally used by Israeli farmers during the second-half of the 20th century and was described by Katan et al. [67]. The authors used a transparent plastic sheet with a thickness of 0.03 mm to reduce the expression of two pathogenic organisms that they had introduced into the soil before spraying in a greenhouse in Israel. The inoculum consisted of *Fusarium oxysporum* f. sp. *lycopersici* and *Verticillum dahliae*. After completing the soil disinfection and carrying out their respective analyses, they obtained a reduction in microbial populations in the first few centimeters of the soil and also a decrease in the severity index of the disease and increased plant growth. This research triggered a series of trials in different parts of the world to test the effect of this technique on crop productivity under different conditions [68]. The technique can be applied both in open-air and greenhouses. In the latter, it can be used directly on natural soil or organic and inorganic substrates [18,39,69–71].

Solarization is a physical soil-disinfection protocol that consists of hydrothermally fumigating the soil through solar energy captured in the surface layer of the soil (0–30 cm) and its humidity. For its execution, the soil to be treated must be conditioned before applying the fumigant treatment. To do so, the land has to be labeled beforehand as it is necessary to uninstall the irrigation system if it exists. After preparing the soil and reinstalling the irrigation system, it is necessary to check the correct emission of water due to the importance of wetting the soil profile for the process. Afterward, the soil is covered with a thin and transparent plastic sheet (about 100 gauges). Once the sheet has been sealed between its perimeter and it is confirmed that there are no air pockets that may reduce the thermal conductivity, irrigation is applied at field capacity. The process should be maintained for 4 to 6 weeks and it is recommended that it be carried out during the summer due to the higher radiation present during this period [18,39,50,67,68,72].

After irrigation, the temperature of the upper-soil profile rises and can exceed 60 °C in greenhouse conditions, provided that there is adequate humidity and heat transfer so that the nocturnal temperature drop is not high [73]. This technique limits the development and expression of pathogens that limit crop productivity. In addition, this control technique is non-hazardous, non-toxic, simple, and economical. Further, it can be combined with chemical fumigants or organic amendments. Ultimately, it enables a reduction in chemical disinfectants, facilitates the expansion of the circular economy within farms, and increases the sustainability of food production [18,30,67,71].

3.1.2. Soil Biodisinfection

This protocol can be subdivided into:

- Biofumigation

Biofumigation is soil disinfection caused by volatile substances generated during the bio-decomposition of organic amendments that have been previously mixed and moistened in the upper-soil profile (0–30 cm). It was initially called biological fumigation and was described by Kirkegaard et al. [74] and renamed biofumigation by the same authors in later works [75]. The researchers observed an increase in the productivity of the evaluated crop after biofumigating the soil due to the mitigation of the disease that limited the crop productivity. Thus, the authors used green manure composed of *Brassicas* plants as an organic amendment and attributed the suppressive effect to the liberation of isothiocyanates produced during the hydrolysis of glucosinolates generated by the enzyme myrosinase. This process has a fumigant effect similar to metam-sodium, a broad-spectrum chemical disinfectant that generates methyl-isothiocyanate [7,76,77]. Another noteworthy feature is that isothiocyanates have variable effectiveness during fumigation since the maximum concentration is produced in the first hours of bio-decomposition [78], specifically in a period ranging from 5 to 8 days after application [79]. A rapid decrease in concentration is observed as a result of isothiocyanates not being able to retain the volatile substances in the soil. In addition, the effectiveness of the technique in arid climates can be reduced by the higher evapotranspiration of the soil, which limits the bio-decomposition of organic nutrients and leads to an increase in the water supply needed to maintain adequate humectation [80]. Soil texture can also harm its efficacy: clay soils are more effective in the hydrolysis of glucosinolates than sandy soils [7].

- Biosolarization

The fact that the biofumigation technique is not suitable for all agricultural fields led to the combination of solarization and biofumigation techniques. The main problem of biofumigation was the retention of gases released during the bio-decomposition of organic amendments and the maintenance of adequate humidity conditions, which was solved with the plastic cover. This led to biosolarization, which allows for the retention of mobile nutrients in the soil, such as nitrates, due to the lower water supply, thus protecting the environment and improving the sustainability of agricultural systems [72,80].

The biosolarization technique can be defined as the hyperthermic fumigant action obtained from the solar energy captured by a plastic film, previously installed on the soil surface and sealed around its entire perimeter, and from the energy released by the metabolism of the microorganisms that participate in the bio-decomposition process of the organic amendments, in combination with the disinfectant action harvested by the volatile substances generated during this bio-decomposition and retained by the plastic film [18,19,72,81–83]. Table 2 shows a comparison of the soil temperature after solarization and biosolarization in a net greenhouse in the Canary Islands. The combination of organic amendments and solarization increases the soil temperature in each parameter evaluated and the number of hours where the soil temperature at 10 cm depth is higher than 40 °C. Another example has been recorded in the protected horticultural system of Almeria, where the biosolarization technique can cause a temperature increase of up to 5 °C compared to solarization [84]. The biosolarization technique has been described as a suitable methodology to reduce the expression of diseases of edaphic origin. A list of diseases controlled via the biosolarization technique is shown in Table 3.

Table 2. Soil temperature reached via solarization and biosolarization vs. control treatment.

	10 cm Depth			30 cm Depth	
Treatment	**Tmax** [1]	**Tmin** [1]	**N. Hours > 40 °C** [1]	**Tmax** [1]	**Tmin** [1]
Control	33.0 ± 1.3	22.6 ± 0.8	0.0 ± 0.0	29.1 ± 1.4	25.5 ± 0.9
Solarization	42.6 ± 1.0	24.8 ± 2.1	107.0 ± 50.4	35.1 ± 1.2	24.9 ± 1.7
Biosolarization	43.6 ± 1.0	26.1 ± 0.4	136.8 ± 56.8	36.4 ± 0.4	26.3 ± 0.51

[1]: average ± standard deviation. Source: own elaboration based on Diaz-Hernandez et al. [83].

Table 3. Soil diseases controlled via soil biosolarization.

Causal Agent	Source
Pythium ultimum	[85]
Phytophthora capsici	[86,87]
Fusarium oxysporum f. sp. *radicis-lycopersici*	[88]
Fusarium oxysporum f. sp. *dianthi*	[70,71,89]
Fusarium oxysporum f. sp. *asparagi*	[90]
Fusarium oxysporum f. sp. *lactucae*	[91]
Fusarium oxysporum f. sp. *cubense*	[92]
Fusarium oxysporum f. sp. *radicis-cucumerinum*	[93]
Pyrenochaeta lycopersici	[83]
Meloidogyne incognita	[86,94]
Meloidogyne spp.	[19,87]
Globodera rostochiensis	[95]
Verticillium dahliae	[69]
Rhizoctonia solani	[69,87]
Macrophomina phaseolina	[96]

Source: own elaboration based on other authors. The source is indicated in the table itself.

The action time of the biosolarization technique is around four weeks in Mediterranean conditions, although it can be up to 12 weeks [18,19,39,72]. During this period, two differentiated phases occur based on the oxygen concentration in the atmosphere generated under plastic. First, an anaerobic phase takes place. During this phase the microorganisms consume the carbon in the organic compounds and the oxygen in the soil while triggering a significant rise in temperature and a large release of carbon dioxide and water. After this release, the expansion of aerobic microorganisms stops and the metabolism of those that develop under anaerobic conditions begins to increase. During the anaerobic phase, compounds such as organic acids (e.g., acetic acid and butyric acid), ammonium, or nitrous acid are generated [81,97–99]. The anaerobic phase of biosolarization has also been named Biological Soil Disinfection, Aerobic Soil Disinfection, and Reductive Soil Disinfection [90,100,101].

- Useful organic amendments in the process

The organic amendment originally used in the biodisinfection process was agricultural biomass obtained from plants of the *Brassicas* genus, as indicated above. The variable presence of glucosinolates in plants of the *Brassicaceae* genus led to the evaluation of different plant species to identify the one that released the greatest amount of the compound. For this purpose, *Brassica campestris, Brassica carinata, Brassica juncea, Brassica napus* [78], *Brassica nigra, Brassica oleracea, Raphanus sativus*, and *Sinapsis alba*, among others, were studied [74,75,78,102–104]. The conclusion was that *Brassica carinata* releases the highest amount of the fumigant molecule, which has led to the formulation of commercial products composed of this plant species.

Different authors have evaluated the disinfection capacity of different organic amendments (manure, agro-industrial waste, plant residue, green manure, etc.) versus the efficacy of *Brassica carinata* pellets and chemical disinfectants. Research results show that manure, agro-industrial residue, plant remains, and green manure achieve similar effects to *Brassica carinata* pellets and chemical disinfectants [71,86,87]. Guerrero et al. [86] obtained this result in a sweet-pepper crop affected by *Phytophthora capsici* and phytopathogenic nematodes of the genus *Meloidogyne*. In this case, their biosolarized treatment with 2.5 kg·m^{-2} of fresh sheep manure showed total commercial productivity similar to biosolarization with *Brassica carinata* pellets and higher than the control group disinfected with methyl bromide for two consecutive crops. This was also shown by García-Ruiz et al. [70] in a carnation crop affected by *Fusarium oxysporum* f. sp. *dianthi*, where biosolarization with 5 kg·m^{-2} of compost consisting of carnation and chrysanthemum residues and 5 kg·m^{-2} of chicken manure resulted in a significantly lower mortality rate compared to the control treatment disinfected with methyl bromide during 672 days of cultivation. These results are due to the reduction in NO_3^-, Mn^{4+}, Fe^{3+}, and SO_4^{2-} during the bio-decomposition of organic amendments, mainly in the anaerobic phase. This process generates various gases (ammonia, hydrogen sulfide, and nitrous oxide) and organic acids (butyric, maleic, lactic, acetic, citric, and propionic), thus resulting in the control of weeds, pathogens, and pests [105–107]. Table 4 identifies a list of organic amendments that have been used in biodesinfection with favorable results. Agricultural residues are suitable for use as an organic amendment, favoring the reuse of by-products within the framework of the circular economy and improving the management of agricultural residues, which has been a major problem in many agricultural systems [18,19,35,51,108].

Table 4. Organic amendments and doses used in biodisinfection.

Organic Amendment	Dose (kg·m^{-2}) [1]	Source
Fresh sheep/goat manure	4.1 ± 0.2	[4,19,69,83,86,87,94,96,109–111]
Mixture of semi-composted horse manure and poultry manure	3.4 ± 2.1	[87]
Chicken manure	1.7 ± 0.9	[4,72,73,87,89,109,112,113]
Tomato plant debris	4.3 ± 1.9	[18,19,39,50]
Sweet-pepper plant waste	6.3 ± 3.3	[114]
Papaya vegetable waste	15.0 ± 5.0	[51]
Beer bagasse	2.0 ± 0.0	[69]
Broccoli plant remains	2.0 ± 0.0	[69]
Brassica carinata pellets	0.6 ± 0.6	[18,39,50,86,89,94,96,113,115]
White mustard (*Sinapsis alba*)	5.9 ± 1.9	[87]
Sugar-beet vinasse	1.5 ± 0.0	[96,113]
Olive residues	2.0 ± 0.7	[89,96]
Carnation and chrysanthemum compost	12.0 ± 0.0	[71]
Dehydrated broccoli (*Brassica oleacea* var. *italica* dehydrated)	0.8 ± 0.0	[72,73]
Glycerin	0.1 ± 0.0	[113]
Feather meal	0.1 ± 0.0	[114]

[1] average; ± standard deviation. Source: own elaboration based on different authors. Note: the source is indicated in the table.

The capacity to control pathogens limiting the productivity of different crops has resulted in biodisinfection techniques, mainly biosolarization (Table 3), to trigger strong increases in crop productivity and plant vigor compared to the non-disinfected plot. However, in some experiments, the increase in productivity has been greater than that achieved in the control disinfected with a chemical fumigant [116]. In this regard, Guerrero-Díaz et al. [94] obtained an increase in the commercial production of a sweet-pepper crop affected by the phyto-parasitic *Meloidogyne incógnita* after biosolarizing with 0.3 kg·m^{-2} of *Brassica carinata* pellets versus plots fumigated with methyl bromide. Guerrero et al. [86] achieved the same increase in productivity during two consecutive years in a sweet-pepper crop affected by *Meloidogyne incognita* and *Phytophthora capsici* via a biosolarization treatment with 0.3 kg·m^{-2} of *Brassica carinata* pellets and via another biosolarization treatment with 2.5 kg·m^{-2} of fresh sheep manure. Table 5 shows a list of crops where biodisinfection has increased productivity compared to the control treatment without disinfection and has equaled or surpassed the control treatment disinfected with a chemical fumigant.

Table 5. List of crops where biodisinfection has increased crop productivity in soils affected by soil-borne pathogens vs. undisinfected treatment.

Crop	Source
Pepper (*Capsicum annuum*) [2] *	[86,94,116–119]
Carnation (*Dianthus caryophyllus*) [2] *	[70,71]
Artichoke (*Cynara scolymus*) [2]	[69]
Tomato (*Solanum lycopersicon*) [2] *	[83,120]
Strawberry (*Fragaria* x *ananassa*) [2]	[96,113,121,122]
Lettuce (*Lactuca sativa*) [2]	[123]
Cucumber (*Cucumis sativus*) [1,2] *	[124]
Asparagus (*Asparagus officinalis*) [2]	[125]
Eggplant (*Solanum melongena*) [2]	[126]
Okra (*Abelmoschus esculentus*) [2]	[126]

[1] reached via biofumigation; [2] reached via biosolarization; * under greenhouse. Source: own elaboration based on other authors.

The benefits of soil biodisinfection on crop productivity also extend to crops with good soil health and not affected by any soil pathology. These results were reported by Mauromicale et al. [127,128] and Marin-Guirao et al. [72]. In this sense, Mauromicale et al. [127,128] recorded an increase in the commercial production of various tomato crops grown in a greenhouse after biosolarizing their experimental plots compared to plots that were only solarized. Furthermore, Marín-Guirao et al. [72] observed an increase in the total and commercial production of a tomato crop grown under greenhouse conditions after applying biosolarization versus biofumigation. The production benefit disappeared in the watermelon crop grown in the same season after the autumn tomato-crop cycle. The authors did not repeat the soil biodisinfection in the following season and obtained the productivity increase in either the autumn tomato or the spring watermelon cycles. Therefore, this benefit could be due to a possible improvement in soil fertility, which benefits the crop grown immediately after the application of biosolarization [39].

The fertilizing power of soil biodisinfection could have been masked in such investigations by the fertigation applied during their production cycles [72,127,128]. In this regard, Castillo-Diaz et al. [18,129] showed the results of a trial in an experimental greenhouse located in the protected agriculture of southeastern Spain (Almería, Spain) where biodesinfection was applied during five consecutive agricultural seasons, the use of tomato agricultural biomass predominated, and soil solarization (combined or not with fertigation) was performed. During the first two years, cycles of up to 170 days after transplanting (DAT) were carried out, which was equivalent to a short production cycle. During the remaining three years, the production cycles were extended up to 217 DAT, which is equivalent to a long production cycle. Their results show that biosolarization without fertigation yielded similar productivity during the five years of cultivation in cycles of up to 170 DAT compared to conventional cultivation composed of solarization and fertigation. Regarding the cycles of up to 217 DAT, they obtained similar productivity in two of the three cycles. However, these authors point out that the production difference between the exclusive fertilization with biosolarization and the combination of solarization and fertigation, in favor of the latter, in one of the campaigns, was due to an uncontrollable epidemic of *Botrytis cinerea* that emerged from 160 DAT and caused the death of up to 80% of the plants in each experimental plot. Figure 1 summarizes the results of the experiments carried out by these authors [18,129].

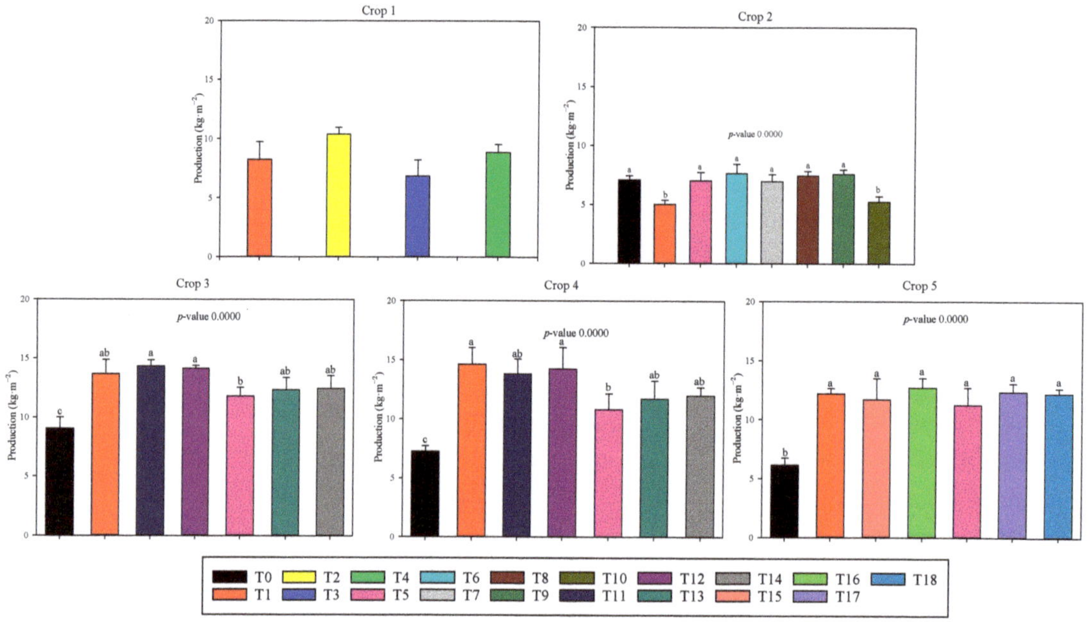

Figure 1. Final production of various tomato crops where biosolarization and solarization were applied with or without fertigation. Note: all treatments were solarized. T0: Without fertilization; T1: Fertigation; T2: 2 kg·m^{-2} of chicken manure; T3: 2 kg·m^{-2} of chicken manure + 5 kg·m^{-2} of vegetable waste; T4: 5 kg·m^{-2} of tomato plant remains; T5: 3.5 kg·m^{-2} of tomato plant remains; T6: 1.6 kg·m^{-2} of radish plant remains; T7: 2 kg·m^{-2} of mustard plant remains; T8: 0.3 kg·m^{-2} of *Brassica carinata* pellets, T9: 0.8 kg·m^{-2} of broccoli; T10: Fertigation + 0.3 kg·m^{-2} of *Brassica carinata* pellets; T11: fertigation + 0.5 kg·m^{-2} of *Brassica carinata* pellets; T12: fertigation + 1.0 kg·m^{-2} of *Brassica carinata* pellets; T12: 3.5 kg·m^{-2} of fresh-tomato plant remains; T13: 3.5 kg·m^{-2} of fresh-tomato plant remains + 0.5 kg/m^{-2} of *Brassica carinata* pellets; T14: 3.5 kg·m^{-2} of fresh-tomato plant remains + 1.0 kg/m^{-2} of *Brassica carinata* pellets; T15: fertigation + 3.5 kg·m^{-2} of fresh-tomato plant remains; T16: fertigation + 5.0 kg·m^{-2} of fresh-tomato plant remains; T17: 5.0 kg·m^{-2} of fresh-tomato plant remains; T18: 6.5 kg·m^{-2} of fresh-tomato plant remains. Different letters indicate significant differences ($p \leq 0.05$; Tukey HDS Test). Source: own elaboration based on other authors [18,129].

Other authors observed similar results in a sweet-pepper crop under greenhouse in southeastern Spanish agriculture (Almeria, Spain) [100]. These authors used papaya residues, combined with coconut fiber, to which they applied the biodisinfection technique, without using any inorganic fertilizer. The results were compared with the final production obtained by the sector in the last six years (2016–2021) in short cycles of 199 DAT (Figure 2). In any case, no statistically significant differences were observed.

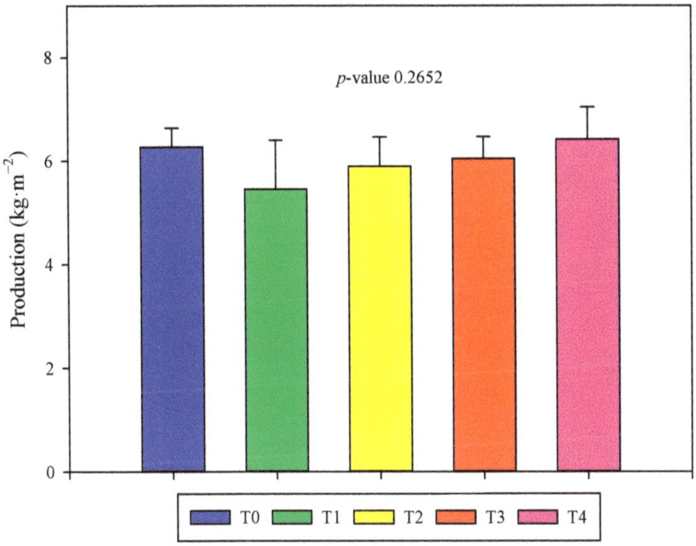

Figure 2. Final production of a sweet-pepper crop fertilized with papaya plant debris versus treatment without fertilization and average productivity of the protected-agriculture sweet pepper in the province of Almeria in production cycles of 199 DAT. T0: average productivity of protected-agriculture sweet pepper of Almeria modulated at 199 DAT from 2016 to 2021. T1: 5 kg·m^{-2} of papaya plant debris; T2: 5 kg·m^{-2} of papaya and 7 kg·m^{-2} of coconut fiber; T3: 10 kg·m^{-2} of papaya and 7 kg·m^{-2} of coconut fiber; T4: 15 kg·m^{-2} of papaya and 7 kg·m^{-2} of coconut fiber. Different letters indicate significant differences ($p \leq 0.05$; Tukey HDS test). Note: The crop was grown in a multitunnel greenhouse and a 59-day biodisinfection procedure was applied between May and June. Source: Own elaboration based on other authors [28,51,130].

The results show the multifunctionality of the soil-biodisinfection technique, which, besides being a useful method to mitigate telluric diseases, affects the fertilization of horticultural crops and allows for a reduction of up to 100% of the synthetic fertilizers normally used in production cycles of up to 217 DAT in intensive-production models, such as protected agriculture of a horticultural nature in southeastern Spain. Moreover, due to the diversity of organic amendments that can be used in the process (Table 4), it facilitates the management of the agricultural biomass recurrently generated at the end of crop cycles. Thus, this technique is a high-value resource for farms because it reduces dependence on fertilizers by applying the principles of the circular economy. This is in line with the EU Farm-to-Table Strategy. In addition, the biodisinfection technique is necessary when reusing agricultural biomass. This organic material is affected by pests and diseases that normally affect plant species during the crop cycle, so these must be inactivated to ensure the plant's health during the next crop, which is achieved with this procedure. Repeated application of the biodisinfection technique does not reduce the density or the biodiversity of agricultural soils. This is because the microorganisms can re-establish themselves at the end of each crop cycle at a similar level to that shown before applying the fumigant treatment, thus respecting the principles indicated in the EU Biodiversity Strategy [16,39].

4. Results and Discussion

Cost Analysis

Table 6 shows the production costs of long-cycle tomato crops, tomato-plus-watermelon short-cycle crops, and sweet-pepper-plus-watermelon crops in the conventional production system. The alternative of self-management of agricultural biomass by means of the biodisinfection technique under the criteria described in the methodology is also presented.

Table 6. Production costs of different vegetable-crop alternatives with and without applying the biodisinfection technique.

Concept (EUR/ha·Year)	Alternative 1								Alternative 2					
	Tomato Long Cycle		Sweet Pepper	Watermelon	Total	Sweet Pepper	Watermelon	Total	Tomato	Watermelon	Total	Tomato	Watermelon	Total
	C	A	C			A			C			A		
Technical assessment	355	355	177	177	355	177	177	355	177	177	355	177	177	355
Soil preparation	4489	3282	4475	4475	8950	3272	3272	6544	0	4475	4475	0	3272	3272
Removal of plant debris	1157	0	1156	1154	2309	0	0	0	0	1154	1154	0	0	0
Incorporation of plant remains	0	1367	0	0	0	1072	1072	2144	0	0	0	682	1072	1754
Solarization	994	1984	496	495	991	991	991	1981	496	495	991	991	991	1981
Water for solarization	132	163	66	66	132	82	82	163	66	66	132	82	82	163
Chemical disinfectant	234	0	117	117	233	0	0	0	117	117	233	0	0	0
Covering and structure	5457	5457	2725	2725	5449	2725	2725	5449	2725	2725	5449	2725	2725	5449
Seeds and seedling production	6592	6592	8044	2323	10,368	8044	2323	10,368	0	2323	2323	0	2323	2323
Labor, supplies, etc.	62,956	62,956	39,178	7270	46,447	39,178	7270	46,447	27,864	7270	35,133	27,864	7270	35,133
Water	2405	1511	920	897	1817	578	563	1141	1216	897	2114	764	563	1327
Fertilizers	4517	1391	2030	2579	4609	0	660	660	2230	2579	4810	1391	660	2051
Total variable cost	89,290	85,059	59,383	22,279	81,661	56,118	19,135	75,252	34,891	22,279	57,170	34,675	19,135	53,809
Soil maintenance	2424	2424	1182	1182	2364	1182	1182	2364	1218	1182	2400	1218	1182	2400
Covering and structure	4852	4852	2366	2366	4732	2366	2366	4732	2438	2366	4804	2438	2366	4804
Energy and fixed supplies	1915	1915	934	934	1869	934	934	1869	962	934	1897	962	934	1897
Insurance, management, and financial services	4219	4219	2058	2058	4115	2058	2058	4115	2120	2058	4177	2120	2058	4177
Equipment and irrigation system	11,625	11,625	4741	4741	9483	4741	4741	9483	5841	4741	10,583	5841	4741	10,583
Total fixed costs	25,034	25,034	11,281	11,281	22,563	11,281	11,281	22,563	12,579	11,281	23,861	12,579	11,281	23,861

C: conventional crop; A: alternative with biodisinfection. Source: own elaboration based on other authors [18,64].

Results suggest that the highest production costs are registered in conventional crops and in the long-cycle tomato crop. Production costs reach up to 114,324 EUR/ha·year, although in the alternatives that combine short cycles, the total cost of cultivation reaches 81,031.0 EUR/ha·year (tomato + watermelon) and 104,224.0 EUR/ha·year (sweet pepper + watermelon) (Figure 3). In recent years, there has been an increase in the cost of inputs and means of production in the protected horticultural system of Almeria, which has accelerated recently due to inflation in the Eurozone. Farmers in this system can see how their economic margins are reduced yearly due to the continued stability of the price at origin of fruit and vegetable products [42,131].

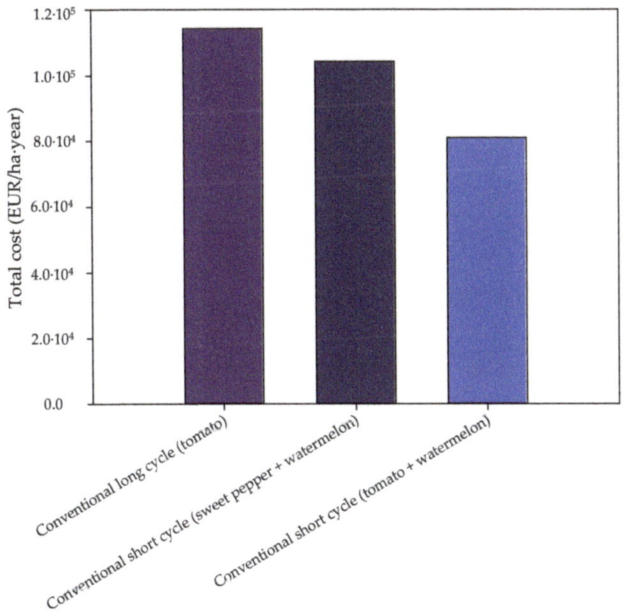

Figure 3. Total cultivation costs of the different production cycles evaluated within the framework of conventional production in Almeria. Source: own elaboration based on other authors [18,64].

The production system that performs self-management of agricultural biomass through the biodisinfection technique shows a decrease in total production costs ranging from 3.7% to 6.1% (Figure 4). Essentially, this decrease takes place in variable costs because fixed operating costs remain at similar levels in the two alternatives: conventional and with biodisinfection technique (Table 6). The reduction in production costs reported by other authors for the main cultivation alternatives in a similar system in the Almeria model is 4.8% in periods of 48 months. This value is very similar to those shown in our work [18]. The long-cycle crop has a lower production-cost-reduction rate than the other two alternatives evaluated, mainly due to the expansion in demand for tomatoes in the long cycle as a result of being the main economic expense (average value of +35.2%). Other research has shown how implementing eco-innovations can expand the economic benefit perceived by agricultural producers either through reduced production costs, expanded crop yields, or increased exports from their marketing channel [131–133].

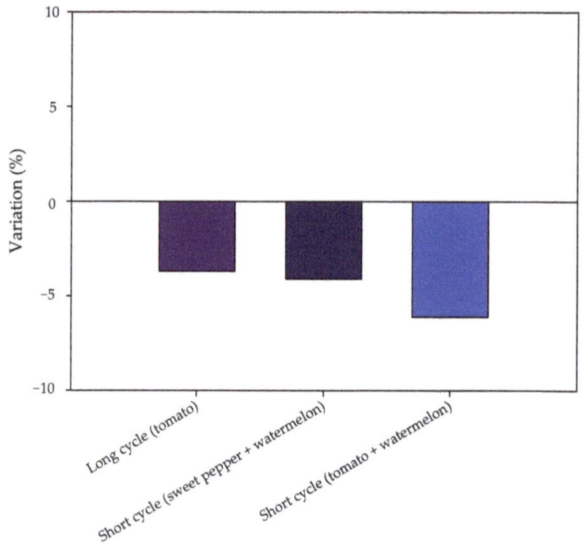

Figure 4. Variation in the total cost of cultivation for the different production cycles evaluated in long and short cycles. Source: own elaboration based on other authors [18,64].

Production costs are reduced because of lower demand for certain inputs, which are cataloged in international literature as environmentally harmful materials. In the first place, there are 100% reductions in the number of chemical soil fumigants because the biodisinfection technique has proved to be a useful procedure to control telluric diseases (Table 3). Also, there has been a decrease in the consumption of synthetic chemical fertilizers. Fertilizer consumption is reduced because added organic amendment provides readily available nutrients for plants. The repeated addition of tomato agricultural biomass can increase the concentration of edaphic organic matter and essential nutrients, such as total nitrogen or readily available potassium. Finally, water consumption is reduced due to the self-management of agricultural biomass as a result of improving soil fertility and hydraulic conductivity [18]. Therefore, this research contributes to showing the existence of real and effective alternatives to reduce agrochemical products, as proposed by the EU, without diminishing the agricultural and economic performance of horticultural producers [17,21].

In addition, if an extra supply of nutrients is needed, this fertilization strategy (agricultural biomass with the biodisinfection technique) could be combined with the application of other types of new-generation fertilizers, such as nano-fertilizers, which offer significant benefits compared to conventional fertilizers when it comes to plant nutrition [18,134].

In this context, the fertilization strategy framed in the principles of the circular economy facilitates both the reduction in inputs and the use of materials that can damage the environment if not properly treated. This material causes serious problems in autumn, and farms far from the productive nucleus continue experiencing difficulties managing it [35].

The results of this research suggest that implementing agricultural policies that promote a fertilization strategy consisting of self-management of agricultural biomass from the previous season and biodisinfection, while reducing synthetic fertilizers, can lower the production costs of greenhouse horticultural crops and sustain crop productivity. However, in the protected agriculture of Almeria, the most important producing area of protected agriculture in southeastern Spain, only 15.3% of the farmers self-manage their crop residues on their farms. Of these farmers, only a few do not use chemical fertilizers on their crops [34,44,135].

Consequently, it is necessary to establish a system of incentives to promote the implementation of this sustainable technique among the operators of horticultural agriculture in Almeria. This incentive system can focus on satisfying the formation needs of agricultural producers and offer economic support as an attractive measure, with each amount segmented for each concept (i.e., formation or economic support) as opposed to the current support that offers a single amount for the incorporation of the crop residues into the soil.

5. Conclusions

This work aimed to identify the availability of alternatives to reduce the dependence on inorganic fertilizers in greenhouse horticultural farms. Thus, the multi-functionality of the biodesinfection technique in its biosolarization modality was identified. This technique was attributed to benefits in the mitigation of diseases of edaphic origin and increasing productivity due to the decrease in the severity index of the disease. However, the utilization of inorganic fertilization in combination with the organic amendments used in the fumigant technique masked its fertilizing power. This fact was demonstrated via research using treatments that do not add this type of synthetic fertilizers compared to conventional cultivation in disease-free soils.

Additionally, given the diversity of organic amendments that can be used in the biodisinfection technique, it is possible to improve the management and treatment of agricultural residues that have generated serious sanitary crises in important horticultural production models (i.e., the Almeria Model). This is achieved thanks to using agricultural biomass, the main organic waste generated in this type of production system, as fumigant material in the biodisinfection technique. It is necessary to apply the fumigant treatment to the agricultural biomass to thermally inactivate the pathogens that may have affected the plant material during the production cycle to ensure the health of the following crop.

Therefore, the biodisinfection technique is a fertilization strategy that makes it possible to significantly reduce the use of inorganic fertilizers in long-term intensive horticultural cycles. This is a result of paramount importance for farmers of this type of production model that require alternatives to frame their production systems within the principles of sustainable development and, in the case of the EU, of the circular economy, where European strategies also call for significant reductions in fertilizers and phytosanitary products. There is a reduction in total costs of up to −6.1% in long-term production cycles in intensive greenhouse agriculture. This strategy also positively affects producer expense accounts because of the reduction in the disbursement of the previously mentioned inputs, the lower consumption of water, and the elimination of the cost of managing the material externally. All of this helps to maintain and promote the socioeconomic development of the territories where economic activity depends on this type of agriculture and also helps to sustain food sovereignty and security. The administration should promote this fertilization technique through an incentive system that includes formation and extra economic support to make its use more attractive.

Although the results of this research expand the knowledge in this field, they are not without limitations. The main limitation derives from using average production costs, since the results do not apply to a particular farmer but to the sector as a whole. However, our study provides a clear overview of the situation and can be the basis for policymakers to make the right decisions. These decisions should lead toward safeguarding the European Union's food sovereignty without reducing the economic benefit of producers by expanding the environmental subcomponent of sustainability.

Future research should identify the adequacy of the fertilization strategy in outdoor horticultural crops and test the fertilization power of agricultural biomass of other horticultural crops, both under greenhouse and outdoor conditions (e.g., watermelon, melon, eggplant, etc.).

Supplementary Materials: The following supporting information can be downloaded at: https://www.mdpi.com/article/10.3390/horticulturae9080859/s1.

Author Contributions: Conceptualization, F.J.C.-D., L.J.B.-U., A.B.-d. and F.C.-F.; methodology, F.J.C.-D., L.J.B.-U., A.B.-d. and F.C.-F.; software, F.J.C.-D., L.J.B.-U., A.B.-d. and F.C.-F.; validation, F.J.C.-D.,

L.J.B.-U., A.B.-d. and F.C.-F.; formal analysis, F.J.C.-D., L.J.B.-U., A.B.-d. and F.C.-F.; investigation, F.J.C.-D., L.J.B.-U., A.B.-d. and F.C.-F.; resources, F.J.C.-D., L.J.B.-U., A.B.-d. and F.C.-F.; data curation, F.J.C.-D., L.J.B.-U., A.B.-d. and F.C.-F.; writing—original draft preparation, F.J.C.-D., L.J.B.-U., A.B.-d. and F.C.-F.; writing—review and editing, F.J.C.-D., L.J.B.-U., A.B.-d. and F.C.-F.; visualization, F.J.C.-D., L.J.B.-U., A.B.-d. and F.C.-F.; supervision, F.J.C.-D., L.J.B.-U., A.B.-d. and F.C.-F.; project administration, F.J.C.-D., L.J.B.-U., A.B.-d. and F.C.-F.; funding acquisition, F.C.-F. All authors have read and agreed to the published version of the manuscript.

Funding: This research received no external funding.

Data Availability Statement: The data used are referenced in the article itself.

Conflicts of Interest: The authors declare no conflict of interest.

References

1. FAO. *Background Notes on Sustainable, Productive and Resilient Agro-Food Systems: Value Chains, Human Capital, and the 2030 Agenda;* FAO: Rome, Italy, 2019; ISBN 9789251316474.
2. Gómez-Tenorio, M.Á.; Magdaleno-González, J.; Tello-Marquina, J.C. *Evaluación e Implementación de Técnicas Regenerativas Para la Mejora de la Fertilidad en el Cultivo del Almendro en las Provincias de Almería y Granada;* Editorial Tecnoagrícola de Spain: Madrid, Spain, 2021; ISBN 978-84-17596-98-9.
3. Jacobs, A.A.; Evans, R.S.; Allison, J.K.; Garner, E.R.; Kingery, W.L.; Mcculley, R.L.; Major, G.; Resource, L.; Soil, A.; Office, S.; et al. Soil & Tillage Research Cover crops and no-tillage reduce crop production costs and soil loss, compensating for lack of short-term soil quality improvement in a maize and soybean production system. *Soil Tillage Res.* **2022**, *218*, 105310. [CrossRef]
4. Martínez-Francés, M.A.; Lacasa-Plasencia, A.; Tello-Marquina, J.C. *Ecología de la Microbiota Fúngica de los Suelos de los Invernaderos de Pimiento y su Interés Agronómico;* Ministerio de Agricultura Pesca y Alimentacion: Madrid, Spain, 2009; ISBN 9788449109874.
5. European Union of Audit. *European Union Audit Preview. Biodiversity in Farming;* European Union of Audit: Luxembourg, 2019.
6. Región de Murcia, C.D.G. *Decreto-Ley 2/2019 de 26 de Diciembre, de Protección Integral del Mar Menor;* Boletín Oficial de la Región de Murcia: Murcia, Spain, 2019; pp. 36008–36089.
7. Díez-Rojo, M.A.; López-Pérez, J.A.; Urbano-Terrón, P.; Bello-Pérez, A. *Biodesinfección de Suelos y Manejo Agronómico Biodesinfección de Suelos y Manejo Agronómico;* Ministerio de Medio Ambiente y Medio Rural y Marino: Madrid, Spain, 2010.
8. UN. *World Population Prospects 2022;* United Nations, Ed.; UN: New York, NY, USA, 2022; ISBN 9789211483734.
9. MAPA; Cajamar. *Diagnóstico Inicial del Estado de la Transformación Digital del Sector Agroalimentario Español;* Ministerio de Agricultura Pesca y Alimentacion y Cajamar: Madrid, Spain, 2022.
10. UN. *Draft Outcome Document of the United Nations Summit for the Adoption of the Post-2015 Development Agenda;* UN: New York, NY, USA, 2015; Volume 13689.
11. Cifuentes-Faura, J. European Union policies and their role in combating climate change over the years. *Air Qual. Atmos. Health* **2022**, *15*, 1333–1340. [CrossRef] [PubMed]
12. European Comission. *The European Green Deal;* Office of the European Union: Brussels, Belgium, 2019; pp. 1–28.
13. European Comission. A New Circular Economy Action Plan. In *For a Cleaner and More Competitive Europe;* Office of the European Union: Brussels, Belgium, 2020; p. 23.
14. Kirchherr, J.; Reike, D.; Hekkert, M. Resources, Conservation & Recycling Conceptualizing the circular economy: An analysis of 114 de fi nitions. *Resour. Conserv. Recycl.* **2017**, *127*, 221–232. [CrossRef]
15. Mazur-Wierzbicka, E. Circular economy: Advancement of European Union countries. *Environ. Sci. Eur.* **2021**, *33*, 111. [CrossRef]
16. European Comission. *EU Biodiversity Strategy for 2030. Bringing Nature Back into Our Lives;* Office of the European Union: Brussels, Belgium, 2020; pp. 1–27.
17. European Comission. *A Farm to Fork Strategy for a Fair, Healthy and Environmentally-Friendly Food System;* Office of the European Union: Brussels, Belgium, 2020; pp. 1–23.
18. Castillo-Díaz, F.J.; Belmonte-Ureña, L.J.; Camacho-Ferre, F.; Tello-Marquina, J.C. Biodisinfection as a Profitable Fertilization Method for Horticultural Crops in the Framework of the Circular Economy. *Agronomy* **2022**, *12*, 521. [CrossRef]
19. Gómez-Tenorio, M.A.; Lupión-Rodríguez, B.; Boix-Ruiz, A.; Ruiz-Olmos, C.; Marín-Guirao, J.I.; Tello-Marquina, J.C.; Camacho-Ferre, F.; De Cara-García, M. Meloidogyne-infested tomato crop residues are a suitable material for biodisinfestation to manage Meloidogyne sp. in greenhouses in Almería (south-east Spain). *Acta Hortic.* **2018**, *1207*, 217–221. [CrossRef]
20. European Commision. *Inception Impact Assessment. EU Nature Restoration Targets;* Office of the European Union: Brussels, Belgium, 2021; p. 4.
21. European Commission. *Proposal for a Regulation of the European Parliament and of the Council on Nature Restoration;* 22.6.2022 COM(2022) 304 Final; European Commission: Brussels, Belgium, 2022; Volume 0195, p. 91.
22. Copa-Cogeca/Europêche. *United Call for Rejection of Unrealistic Nature Legislation That Endangers Farmers and Fishers' Livelihoods and Food Production in the EU;* Copa-Cogeca/Europêche: Bruxelles, Belgium, 2023; p. 2.
23. Rabobank. *World Vegetable Map;* Rabobank Gr.: Utrecht, The Netherlands, 2017; Volume 2017.

24. MAPA. *Encuesta Sobre Superficies y Rendimientos de Cultivos*; Ministerio de Agricultura, Pesca y Alimentación: Madrid, Spain, 2023; p. 45.
25. Junta de Andalucía. Cartografía de Invernaderos en Almería, Granada y Málaga; Año 2021. Cons. Agric. Ganad. Pesca y Desarro. Sosten.; la Junta Andalucía. 2022, pp. 1–27. Available online: https://www.juntadeandalucia.es/sites/default/files/2021-11/Cartografia%20_inv_AL_GR_MA_v210928%20_1.pdf (accessed on 3 December 2022).
26. EUROSTAT. Crop Production in EU Standard Humidity (from 2000 Onwards). [V0000_S0000] Fresh Vegetables (Including Melons) and Strawberries. Available online: https://ec.europa.eu/eurostat/databrowser/view/APRO_CPSH1__custom_4031262/default/table?lang=en (accessed on 3 December 2022).
27. Gutiérrez-Gutiérrez, C. *Anuario Agrícola. 2021. Fhmurcia*; Fhmurcia: Murcia, Spain, 2022.
28. Junta de Andalucía. Anuario. Available online: https://www.juntadeandalucia.es/organismos/agriculturaganaderiapescaydesarrollosostenible/servicios/estadistica-cartografia/anuarios.html (accessed on 15 June 2022).
29. Cajamar. *Análisis de la Campaña Hortofrutícola. Campaña 2021/2022*; Cajamar Caja Rural: Barcelona, Spain, 2022.
30. Valera-Martínez, D.L.; Belmonte-Ureña, L.J.; Molina Aiz, F.D.; López MArtínez, A. *Los Invernaderos de Almería. Análisis de su Tecnología y Rentabilidad*; Cajamar Caja Rural: Almeria, Spain, 2014; p. 504.
31. Vanthoor, B.H.E.; Stigter, J.D.; van Henten, E.J.; Stanghellini, C.; de Visser, P.H.B.; Hemming, S. A methodology for model-based greenhouse design: Part 5, greenhouse design optimisation for southern-Spanish and Dutch conditions. *Biosyst. Eng.* **2012**, *111*, 350–368. [CrossRef]
32. Junta de Andalucía. *Caracterización de los Invernaderos de Andalucía*; Junta de Andalucía: Sevilla, Spain, 2015.
33. Duque-Acevedo, M.; Lancellotti, I.; Andreola, F.; Barbieri, L.; Belmonte-ureña, L.J.; Camacho-ferre, F. Management of Agricultural Waste Biomass as raw material for the construction sector. An analysis of sustainable and circular alternatives. *Environ. Sci. Eur.* **2022**, *34*, 70. [CrossRef]
34. Junta de Andalucía. *Líneas de Actuación en Materia de Gestión de Restos Vegetales en la Horticultura de Andalucía*; Cons. Agric. Ganad. Pesca y Desarro. Sosten.; la Junta Andalucía. 2016, pp. 1–45. Available online: https://www.juntadeandalucia.es/export/drupaljda/Lineas_actuacion_materia_gestion_restos_vegetales_horticultura_Andalucia.pdf (accessed on 3 December 2022).
35. Duque-Acevedo, M.; Belmonte-Ureña, L.J.; Plaza-Úbeda, J.A.; Camacho-Ferre, F. The management of agricultural waste biomass in the framework of circular economy and bioeconomy: An opportunity for greenhouse agriculture in Southeast Spain. *Agronomy* **2020**, *10*, 489. [CrossRef]
36. Camacho-Ferre, F. Diferentes alternativas para la gestión del residuo biomasa procedente de cultivos de invernadero. In *Innovaciones Tecnológicas en Cultivos de Invernadero*; Rodríguez, E.J.F., Ed.; Universidad de Almería—Junta de Andalucía: Almería, Spain, 2003; pp. 211–237. ISBN 84-8748-052-7.
37. European Commission. *A Sustainable Bioeconomy for Europe: Strengthening the Connection between Economy, Society and the Environment. Updated Bioeconomy Strategy*; European Commission: Brussels, Belgium, 2018; ISBN 978-92-79-94144-3.
38. Sayadi-Gmada, S.; Roc, C.; Rojas-Serrano, F.; Garc, R.; Lorbach-Kelle, M.B.; Manrique-Gordillo, T. Inorganic Waste Management in Greenhouse Agriculture in Almeria (SE Spain): Towards a Circular System in Intensive Horticultural Production. *Sustainability* **2019**, *11*, 3782. [CrossRef]
39. Castillo Díaz, F.J.; Marín-Guirao, J.I.; Belmonte-Ureña, L.J.; Tello-marquina, J.C. Effect of Repeated Plant Debris Reutilization as Organic Amendment on Greenhouse Soil Fertility. *Int. J. Environ. Res. Public Health* **2021**, *18*, 11544. [CrossRef] [PubMed]
40. Camacho-Ferre, F. *Técnicas de Producción en Cultivos Protegidos*; Instituto de Publicaciones de Cajamar: Almería, Spain, 2004; ISBN 84-95531-17-8.
41. García-García, M.C.; Céspedes-López, A.J.; Cano-Banderas, M.; Martín-Expósito, E.; Cánovas-Fernández, G.; Parra-Gómez, S. Trends in soil disinfection in greenhouses in south-eastern Spain. In Proceedings of the X International Symposium on SOIL and Substrate Disinfestation, Almeria, Spain, 6–8 June 2023.
42. Honoré, M.N.; Belmonte-Ureña, L.J.; Navarro-Velasco, A.; Camacho-Ferre, F. Profit analysis of papaya crops under greenhouses as an alternative to traditional intensive horticulture in Southeast Spain. *Int. J. Environ. Res. Public Health* **2019**, *16*, 2908. [CrossRef] [PubMed]
43. Acebedo, M.M.; Diánez, F.; Santos, M. Almeria's Green Pest Management Revolution: An Opportunity That Arose from a Food Safety Alert. *Agronomy* **2022**, *12*, 619. [CrossRef]
44. COEXPHAL. *Memoria Coexphal 2020/2021*; Coexphal: Almeria, Spain, 2022.
45. Valera-Martínez, D.L.; Belmonte-Ureña, L.J.; Molina Aiz, F.D.; Camacho-Ferre, F. The greenhouses of Almería, Spain: Technological analysis and profitability. *Acta Hortic.* **2017**, *1170*, 2019–2226. [CrossRef]
46. Kitchenham, B. *Procedures for Performing Systematic Reviews*; Keele University Technical Report TR/SE-0401; Keele University: Newcastle, UK, 2004; pp. 1–34.
47. Tranfield, D.; Denyer, D.; Smart, P. Towards a Methodology for Developing Evidence-Informed Management Knowledge by Means of Systematic Review. *Br. J. Manag.* **2003**, *14*, 207–222. [CrossRef]
48. Batlles-delaFuente, A.; Abad-segura, E.; González, M.; Cortés-García, F.J. An Evolutionary Approach on the Framework of Circular Economy Applied to Agriculture. *Agronomy* **2022**, *12*, 620. [CrossRef]
49. Elsevier Scopus. Available online: https://www.elsevier.com/es-es/solutions/scopus (accessed on 14 August 2022).

50. García-Raya, P.; Ruiz-Olmos, C.; Marín-Guirao, J.I.; Asensio-Grima, C.; Tello-Marquina, J.C.; de Cara-García, M. Greenhouse Soil Biosolarization with Tomato Plant Debris as a Unique Fertilizer for Tomato Crops Greenhouse Soil Biosolarization with Tomato Plant Debris as a Unique Fertilizer for Tomato Crops. *Int. J. Environ. Res. Public Health* **2019**, *16*, 279. [CrossRef]
51. Ruiz-Olmos, C.A.; Gómez-Tenorio, M.Á.; Camacho-ferre, F.; Belmonte-Ureña, L.J.; Tello-Marquina, J.C. Control de nematodos del género Meloidogyne en un suelo de invernadero cultivado con papaya utilizando la técnica de biosolarización de suelos. *Terralia* **2018**, *116*, 53–62.
52. Batlles-delaFuente, A.; Belmonte-Ureña, L.J.; Duque-Acevedo, M.; Camacho-Ferre, F. A Profitable Alternative for the Spanish Southeast: The Case of Production of Figs in Greenhouses. *Agronomy* **2022**, *12*, 2577. [CrossRef]
53. Subhashree, S.N.; Igathinathane, C.; Hendrickson, J.; Archer, D.; Liebig, M. Forage economics calculator web tool: A decision support system for forage management. *Comput. Electron. Agric.* **2023**, *208*, 107775. [CrossRef]
54. Xu, Z.; Sarmento, L.; Ito, L.; Tokai, A. Cost and health benefit analysis of remediation alternatives for the heavy-metal-contaminated agricultural land in a Pb—Zn mining town in China. *J. Clean. Prod.* **2023**, *397*, 136503. [CrossRef]
55. Zhuo, C.; Xueqin, L.; Zhiwei, W.; Yantao, Y.; Tanglei, S.; Taoli, H. Industrial Crops & Products Techno-economic and whole life cycle assessment of ester fuels production from agricultural waste via hydrothermal liquefaction. *Ind. Crop. Prod.* **2023**, *192*, 116096. [CrossRef]
56. Castillo-Díaz, F.J.; Belmonte-Ureña, L.J.; Batlles-delaFuente, A.; Camacho-Ferre, F. Impact of the new measures related to the circular economy on the management of agrochemical packaging in Spanish agriculture and the use of biodegradable plastics. *Environ. Sci. Eur.* **2022**, *34*, 94. [CrossRef]
57. Celiktopuz, E.; Kapur, B.; Saridas, M.A.; Güney, O.; Aksoy, F. Plant Physiology and Biochemistry Yield, quality and physiological variation of strawberry in response to irrigation regimes and exogenous proline with a cost benefit analysis. *Plant Physiol. Biochem.* **2023**, *195*, 238–246. [CrossRef] [PubMed]
58. Colla, G. Foliar Applications of Protein Hydrolysate, Plant and Seaweed Extracts Increase Yield but Differentially Modulate Fruit Quality of Greenhouse Tomato. *HortScience* **2017**, *52*, 1214–1220. [CrossRef]
59. Heinrich, T.; Park, H.; Orozco, R.; Ding, Z.; Álvarez-López, V.; Mosquera-losada, M.R.; Steinbeis, L.; Hoffmann, T. Biochar production from late-harvest grass—Challenges and potential for farm-scale implementation. *Sustain. Prod. Consum.* **2023**, *37*, 256–267. [CrossRef]
60. López-Serrano, M.A.; Velasco-Muñoz, J.F.; Aznar-Sánchez, J.A.; Román-Sánchez, I.M. Financial Evaluation of the Use of Reclaimed Water in Agriculture in Southeastern Spain, A Mediterranean Region. *Agronomy* **2021**, *11*, 2218. [CrossRef]
61. Morselli, N.; Boccaletti, S.; Meglioraldi, S.; Puglia, M.; Pedrazzi, S.; Allesina, G. Biomass-powered thermal weeding in wine farms: An environmental and economic assessment. *J. Clean. Prod.* **2023**, *385*, 135684. [CrossRef]
62. Roberts, M.; Hawes, C.; Young, M. Environmental management on agricultural land: Cost benefit analysis of an integrated cropping system for provision of environmental public goods. *J. Environ. Manag.* **2023**, *331*, 117306. [CrossRef]
63. Streimikis, J.; Baležentis, T. Agricultural sustainability assessment framework integrating sustainable development goals and interlinked priorities of environmental, climate and agriculture policies. *Sustain. Dev.* **2020**, *28*, 1702–1712. [CrossRef]
64. Torresano, F.; Camacho-Ferre, F. *Valoración de las Diferentes Labores Culturales en los Cultivos de Tomate, Pimiento, Calabacín, Pepino, Sandía, Melón, Judía y Berenjena*; Agrupación Española de Entidades Aseguradoras de los Seguros Agrarios Combinados (Agroseguro); Universidad de Almería: Almería, Spain, 2012.
65. Agroseguro. Agroseguro. Available online: https://agroseguro.es/conocenos/quienes-somos/ (accessed on 14 April 2023).
66. INE. Índices Nacionales: General y de Grupos ECOICOP. Available online: https://www.ine.es/jaxiT3/Tabla.htm?t=22344 (accessed on 6 April 2023).
67. Katan, J.; Greenberger, A.; Alon, H.; Grinstein, A. Solar Heating by Polyethylene Mulching for the Control of Diseases caused by Soil-Borne Pathogens. *Phytopathology* **1976**, *66*, 683–688. [CrossRef]
68. Katan, J. Soil solarization: The idea, the research and its development. *Phytoparasitica* **2015**, *43*, 1–4. [CrossRef]
69. Guerrero, M.M.; Lacasa, C.M.; Martínez, V.; Martínez-Lluch, M.C.; Larregla, S.; Lacasa, A. Soil biosolarization for Verticillium dahliae and Rhizoctonia solani control in artichoke crops in southeastern Spain. *Span. J. Agric. Res.* **2019**, *17*, e1002. [CrossRef]
70. García-Ruiz, A.; Tello-Marquina, J.C.; Avilés-Guerrero, M.; Ordovás-Ascaso, J. *Fusariosis del Clavel (Fusarium oxysporum f. sp. dianthi) Últimos Avances en su Control*; Ediciones Agrotécnicas: Madrid, Spain, 2009.
71. Palmero, D.; de Cara-García, M.; Santos, M.; Tello-Marquina, J.C. Control of diseases from forma especiales of Fusarium oxysporum causing wilt in intensive horticultural crops. *Res. Signpost* **2011**, *661*, 209–228.
72. Marín-Guirao, J.I.; Tello-Marquina, J.C.; Díaz, M.; Boix, A.; Ruiz-Olmos, C.A.; Camacho-Ferre, F. Effect of greenhouse soil bio-disinfection on soil nitrate content and tomato fruit yield and quality. *Soil Res.* **2016**, *54*, 200–206. [CrossRef]
73. Marín-Guirao, J.I.; de Cara-García, M.; Tello-Marquina, J.C. Effect of soil biodisinfection on soil fungal communities associated to horticultural crops. *Ecosistemas* **2019**, *28*, 63–72. [CrossRef]
74. Kirkegaard, J.A.; Gardner, P.A.; Desmarchelier, J.M.; Angus, J.F. Benefits of brassica break crops in the Southeast wheatbelt. In Proceedings of the 7th Australian Agronomy Conference, Wagga Wagga, Australia, 19–24 September 1993; pp. 282–285.
75. Kirkegaard, J.A.; Gardner, J.; Desmarcherlier, J.M.; Angus, J.F. Biofumigation Using Brassica species to Control Pest and Diseases in Horticulture and Agriculture. In Proceedings of the 9th Australian Research Assembly on Brassicas, Waga Wagga, Australia, 5–7 October 1993; Wrather, N., Mailes, R.J., Eds.; pp. 77–82.
76. Lazzeri, L.; Leoni, O.; Manici, L.M. Biocidal plant dried pellets for biofumigation. *Ind. Crop. Prod.* **2004**, *20*, 59–65. [CrossRef]

77. Lazzeri, L.; Manici, M. The glucosinolate-myrosinase system: A natural and practical tool for biofumigation. *Acta Hortic.* **2000**, *532*, 89–95. [CrossRef]
78. Gardiner, J.B.; Morra, M.J.; Eberlein, C.V.; Brown, P.D.; Borek, V. Allelochemicals released in soil following incorporation of rapeseed (Brassica napus) green manures. *J. Agric. Food Chem.* **1999**, *47*, 3837–3842. [CrossRef] [PubMed]
79. Gimsing, A.L.; Kirkegaard, J.A. Glucosinolate and isothiocyanate concentration in soil following incorporation of Brassica biofumigants. *Soil Biol. Biochem.* **2006**, *38*, 2255–2264. [CrossRef]
80. Marín-guirao, J.I. *Evaluación del Efecto de Distintas Materias Orgánicas Sobre la Microbiota Edáfica, Contenido de Nitratos en Suelo, y Producción y Calidad de Cultivos de Tomate y Sandía en Suelo Arenado Bajo Plástico*; Universidad de Almería: Almería, Spain, 2016.
81. Achmon, Y.; Harrold, D.R.; Claypool, J.T.; Stapleton, J.J.; Vandergheynst, J.S.; Simmons, C.W. Assessment of tomato and wine processing solid wastes as soil amendments for biosolarization. *Waste Manag.* **2016**, *48*, 156–164. [CrossRef]
82. Seo, M.W.; Lee, S.W.; Lee, S.H.; Jang, I.B.; Heo, H.J. Effect of Green Manure Incorporation and Solarization on Root Rot Disease of 3-year-old Ginseng in Soil of Continuous Cropping Ginseng. *Korean J. Med. Crop Sci.* **2019**, *27*, 284–291. [CrossRef]
83. Díaz-Hernández, S.; Gallo-Llobet, L.; Domínguez-Correa, P.; Rodríguez, A. Effect of repeated cycles of soil solarization and biosolarization on corky root, weeds and fruit yield in screen-house tomatoes under subtropical climate conditions in the Canary Islands. *Crop Prot.* **2017**, *94*, 20–27. [CrossRef]
84. García-Raya, P. *Efecto de la Adición de Restos de Cosecha de Tomate Mediante Biosolarización en la Producción, Calidad del Tomate y Morfología de la Planta*; Universidad de Almería: Almeria, Spain, 2018.
85. Gamliel, A.; Stapleton, J.J. Effect of Chicken Compost or Ammonium Phosphate and Solarization on Pathogen Control, Rhizosphere Microorganisms, and Lettuce Growth. *Plant Dis.* **1993**, *77*, 886. [CrossRef]
86. Guerrero, M.M.; Ros, C.; Lacasa, C.M.; Martínez, V.; Lacasab, A.; Fernández, P.; Núñez-Zofío, M.; Larreglac, S.; Martíneza, M.A.; Díez-Rojo, M.A.; et al. Effect of biosolarization using pellets of brassica carinata on soil-borne pathogens in protected pepper crops. *Acta Hortic.* **2010**, *883*, 337–344. [CrossRef]
87. Núñez-Zofio, M.; Larregla, S.; Garbisu, C. Repeated biodisinfection controls the incidence of Phytophthora root and crown rot of pepper while improving soil quality. *Span. J. Agric. Res.* **2012**, *10*, 794–805. [CrossRef]
88. Klein, E.; Katan, J.; Austerweil, M.; Gamliel, A. Laboratory System to Study Soil Solarization and Organic Amendment Effects on Plant Pathogens. *Phytopathology* **2007**, *97*, 1476–1983. [CrossRef] [PubMed]
89. Basallote-Ureba, M.J.; Vela-Delgado, M.D.; Capote, N.; Melero-Vara, J.M.; López-Herrera, C.J.; Prados-Ligero, A.M.; Talavera-Rubia, M.F. Control of Fusarium wilt of carnation using organic amendments combined with soil solarization, and report of associated Fusarium species in southern Spain. *Crop Prot.* **2016**, *89*, 184–192. [CrossRef]
90. Blok, W.J.; Lamers, J.G.; Termorshuizen, A.J.; Bollen, G.J. Control of soilborne plant pathogens by incorporating fresh organic amendments followed by tarping. *Phytopathology* **2000**, *90*, 253–259. [CrossRef] [PubMed]
91. Gilardi, G.; Pugliese, M.; Gullino, M.L.; Garibaldi, A. Effect of different organic amendments on lettuce fusarium wilt and on selected soilborne microorganisms. *Plant Pathol.* **2016**, *65*, 704–712. [CrossRef]
92. Huang, X.Q.; Wen, T.; Zhang, J.B.; Meng, L.; Zhu, T.B.; Liu, L.L.; Cai, Z.C. Control of soil-borne pathogen Fusarium oxysporum by biological soil disinfestation with incorporation of various organic matters. *Eur. J. Plant Pathol.* **2015**, *143*, 223–235. [CrossRef]
93. Marín-Guirao, J.I.; Tello-Marquina, J.C. Microbiota edáfica y fatiga de suelo en invernaderos de la provincia de Gran. In *I Jornadas de Transferencia Hortofrutícola de Ciambital*; Camacho-Ferre, F., Valera-Martínez, D.L., Belmonte-Ureña, L., Herrero-Sánchez, C., Reca-Cardeña, J., Marín-Membrive, P., del Pino-Gracia, A., Casa-Fernández, M., Eds.; Universidad de Almería: Almería, Spain, 2017; pp. 17–36, ISBN 978-84-16389-98-8.
94. Guerrero-Díaz, M.M.; Lacasa-Martínez, M.; Hernández-Piñera, A.; Martínez-Alarcon, V.; Lacasa-Plasencia, A. Evaluation of repeated biodisinfestation using *Brassica carinata* pellets to control Meloidogyne incognita in protected pepper crops. *Span. J. Agric. Res.* **2013**, *11*, 485–493. [CrossRef]
95. Valdes, Y.; Viaene, N.; Moens, M. Effects of yellow mustard amendments on the soil nematode community in a potato field with focus on Globodera rostochiensis. *Appl. Soil Ecol.* **2012**, *59*, 39–47. [CrossRef]
96. Chamorro, M.; Miranda, L.; Domínguez, P.; Medina, J.J.; Soria, C.; Romero, F.; Santos, B.D.L. Evaluation of biosolarization for the control of charcoal rot disease (Macrophomina phaseolina) in strawberry. *Crop Prot.* **2015**, *67*, 279–286. [CrossRef]
97. Epstein, E. *The Science of Composting*; CRC Press: New York, NY, USA, 1997.
98. Simmons, C.W.; Guo, H.; Claypool, J.T.; Marshall, M.N.; Perano, K.M.; Stapleton, J.J.; VanderGheynst, J.S. Managing compost stability and amendment to soil to enhance soil heating during soil solarization. *Waste Manag.* **2013**, *33*, 1090–1096. [CrossRef]
99. Lazarovits, G. Management of soil-borne plant pathogens with organic soil amendments: A disease control strategy salvaged from the past. *Can. J. Plant Pathol.* **2001**, *23*, 1–7. [CrossRef]
100. Momma, N.; Kobara, Y.; Uematsu, S.; Kita, N.; Shinmura, A. Development of biological soil disinfestations in Japan. *Appl. Microbiol. Biotechnol.* **2013**, *97*, 3801–3809. [CrossRef] [PubMed]
101. Katase, M.; Kubo, C.; Ushio, S.; Ootsuka, E.; Takeuchi, T.; Mizukubo, T. Nematicidal activity of volatile fatty acids generated from wheat bran in reductive soil disinfestation. *Jpn. J. Nematol.* **2009**, *39*, 53–62. [CrossRef]
102. Kirkegaard, J.A.; Sarwar, M. Biofumigation potential of brassicas: I. Variation in glucosinolate profiles of diverse field-grown brassicas. *Plant Soil* **1998**, *201*, 71–89. [CrossRef]
103. Hansen, Z.R.; Keinath, A.P. Increased pepper yields following incorporation of biofumigation cover crops and the effects on soilborne pathogen populations and pepper diseases. *Appl. Soil Ecol.* **2013**, *63*, 67–77. [CrossRef]

104. Ros, C.; Sánchez, F.; Martínez, V.; Lacasa, C.M.; Hernández, A.; Torres, J.; Guerrero, M.M.; Lacasa, A. El cultivo de brásicas para biosolarización reduce las poblaciones de Meloidogyne incognita en los invernaderos de pimiento del sudeste de Spain. *ITEA Inf. Técnica Económica Agrar.* **2016**, *112*, 109–126. [CrossRef]
105. Lamers, J.G.; Runia, W.T.; Molendijk, L.P.G.; Bleeker, P.O. Perspectives of anaerobic soil disinfestation. *Acta Hortic.* **2010**, *883*, 277–284. [CrossRef]
106. Rosskopf, E.N.; Kokalis-Burelle, N.; Hong, J. Status of ADS Development in Florida. In Proceedings of the Annual International Research Conference on Methyl Bromide. Alternatives and Emissions Reductions, Orlando, FL, USA, 30 October–2 November 2011; Volume 42, pp. 103–112.
107. Fernandez-Bayo, J.D.; Shea, E.A.; Parr, A.E.; Achmon, Y.; Stapleton, J.J.; VanderGheynst, J.S.; Hodson, A.K.; Simmons, C.W. Almond processing residues as a source of organic acid biopesticides during biosolarization. *Waste Manag.* **2020**, *101*, 74–82. [CrossRef]
108. MAGRAMA. *Plan Estatal Marco de Gestión de Residuos PEMAR (2016–2022)*; Ministerio de Agricultura, Pesca y Alimentación: Madrid, Spain, 2015; pp. 1–182.
109. Guerrero, M.M.; Guirao, P.; Martinez-lluch, M.C.; Tello, J.C.; Lacasa, A. Soil fatigue and its specificity towards pepper plants in greenhouses. *Span. J. Agric. Res.* **2014**, *12*, 644–652. [CrossRef]
110. Fernández, P.; Lacasa, A.; Guirao, P.; Larregla, S. Effects of Biosolarization with fresh sheep manure on soil physical properties of pepper greenhouses in Campo de Cartagena. In Proceedings of the 6th Workshop on Agri-Food Research, Cartagena, Spain, 8–9 May 2018; Artés-Hernández, F., Cos, J.E., Fernández-Hernández, J.A., Calatrava, J.A., Aguayo, E., Alarcón, J.J., Guitiérrez-Cortines, M.E., Eds.; pp. 97–100, ISBN 9788416325641.
111. Ros, C.; Martínez, V.; Sánchez-Solana, F.; López-Marín, J.; Lacasa, C.M.; Guerrero, M.d.M.; Lacasa, A. Combination of biosolarization and grafting to control Meloidogyne incognita in greenhouse pepper crops. *Crop Prot.* **2018**, *113*, 33–39. [CrossRef]
112. Medina, J.J.; Miranda, L.; Soria, C.; Palencia, P. Non-Chemical Alternatives to Methyl Bromide for Strawberry: Biosolarization as Case-Study in Huelva (Spain). *Acta Hortic.* **2009**, *842*, 961–964. [CrossRef]
113. Miranda, L.; Domínguez, P.; Soria, C.; Zea, T.; Talavera, M.; Velasco, L.; Romero, F.; Santos, B.D.L.; Newton, A.I. Soil Biosolarization for Strawberry Cultivation. *Acta Hortic.* **2012**, *926*, 407–414.
114. Marín-guirao, J.I.; Gómez-Tenorio, M.Á.; Castillo-Díaz, F.J.; Magdaleno-González, J.; Tello-Marquina, J.C. Incidencia de la biodesinfección de suelos de invernadero usando restos vegetales del cultivo de pimiento sobre la microbiota edáfica (*Hongos y bacterias*). In Proceedings of the I Congreso Internacional de Sistemas Agrarios Tradicionales, V Encuentro InterVegas, Tenerife, Spain, 24–26 October 2019.
115. Serrano-Pérez, P.; De Santiago, A.; Rodríguez-Molina, M.d.C. Biofumigation with Pellets of Defatted Seed Meal of *Brassica carinata*: Factors Affecting Performance against Phytophthora nicotianae in Pepper Crops. *Front. Sustain. Food Syst.* **2021**, *5*, 664531. [CrossRef]
116. Ros, M.; Garcia, C.; Hernandez, M.T.; Lacasa, A.; Fernandez, P.; Pascual, J.A. Effects of biosolarization as methyl bromide alternative for Meloidogyne incognita control on quality of soil under pepper. *Biol. Fertil. Soils* **2008**, *45*, 37–44. [CrossRef]
117. Guerrero, M.; Martínez, M.; Ros, C.; Martínez, M.; Bello, A.; Lacasa, A. Biosolarización y biofumigación para la producción de pimiento ecológico en invernadero. In Proceedings of the VII Congreso SEAE, Zaragoza, Spain, 18–23 September 2006; p. 103.
118. Martínez, M.A.; Guerrero, M.M.; Martínez, M.C.; Ros, C.; Lacasa, A.; Tello-Marquina, J.C. Efecto de la biosolarización reiterada sobre la microbiota fúngica en cultivos de pimiento. In Proceedings of the VII Congreso SEAE, Zaragoza, Spain, 18–23 September 2006; p. 212.
119. Guerrero-Díaz, M.M.; Lacasa-Martinez, C.M.; Hernández, A.; Martínez, V.; Martínez, M.C.; Fernández-Molina, P.; Lacasa, A. Biosolarization with agroindustrial byproduct for the control of soilborne pathogens in protected pepper crops in southeast Spain. *Acta Hortic.* **2014**, *1044*, 157–161. [CrossRef]
120. Mitidieri, M.S.; Brambilla, V.; Barbieri, M.; Piris, E.; Celié, R. Tomato Crop Health, Yield, and Greenhouse Soil Conditions after 17 Years of Repeated Treatments of Biofumigation and Solarization. *Glob. J. Agric. Innov. Res. Dev.* **2021**, *8*, 123–139. [CrossRef]
121. Domínguez, P.; Miranda, L.; Soria, C.; de los Santos, B.; Chamorro, M.; Romero, F.; Daugovish, O.; López-Aranda, J.M.; Medina, J.J. Soil biosolarization for sustainable strawberry production. *Agron. Sustain. Dev.* **2014**, *34*, 821–829. [CrossRef]
122. Domínguez, P.; Miranda, L.; Medina, J.J.; Santos, B.D.L.; Daugovish, O.; Soria, C.; Chamorro, M.; López-, J.M.; Domínguez, P.; Miranda, L.; et al. Evaluation of Non-Fumigant Alternative Soil Treatments for Strawberry Production in Huelva Evaluation of Non-Fumigant Alternative Soil Treatments for Strawberry Production in Huelva (Spain). *Int. J. Fruit Sci.* **2016**, *16*, 28–36. [CrossRef]
123. Sánchez-Navarro, A.; Jiménez-Ballesta, R.; Girona-Ruiz, A.; Alarcón-Vera, I.; Delgado-Iniesta, M.J. Rapid Response Indicators for Predicting Changes in Soil Properties Due to Solarization or Biosolarization on an Intensive Horticultural Crop in Semiarid Regions. *Land* **2022**, *11*, 64. [CrossRef]
124. López-Elías, J.; Jiménez, L.J.; Huez, L.M.A.; Garza, O.S.; Cruz, B.F.; Bautista, O.A.L. Medidas de control biológico en la producción de pepino, bajo condiciones de invernadero. *Idesia (Arica)* **2017**, *35*, 7–12. [CrossRef]
125. De la Lastra, E.; Marín-Guirao, J.I.; López-Moreno, F.J.; Soriano, T.; de Cara-García, M.; Capote, N. Potential inoculum sources of Fusarium species involved in asparagus decline syndrome and evaluation of soil disinfestation methods by qPCR protocols. *Pest Manag. Sci.* **2021**, *77*, 4749–4757. [CrossRef]

126. Khadka, R.B.; Marasini, M.; Rawal, R.; Testen, A.L.; Miller, S.A. Effects of anaerobic soil disinfestation carbon sources on soilborne diseases and weeds of okra and eggplant in Nepal. *Crop Prot.* **2020**, *135*, 104846. [CrossRef]
127. Mauromicale, G.; Lo Monaco, A.; Longo, A.M.G. Improved efficiency of soil solarization for growth and yield of greenhouse tomatoes. *Agron. Sustain. Dev.* **2010**, *30*, 753–761. [CrossRef]
128. Mauromicale, G.; Longo, A.M.G.; Lo Monaco, A. The effect of organic supplementation of solarized soil on the quality of tomato fruit. *Sci. Hortic. (Amsterdam)* **2011**, *129*, 189–196. [CrossRef]
129. Castillo-Díaz, F.J.; Ruiz-Olmos, C.A.; Gómez-Tenorio, M.Á.; Tello-Marquina, J.C. Efecto de la biosolarización sobre la producción de tomate cultivado bajo invernadero en Almería. Parte I: Evaluación de diferentes restos vegetales. *Agrícola Vergel* **2021**, *432*, 103–112.
130. Junta de Andalucía. Observatorio de Precios y Mercados. Costes. Available online: https://www.juntadeandalucia.es/agriculturaypesca/observatorio/servlet/FrontController?action=List&table=11200&subsector=&page=1&ec=default (accessed on 12 April 2023).
131. Castillo-Díaz, F.J.; Batlles-delafuente, A.; Belmonte-Ureña, L.J.; Camacho-ferre, F. Impact of environmental policies on the profitability of greenhouse agriculture in southeastern Spain. *Sustain. Dev.* **2023**, 1–18. [CrossRef]
132. Martos-Pedrero, A.; Jiménez-Castillo, D.; Ferrón-Vílchez, V.; Cortés-García, F.J. Corporate social responsibility and export performance under stakeholder view: The mediation of innovation and the moderation of the legal form. *Corp. Soc. Responsib. Environ. Manag.* **2022**, *30*, 248–266. [CrossRef]
133. Plaza-Úbeda, J.A.; Abad-Segura, E.; de Burgos-Jiménez, J.; Boteva-Asenova, A.; Belmonte-Ureña, L.J. Trends and New Challenges in the Green Supply Chain: The Reverse Logistics. *Sustainability.* **2020**, *13*, 331. [CrossRef]
134. Sidorowicz, A.; Maqbool, Q.; Nazar, M. Future of Nanofertilizer. In *Nanotechnology for Agriculture: Crop Production & Protection*; Springer Nature: Berlin/Heidelberg, Germany, 2019; Volume 1, pp. 143–152.
135. European Union. Regulation (eu) 2018/848 of the European Parliament and of the Council of 30 May 2018 on Organic Production and Labelling of Organic Products and Repealing Council Regulation (EC) No 834/2007. *Off. J. Eur. Union* **2018**, *150*, 1–92.

horticulturae

Article

Effects of Using Plasma-Activated Water as a Nitrate Source on the Growth and Nutritional Quality of Hydroponically Grown Green Oak Lettuces

Soraya Ruamrungsri [1,2], Choncharoen Sawangrat [3,4], Kanokwan Panjama [1,2], Phanumas Sojithamporn [4], Suchanuch Jaipinta [3], Wimada Srisuwan [3], Malinee Intanoo [3], Chaiartid Inkham [2,5,*] and Sa-nguansak Thanapornpoonpong [1,3,*]

1 Department of Plant and Soil Science, Faculty of Agriculture, Chiang Mai University, Chiang Mai 50200, Thailand
2 H.M. the King's Initiative Centre for Flower and Fruit Propagation, Chiang Mai University, Chiang Mai 50200, Thailand
3 Agriculture and Bio Plasma Technology Center (ABPlas), Thai—Korean Research Collaboration Center (TKRCC), Science and Technology Park, Chiang Mai University, Chiang Mai 50200, Thailand
4 Department of Industrial Engineering, Faculty of Engineering, Chiang Mai University, Chiang Mai 50200, Thailand
5 Science and Technology Research Institute, Chiang Mai University, Chiang Mai 50200, Thailand
* Correspondence: sunwins111@hotmail.com (C.I.); sa-nguansak.t@cmu.ac.th (S.-n.T.)

Abstract: Nitrate is a major source of the inorganic nitrogen taken up by the roots of plants. Nitrate sources are generally derived from inorganic minerals by an energy-consuming chemical process; as a result, the price of chemical fertilizers is gradually increasing year by year. NO_3-N, generated from N_2 using the plasma technique, is an alternative method of producing nitrate from the air. Therefore, in this research, we aimed to determine the efficiency of generating NO_3-N using plasma-activated water (PAW) to replace nitrates from chemical fertilizer in a nutrient solution. Green oak lettuce (*Lactuca sativa* L.) was grown in a hydroponics system using the double-pot technique. The plants were supplied with three different nutrient solutions (based on Hoagland's solution), i.e., T1, no nitrate in the nutrient solution ($NO_3^- = 0$); T2, using nitrate sourced from a commercial chemical fertilizer (normal nitrate); and T3, using a nitrate source generated using the pinhole plasma jet technique (plasma nitrate). The other macronutrients and micronutrients in each treatment were equally supplied. The results show that, at the harvested stage (21 days after the plants received treatment), the no-nitrate (T1) treatment provided lower growth and yields. Moreover, compared with the normal nitrate (T2) and plasma nitrate (T3), the results indicate that most growth and yields showed no statistical differences. In terms of nitrate accumulation within plants, it was found that the normal nitrate treatment (T2) had the highest levels of nitrate accumulation, in both the underground and aboveground parts of green oak lettuce. These results confirmed that plasma nitrate could be an alternative source of nitrate N which provided a safer way for the environment and human health in terms of nitrate accumulation. In addition, data related to the chemical analysis of free amino acid concentrations in each treatment are discussed in this research.

Keywords: plasma-activated water; pinhole plasma jet; hydroponics system; green oak lettuce; nitrogen fertilizer; plasma nitrate

1. Introduction

Globally, the production levels of lettuces such as Batavia, butterhead, red oak, red coral, and green oak have risen due to population increases and growing health concerns [1]. According to the Food and Agricultural Organization of the United Nations [2], the global production of lettuce and chicory increased from 2015 to 2020 from approximately 26 to 28 million tons. Green oak (*Lactuca sativa* L.) is a leafy vegetable that is rich in nutrients such as fiber, folate, carotenoids, phenolic and antioxidant compounds, minerals, and various vitamins [3]. Although soil-grown lettuce is consumed worldwide, the risk of exposure to inorganic fertilizers and pesticides, which cause pollution and contamination, is high [4]. Hydroponics, an alternative method of growing lettuce, is an innovative agricultural technology that has been adopted in several countries. Due to the absence of soil, water is responsible for providing nutrients, hydration, and oxygen to plants or vegetables. Therefore, the risk of disease occurring due to organisms in the soil is low in such systems [5]. Furthermore, hydroponic systems are a highly popular technique for cultivating plants because they lead to improvements in growth, yields, income, profit, and the number of crops produced per year [6]. The fertilizer solution is an important factor in the functioning of hydroponics systems, and NO_3-N is one of the main mineral elements that is usually derived from chemical fertilizers.

In agriculture, nitrogen is an essential macronutrient for plant growth and development. It is generally used to increase crop productivity, and the level of nitrogen available to plants is one of the factors that most strongly limit crop productivity. Plants normally absorb nitrogen in the form of NO_3^- and NH_4^+. Nitrate, as a signal molecule involved in plant development and metabolism, is the nutrient that leads to the production of amino acids and nitrogen compounds [7]. Therefore, NO_3^- accumulation in plant tissues can be influenced by many preharvest factors, including plant species, the photoperiod, temperature, light intensity, and the type of fertilizer applied; this can occur when the nitrate uptake exceeds plant assimilation [8].

Ammonia, nitrate, and urea are the three main nitrogen-based fertilizers. The production of nitrogen-based fertilizers has always been a challenge for horticulturists due to the increasing demand for such products [9]. Traditionally, ammonia was produced through the Haber–Bosch (HB) ammonia synthesis process, which consumes large amounts of energy and thus has a substantial carbon footprint [10]. The annual global energy consumption and CO_2 emissions of ammonia production accounted for 1–2% and 300 million tons per year, respectively [11]. Consequently, it is imperative to ensure the availability of fertilizers that are safe for human health, sustainable, and environmentally friendly. Moreover, Russia is the second-largest exporter of fertilizers worldwide but terminated fertilizer exports during the 2022 Russia–Ukraine war, causing shortages around the world, in countries including Thailand [12,13]. Therefore, the cost of fertilizer is undergoing a continuous increase.

Plasma, which is the fourth state of matter, is a source of electrons, positively charged ions, radicals, gas atoms, molecules (in excited or basic states), and photons with a range of energies [14]. Plasma can occur in both natural phenomena on Earth (e.g., the aurora borealis and fire) and under laboratory conditions, by providing sufficient energy to excite gaseous molecules [15]. Plasma falls into the two categories of nonthermal plasma and thermal plasma. Nonthermal plasma has received considerable attention for applications in the agricultural sector [16]. Reactive oxygen and nitrogen species (RONS) generated from plasma are important factors that can enhance plant physiology and growth [17]. However, the nonthermal plasma technology in the form of gas is not sufficiently flexible for use in the agricultural sector due to the risk of plant surface damage [18]. Therefore, another form of plasma, which is more flexible than the gas form, is generated by exposing plasma to water; this is called plasma-activated water (PAW). Several researchers have proven the potential of PAW in various applications, including microbial inactivation [19,20], enhancement of seed germination [21,22], and plant growth improvement [23,24]. In addition to these applications in the agricultural sector, some research has used plasma-activated water as a source of nitrogen species. Graves et al. [25] indicated that using air plasma to activate liquid solutions of organic material resulted in a reduction in pH and the addition of NO_2^-, NO_3^-, and H_2O_2. Therefore, the generated NO_3^- provides plants with accessible nitrate, and the nitrate increases linearly with the increase in power usage under all tested flow rates (i.e., 0.5, 1.25, and 2.0 L/min) [26]. In addition, Carmassi et al. [27] investigated the effect of nonthermal plasma (NTP) technologies on hydroponic nutrient solutions and determined their effects on the growth and quality of baby-leaf lettuce (*Lactuca sativa* var. *acephala* Alef.). Their results prove that NTP treatment of the nutrient solution could improve the production and quality of baby-leaf lettuce grown in a hydroponic growing system (HGS). In hydroponics systems, the occurrence of algae leads to major problems, including intensive odors, a reduction in dissolved oxygen, nutrient deprivation, and a pH reduction in the system [28,29]. However, Date et al. [30] showed that plasma-activated nutrient solutions could reduce algae growth to a greater degree than standard nutrient solutions.

Some previous studies have indicated that nitrate in plasma-activated water effectively enhances plant growth, especially leafy plants, and little is known about the potential of nitrate from different nutrient solutions. Additionally, a pinhole plasma jet has not been applied to nitrate fertilizer generation before. Therefore, in the present study, we aimed to investigate the effects of nitrate synthesized from plasma-activated water on the growth performance, yields, and nutritional value of green oak lettuce in a hydroponics system. Plant morphological characteristics, including plant height, root length, canopy, leaf width, number of leaves, and leaf area were compared between three treatment groups (i.e., no nitrate, normal nitrate, and plasma nitrate). Finally, the chemical compositions of the three treatment groups of green oak were investigated and compared.

2. Materials and Methods

2.1. Plasma Fertilizer Preparation

A pinhole plasma jet was developed to generate plasma-activated water (PAW) at the Agriculture and Bio Plasma Technology Laboratory, Science and Technology Park, Chiang Mai University, Thailand. The schematic of the pinhole plasma jet discharge system is shown in Figure 1. This system consists of a pinhole plasma jet discharge system with a 125-watt neon transformer, a high-voltage power supply, a gas transport system to provide controlled flows of air and oxygen gas, control devices, and a tank containing tap water. Tap water (40 L) was used as a solution because it is convenient and cost-effective. The working gases were mixed from air and oxygen with flow rates of 1.5 and 1.5 L/min, respectively. After discharging the plasma for 16 h the nitrate and nitrite concentrations were set to approximately 800 mg/L, which is comparable with Hoagland fertilizer. The chemical composition of the PAW (i.e., NO_3^- and NO_2^-) was determined using a commercial kit containing Griess reagents (Cayman Chemicals, Ann Arbor, MI, USA), and a pink and brown azo-product was formed. The NO_3^- and NO_2^- concentrations were measured using established UV/Vis spectrophotometric methods (Shimadzu UV–1800, Kyoto, Japan) with maximum absorbance peaks at 372 and 507 nm, respectively. The concentration of H_2O_2 was determined using the iodometric titration method. Then, 1 mL of 2% potassium iodide solution and 2 M HCl was added to the PAW. The solution was kept in the dark for 15 min, resulting in the solution turning yellow in color. Then, 0.1 M sodium thiosulfate solution was added until the solution turned a lighter yellow. Dropping a starch indicator into the solution turned the solution blue in color. Then, the blue solution was titrated with a 0.1 M sodium thiosulfate solution until the solution turned clear. The concentrations of H_2O_2, NO_3^-, and NO_2^- in the PAW were measured immediately after plasma activation, and their values were 102.99, 883.59, and 31.56 mg/L, respectively.

2.2. Plant Materials and Growth Conditions

The experiment was conducted in an evaporative greenhouse at H.M. The King Initiative Centre for Flower and Fruit Propagation, Chiang Mai University, Thailand. The temperature and relative humidity (RH) inside the evaporative greenhouse were 27.8 °C and 89.28%, respectively. The green oak lettuce (*Lactuca sativa*) seeds were germinated on a sponge in a plastic tray floating on a half-strength Hoagland fertilizer formula (Electrical conductivity (EC) = 1.4 µS/cm pH = 5.5) for 21 days. Then, the plants were transferred to the hydroponics system and supplied with three different nutrient solutions (based on Hoagland's solution), i.e., T1 = no nitrate in the nutrient solution (NO_3-N-free solution), T2 = using a nitrate source from a commercial chemical fertilizer (chemical fertilizer NO_3-

N), and T3 = using a nitrate source generated using the pinhole plasma jet technique (PAW-NO_3-N-). The double-pot system used in this research is illustrated in Figure 2. The other macronutrients and micronutrients in each nutrient solution were equally supplied, as shown in Table 1.

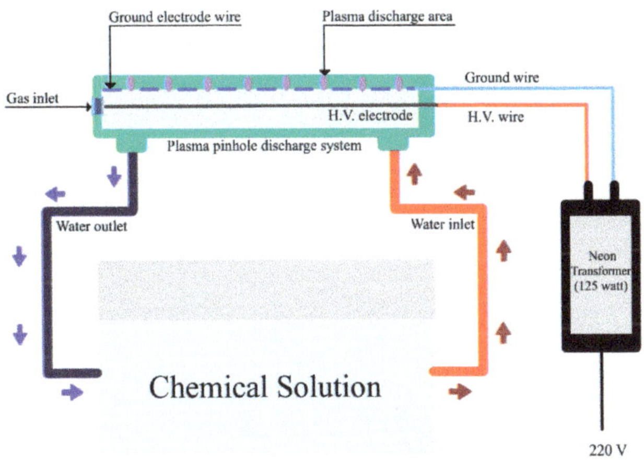

Figure 1. Pinhole discharge system for generating plasma-activated water.

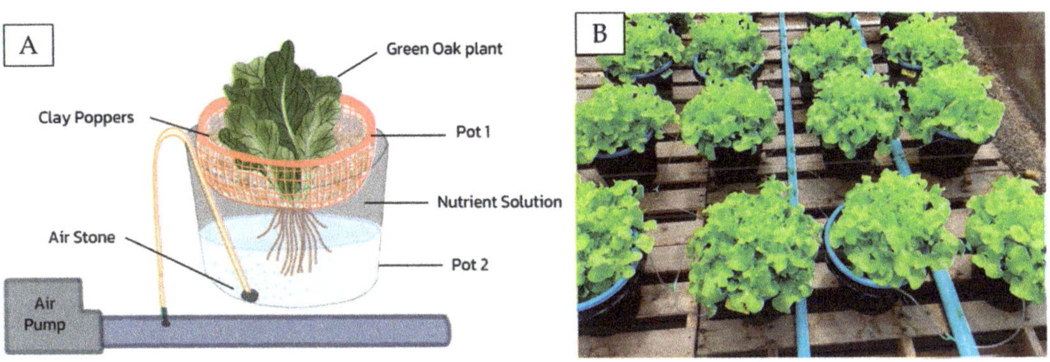

Figure 2. (**A**) Diagram and (**B**) actual experimental setup of the double-pot system used in this experiment.

Table 1. Macronutrient and micronutrient concentration (mg/L) in each nutrient solution.

Treatment	Macronutrients (mg/L)							Micronutrients Concentration (mg/L)					
	N Source			P	K	Ca	Mg	B	Mn	Zn	Cu	Mo	Fe
	NO_3^-	Plasma-NO_3^-	NH_4										
T1	0	0	234.97	161.15	48.86	0.50	0.50	0.05	0.02	0.01	5.05	0.01	5.05
T2	871.11	0	234.75	161.20	48.86	0.50	0.50	0.05	0.02	0.01	5.05	0.01	5.05
T3	0	883.59	234.97	161.15	48.86	0.50	0.50	0.05	0.02	0.01	5.05	0.01	5.05

T1 = No nitrate, T2 = Normal nitrate, and T3 = Plasma nitrate.

The pH of each nutrient solution was adjusted to 5.5–6 by the pH-down adjuster and pH-up adjuster (KOH and diluted HNO_3, respectively). The EC of plasma-activated water was 1.5 µS/cm. The pH, EC, and temperature were monitored using a SevenCompactTM Duo pH/Conductivity meter S213 (Mettler Toledo International Inc., Greifenseem, Switzerland). The solution was completely renewed every 7 days.

2.3. Measurement of Plant Morphology and Growth Characteristics

Four plant samples were randomly chosen from each replicate. Plant morphological characteristics, including plant height, root length, canopy width, leaf width, number of leaves, leaf area (LI-3100C, LI-COR Biosciences, Lincoln, OR, USA), and leaf color intensity (SPAD-502 Plus, Spectrum Technologies Inc., Aurora, IL, USA) were measured for plants from each treatment every seven days. The roots were rinsed with double-deionized water to remove visible hydroponic planting material (sponge) from the surface. At the same time, fresh and dry weights were also determined. To record the plants' dry weights, the aboveground part and the underground part were washed with tap water, dried, and placed in an oven (UN55, MEMMERT, Büchenbach, Germany) at 60 °C for 1 week. The photosynthesis rate and stomatal conductance were measured at the harvest stage (when the plants had been cultivated for 21 days after treatment (DAT)) with a portable LCpro-SD (ADC BioScientific Ltd., Hoddesdon, England). The leaves' greenness was evaluated using a SPAD-502 Plus chlorophyll meter (Spectrum Technologies Inc., Aurora, IL, USA).

2.4. Determination of Nitrate Contents in the Plants

After the lettuce samples were dried (~7 days in the oven), 10 mg of lettuce was immersed in 10 mL of distilled water in a centrifuge tube and incubated at room temperature for 1 night. The mixture was centrifuged at 15,000 rpm for 20 min. The supernatant was collected at 0.1 mL and added to 0.4 mL 5% (*w*/*v*) salicylic acid–sulfuric acid (5 g salicylic acid in 100 mL sulfuric acid) in each tube; the sample was mixed well, and then the reactants were incubated for 20 min at room temperature until they had cooled down. Then, 9.5 mL of 8% (*w*/*v*) NaOH solution was added to each tube and the tubes were cooled to room temperature (about 20–30 min). The absorbance measurement was made at 410 nm using established UV/Vis spectrophotometric methods (Shimadzu UV–1800, Kyoto, Japan). To generate the standard curve, the concentrations of the series of standard solutions were 0, 62.5, 125, 250, 500, and 1,000 mg/L KNO_3. Additionally, the concentration of the KNO_3 standard solution was plotted, and the R2 value was determined [31].

2.5. Determination of Total Phenolic Compounds

The total phenolic compounds were determined using a modified Folin–Ciolcateu method [32]. In brief, 1.0 g of fresh lettuce was extracted in 10 mL of 80% (*v*/*v*) methanol, sonicated for 5 min at 37 °C, and centrifuged at 4000 rpm for 10 min, followed by the collection of supernatants. An aliquot (500 µL) of supernatant was reacted with 2.5 mL 10% Folin–Ciolcalteu reagent, followed by the addition of 2.0 mL 7.5% Na_2CO_3 solution, and vortexed. The mixture was incubated at room temperature for 30 min. Absorbance was read at 765 nm using established UV/Vis spectrophotometric methods (Shimadzu UV–1800, Kyoto, Japan). Total phenolic content was calculated using the standard curves of 0, 5, 10, 20, 40, and 80 mg/L solutions of gallic acid (mg GAE/g FW) following the formula T = CV/M, where T = the total phenolic content in mg GAE/g FW, C = the concentration of gallic acid established from the calibration curve in mg/mL, V = the volume of the extract in mL, and M = the fresh weight of the plant extract in g.

2.6. Determination of Phosphorus, Potassium, Calcium, and Magnesium

The determination of exchangeable potassium and phosphorus in the fertilizer was conducted using an atomic emission spectrophotometer (AES) [33]; an atomic absorption spectrophotometer (AAS) was used to analyze calcium and magnesium [34].

2.7. Determination of Free Amino Acids

The amino acids in green oak leaves were identified using a high-speed amino acid analyzer (L-8900, Hitachi High-Technologies Corporation, Japan) equipped with a Hitachi custom ion-exchange column (4.6 × 60 mm) (Hitachi High-Technologies Corporation, Tokyo, Japan). For the analysis, a dry green oak powder was prepared using the methods described by Wang et al. [35].

2.8. Statistical Analysis

The data are presented as mean values ± standard deviation and were obtained using SPSS software (IBM Corp., Armonk, NY, USA). The one-way analysis of variance (ANOVA) and subsequent multiple range test using the least significant difference method (LSD) were performed to judge the differences between the groups. The different lowercase letters represent statistically significant differences at probability $p < 0.05$.

3. Results and Discussions

3.1. Changes in the NO_3 Concentration in the Solution

According to several studies, reactive nitrogen species including NO_2^- and NO_3^- are the main long-living species produced during the plasma–water interaction [36]. Various studies have examined the chemical reaction for the generation of nitrate. Al-Sharify et al. [37] proposed one of the possible nitrate-generation paths, as presented in Equation (1). In the present study, the pinhole plasma jet was directly discharged into water, in which the obtained concentrations of nitrate and nitrite were higher than when the jet was discharged above the water's surface [38].

$$NO_2^- + H_2O_2 + H^+ \rightarrow NO_3^- + H_2O + H^+ \quad (1)$$

Nitrate is a general form of nitrogen taken up and assimilated by plants. Andrews and Raven [39] explained that nitrate that is taken up by plants is reduced to nitrite and ammonium (NH_4^+) by the enzymes nitrate reductase (NR) and nitrite reductase (NiR), respectively. Table 2 shows the remaining nitrate concentration in each nutrient solution (i.e., T1, T2, and T3) at each sampling time (i.e., 7, 14, and 21 DAT). Note that each nutrient solution was mixed and renewed every seven days. Therefore, the initial concentration of nitrate was the same as that on day 1. As shown in Table 2, the nitrate concentration in T3 decreased by approximately 33.75%, 35.88%, and 49.09% after 7 DAT, 14 DAT, and 21 DAT, respectively. The reduction of nitrate in T3 was more than that in T1 and T2 at every sampling time. The concentration of nitrate generated in PAW gradually decreased with the increased storage time [38]. This indicates that plants can easily take in NO_3-N generated from PAW.

Table 2. Nitrate content remaining in each treatment at each sampling time.

Treatments	Nitrate Content in Nutrient Solution (mg/L)			
	Day 1	7 DAT	14 DAT	21 DAT
T1	0	0	0	0
T2	871.11	780.08	752.45	572.19
T3	883.59	585.35	566.49	449.82
%CV		17.44	3.02	10.97
LSD0.05		158.57	26.52	74.69

T1 = No nitrate, T2 = Normal nitrate, and T3 = Plasma nitrate.

3.2. Growth Characteristics and Yields

Visualizations from above and from the side of the plants supplied with the nutrient solutions without nitrate (T1), with normal nitrate (T2), and with plasma nitrate (T3) are shown in Figure 3. T1 plants exhibited smaller heads and longer roots than T2 and T3. Jia et al. [40] indicated that the root foraging response prompted by increasing root length results from a lack of nitrogen. They explained that variation in root elongation under low N results from variation in the brassinosteroid (BR) biosynthesis gene DWARF1 and from the overexpression of DWF1. The average morphological features including plant height, root length, canopy width, leaf width, number of leaves, and leaf area after the plants were harvested at 21 days after treatment (DAT) are presented in Table 3. Plant height, leaf width, canopy width, and the number of leaves in the T2 and T3 groups were not statistically different ($p > 0.05$). Similar results were obtained by Noh et al. [41], which indicates that growth parameters are not significantly different between lettuces grown with and without plasma-activated water. In contrast, lettuce cultivated under T3 had a leaf area larger than T1 and T2 by 75.41% and 27.77%, respectively. Even though the number of leaves was not significantly different between T2 and T3, this difference can lead to a significantly larger leaf area. A larger leaf area could result in a better interception of light and a higher photosynthetic rate [42]. RONS generated in plasma-activated water play a key role in stimulating plant growth and responses to stress [27].

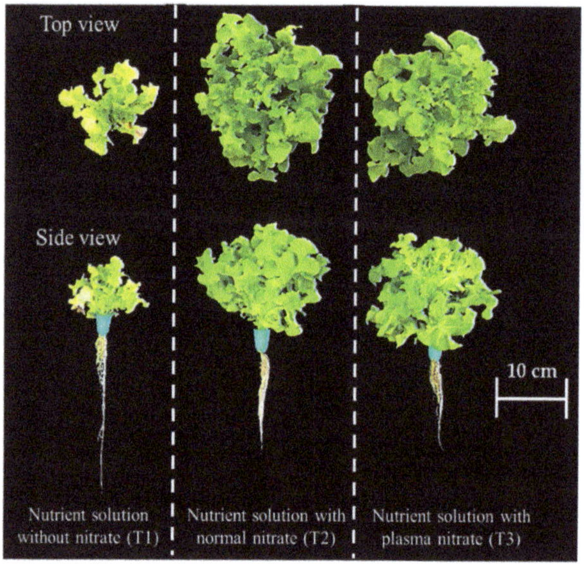

Figure 3. Growth of green oak lettuce grown in a hydroponics system and supplied with different NO_3^- sources at harvest stage (21 days after treatment).

Table 3. The morphologies of hydroponically grown lettuce cultivated in nutrient solutions without nitrate (T1), normal nitrate (T2), and plasma nitrate (T3) at harvest stage (21 DAT).

Treatments	Plant Height (cm)	Root Length (cm)	Canopy Width (cm)	Leaf Width (cm)	Leaves Number	Leaves Area (cm^2)
T1	13.00 ± 2.90 b	24.57 ± 3.65 a	15.67 ± 5.39 b	7.17 ± 2.50 b	6.67 ± 1.03 b	548.05 ± 37.27 c
T2	19.42 ± 1.50 a	18.55 ± 3.93 b	29.17 ± 2.16 a	14.50 ± 0.89 a	14.50 ± 2.26 a	1631.87 ± 135.74 b
T3	18.00 ± 2.00 a	16.37 ± 2.89 b	28.33 ± 2.66 a	14.75 ± 4.00 a	17.00 ± 2.68 a	2228.44 ± 167.40 a
%CV	13.15	17.75	15.12	22.83	16.59	8.59
LSD0.05	2.72	4.33	4.53	3.41	2.60	155.40

Means in the same column followed by different letters are significantly different ($p < 0.05$) by LSD.

Chlorophyll, which gives plants their green color, has nitrogen as a part of its molecular makeup. The samples cultivated in the T3 group had higher levels of nitrate uptake than the T2 group, as presented in Table 2. In addition, Table 4 shows that plants cultivated with a nutrient solution containing plasma nitrate (T3) had significantly higher ($p < 0.05$) greenness of leaves than T1 and T2, by 5.5 and 3.45 SPAD units, respectively. This indicates that plasma nitrate could be assimilated into chlorophyll molecules to a greater extent than in the other treatments. Nitrogen deficiency leads to a reduction in chlorophyll in leaves, which means that the leaves are less green [43,44], as shown in T1. Chlorophyll is an important photosynthetic pigment and the leaves' appearance represents the rate of photosynthesis. Furthermore, the chlorophyll content directly relates to the leaves' area. A larger leaf area represents a higher photosynthetic rate as there is a larger area for absorbing light [44]. The

T2 and T3 groups presented significantly higher photosynthetic rates than the T1 group, and a similar trend was seen for the leaf area. Kučerová et al. [45] demonstrated that both nitrate concentration and hydrogen peroxide are responsible for higher photosynthesis rates. Nitrogen is a fundamental constituent of chlorophyll, which correlates with the photosynthetic rate, carboxylation enzymes, and protein membranes [46]. Correspondingly, the lowest photosynthesis rate was observed in the T1 group. The stomatal conductance of the green oak cultivated in the T2 and T3 groups was significantly higher than in the T1 group. Many studies have indicated that stomatal conductance is a physiological plant mechanism that plays an important role in photosynthetic processes [47,48]. Saberi et al. [49] also showed the positive effects of cold plasma technology under haze conditions on the photosynthetic rate, chlorophyll content, and stomatal conductance of wheat, which increased by 34%, 32%, and 93%, respectively. As a result, nitrate content from any source (e.g., normal nitrate nutrient solutions and plasma nitrate) is essential to most growth parameters, including plant height, root length, leaf area, photosynthetic rate, and dry fresh weight.

Table 4. Leaves color intensity, photosynthesis rate, and stomatal conductance of hydroponically grown lettuce nutrient solution without nitrate (T1), normal nitrate (T2), and plasma nitrate (T3) at 42 days.

Treatments	Leaves Color Intensity (SPAD Unit)	Photosynthesis Rate ($\mu mol/m^2/s$)	Stomatal Conductance ($\mu mol/m^2/s$)
T1	19.93 ± 1.35 c	0.72 ± 0.33 b	0.10 ± 0.02 b
T2	21.98 ± 0.88 b	2.09 ± 0.92 a	0.20 ± 0.05 a
T3	25.43 ± 0.94 a	3.02 ± 0.56 a	0.14 ± 0.03 a
%CV	4.81	33.67	24.40
LSD0.05	1.33	0.80	0.04

Means in the same column followed by different letters are significantly different ($p < 0.05$) by LSD.

Non-significant ($p > 0.05$) differences in the fresh- and dry-weight yields were observed between T2 and T3 (Table 5). However, plants cultivated with a nutrient solution without nitrate had significantly lower fresh and dry weights compared to the T2 and T3 groups ($p < 0.05$). Nitrogen is known to be the mineral element that is most necessary for plant growth, and nitrate (NO_3^-) is the major form in which plants absorb nitrogen. Several studies indicate that plasma-activated water can be used as a nitrogen fertilizer as it generates nitrogen species (e.g., NO_2^- and NO_3^-) [50,51]. The effects of plasma-activated water on the germination rate, growth characteristics, and total phenolic and flavonoid contents of mung bean sprouts were investigated by Fan et al. [52]. Their results show a remarkable increase in growth characteristics (i.e., stem length and average weight) with PAW15 (with a discharge time of 15 s). The positive effects of PAW treatment on mass accumulation could arise due to the stimulation of protein synthesis [53]. However, in our research, we found non-significant differences in the effects on plant morphology and growth characteristics between T2 and T3. The reason for this could be that the nitrate concentration is equivalent to the normal nitrate nutrient solution and the plasma nitrate.

Table 5. Yields of hydroponically grown lettuce cultivated in nutrient solution without nitrate (T1), normal nitrate (T2), and plasma nitrate (T3) at 21 DAT.

Treatments	Total Plant Fresh Weight (g)	Total Plant Dry Weight (g)
T1	12.14 ± 2.92 b	0.45 ± 0.04 b
T2	77.67 ± 4.72 a	3.62 ± 0.06 a
T3	81.20 ± 13.54 a	3.60 ± 0.60 a
%CV	14.50	15.20
LSD0.05	16.52	0.78

Means in the same column followed by different letters are significantly different ($p < 0.05$) by LSD.

3.3. Nutritional Quality Analysis

Leafy vegetables, especially lettuce, are considered to be a notable nitrate-accumulating species [54]. By analyzing the NO_3^- content in the aboveground and underground parts of green oak lettuce, the plants cultivated in normal nitrate nutrient solution (T2) were shown to exhibit the highest accumulation of NO_3^- content ($p < 0.05$), as shown in Table 6. However, high nitrate content poses a risk to human health and the surrounding environment [54]. Interestingly, lettuce cultivated in the T3 group showed significantly lower nitrate concentrations than those of the T2 group, even though the initial nitrate concentrations of the fertilizers were not significantly different. This result could indicate that NO_3-N produced using cold plasma technology (T3) is rapidly assimilated into an organic form such as free amino acids, as shown in Table 7; therefore, the NO_3-N accumulation in the vacuole was lower than in the T2 group. This result has implications that may be beneficial to the environment and to human health.

Table 6. Nitrate concentration, total nitrogen, and total phenolic in grown lettuce at 42 days.

Treatments	Nitrate Concentration in Plant (mg/kg) FW		Total Phenolic (mg GAE/g FW)
	Aboveground	Underground	
T1	97.27 ± 6.43 c	83.41 ± 12.86 c	3.16 c ± 0.76 c
T2	5370.00 ± 109.28 a	7256.4 ± 604.25 a	27.67 a ± 0.42 a
T3	1674.54 ± 205.70 b	3456.4 ± 154.28 b	22.74 ± 3.79 b
%CV	5.22	10.01	12.58
LSD0.05	44.40	1146.10	4.49

Means in the same column followed by different letters are significantly different ($p < 0.05$) by LSD.

Table 7. Concentration of amino acids (nmol/mg) in green oak lettuce after being cultivated in nutrient solution without nitrate (T1), normal nitrate (T2), and plasma nitrate (T3).

Amino Acid	Treatments			%CV	$LSD_{0.05}$
	No Nitrate (T1)	Normal Nitrate (T2)	Plasma Nitrate (T3)		
Aspartic acid (Asp)	17.06 ± 0.15 b	47.43 ± 1.22 a	50.17 ± 0.21 a	1.88	2.28
Threonine (Thr)	9.65 ± 0.18 c	20.40 ± 1.22 b	28.97 ± 0.32 a	4.15	2.60
Serine (Ser)	10.93 ± 0.04 c	24.5 ± 0.79 b	31.41 ± 0.18 a	2.11	1.50
Glutamic acid (Glu)	22.00 ± 0.06 c	58.20 ± 0.46 b	69.82 ± 0.01 a	0.54	0.85
Proline (Pro)	9.65 ± 0.37 c	31.74 ± 0.52 a	29.16 ± 0.23 b	1.66	1.24
Glycine (Gly)	19.02 ± 0.04 c	58.50 ± 0.11 a	57.56 ± 0.09 b	0.18	0.26
Alanine (Ala)	18.96 ± 0.02 c	52.07 ± 0.27 b	54.30 ± 0.01 a	0.37	0.50
Cysteine (Cys)	1.04 ± 1.26 a	0.82 ± 0.91 a	0.31 ± 0.05 a	125.15	2.87
Valine (Val)	13.09 ± 0.03 b	37.96 ± 0.28 a	37.44 ± 0.06 a	0.55	0.51
Methionine (Met)	0.58 ± 0.06 c	2.88 ± 0.10 a	2.14 ± 0.00 b	3.35	0.20
Isoleucine (Lie)	9.17 ± 0.01 c	27.15 ± 0.20 a	25.29 ± 0.01 b	0.56	0.37
Leucine (Leu)	16.99 ± 0.09 c	47.90 ± 0.20 a	47.24 ± 0.06 b	0.35	0.41
Tyrosine (Tyr)	3.66 ± 0.07 c	5.05 ± 0.46 b	10.99 ± 0.06 a	4.18	0.87
Phenylalanine (Phe)	7.63 ± 0.03 b	23.56 ± 0.46 a	23.78 ± 0.08 a	1.47	0.86
Lysine (Lys)	7.82 ± 0.00 c	23.66 ± 0.23 b	24.96 ± 0.08 a	0.76	0.45
Histidine (His)	2.69 ± 0.03 c	9.32 ± 0.05 a	9.10 ± 0.03 b	0.52	0.12
Arginine (Arg)	5.98 ± 0.07 c	17.67 ± 0.06 b	18.54 ± 0.13 a	0.67	0.30

Means in the same row followed by different letters are significantly different ($p < 0.05$) by LSD.

The results of the present study show that the lettuce cultivated with T2 had levels of total phenolic compounds 0.35 ± 0.05 mg GAE/g FW higher than that cultivated with T3 by 0.28 ± 0.05 mg GAE/g FW ($p < 0.05$), as shown in Table 6. Phenolic compounds, one of the most abundant groups of secondary metabolites found in vegetables, contain several thousand compounds. Several environmental factors, including temperature, light, and nutrient management, are responsible for increasing the biosynthesis of phenolic compounds in plants [55]. Nitrogen, an important nutrient in plants, plays a critical role in regulating phenolic accumulation in plants [56]. Previous research has studied the effect of PAW on the total phenolic compounds in apples [57], mung bean sprouts [58], and rocket leaves [59]. Most of this research indicates that phenolic accumulation depends on plasma exposure time, whereby longer treatment times lead to lower levels of phenolic compounds [50,57]. Since PAW has stronger oxidation abilities with increasing exposure time, the rapid oxidation of phenols causes an immediate reduction in the phenolic content levels [60]. In the present work, plasma nitrate did not induce phenolic compounds in plants to the same extent as commercial fertilizers do (i.e., normal nitrate nutrient solution); however, our results still indicate an effective growth enhancement. Therefore, further research should investigate the suitable conditions for generating plasma nitrate to enhance total phenolic accumulation in plants.

The effects of plasma nitrate (T3), especially the role of NO_3^- species and reactive oxygen species, the nutrient solution without nitrate (T1), and normal nitrate (T2) on the concentration of 17 free amino acids were determined and are shown in Table 7. The results illustrate that threonine, serine, glutamic acid, alanine, tyrosine, phenylalanine, lysine, and arginine have significantly higher concentrations in T3 than in T2 ($p < 0.05$). Zhou et al. [61] showed that RNS generated from PAW, especially NO_3^-, was assimilated by plant enzymes, leading to the generation of organic nitrogen compounds, such as free amino acids, in accordance with the assimilation pathway of nitrate. Therefore, nitrate content generated from plasma nitrate might more easily transform into amino acids. However, green oak lettuces cultivated in the T3 group showed significantly lower levels of proline, glycine, methionine, isoleucine, leucine, and histidine than those in the T2 group ($p < 0.05$). Han et al. [62] explained that cold plasma treatment causes the chemical modification of amino acids through hydroxylation, dehydrogenation, nitration, and dimerization, leading to the degradation of sulfur-containing and aromatic amino acids. Interestingly, very small amounts of methionine were present in the green oak cultivated with T1 (Table 7). Methionine, a sulfur-containing amino acid, is an essential component for the synthesis of all proteins. Wilson et al. [63] also indicated that both nitrogen and sulfur affect the synthesis of amino acids.

4. Conclusions

Air and oxygen were applied using a pinhole plasma jet with a flow rate of 1.5 L/min and discharged for 16 h to generate a nitrogen fertilizer with a nitrate concentration of approximately 800 mg/L. Plant height, root length, canopy width, leaf width, number of leaves, photosynthesis rate, and stomatal conductance of green oak lettuce cultivated in plasma nitrate solution compared to those grown in the normal nitrate nutrient solution were not significantly different, except in respect of leaf area and leaf greenness. Similarly, the yields, including the fresh and dry weights of plants cultivated in plasma nitrate, were the same as those of the plants cultivated in commercial nitrate. Significantly lower levels of chemical compositions (including nitrate concentrations in the underground and aboveground parts and the total phenolic compounds) were found in green oak irrigated with plasma nitrate than in that grown with a normal nitrate nutrient solution ($p < 0.05$). The presence of fewer nitrate residues in the green oak samples indicates a reduced threat to human health and to the environment. Moreover, plasma nitrate was rapidly assimilated to free amino acids, resulting in significantly higher concentrations of amino acids including Thr, Ser, Glu, Ala, Tyr, Phe, Lys, and Arg than in the normal nitrate nutrient solution. These results could indicate that plasma-activated water can be used as an alternative nitrogen fertilizer, as well as an innovative environmentally friendly technology. However, the nitrogen assimilation pathway from plasma fertilizers should be further explored in order to understand why the nitrate levels of green oak cultivated in plasma nitrate are lower than those of the normal nitrate nutrient solution. Moreover, the economic effectiveness of this method should be determined to prepare for the adoption of this technology in the commercial sector.

Author Contributions: Conceptualization, S.R., C.I. and S.-n.T.; methodology, S.J., M.I. and S.-n.T.; software, P.S. and S.J.; validation, P.S., S.J. and S.-n.T.; formal analysis, P.S. and S.J.; investigation, P.S. and S.J.; resources, S.R., K.P. and C.S.; data curation, S.J. and M.I.; writing—original draft preparation, P.S., S.J. and M.I.; writing—review and editing, S.R., K.P. and C.I.; visualization, P.S., S.J. and M.I.; supervision, S.R., C.S. and S.-n.T.; project administration, W.S.; funding acquisition, C.S. and S.-n.T. All authors have read and agreed to the published version of the manuscript.

Funding: This research received external funding from Fundamental Fund 2022, Chiang Mai University (FF65/117).

Institutional Review Board Statement: Not applicable.

Informed Consent Statement: Not applicable.

Data Availability Statement: Data is contained within the article.

Acknowledgments: This work was supported by H.M. The King Initiative Centre for Flower and Fruit Propagation, Chiang Mai with the research location. The authors also thank all staff members at the Agriculture and Bio Plasma Technology Center (ABPlas), Science and Technology Park, Chiang Mai University, Thailand, (CMU SteP), for their efforts and research support.

Conflicts of Interest: The authors declare no conflict of interest.

References

1. Pornsuriya, C.; Ito, S.; Sunpapao, A. First report of leaf spot on lettuce caused by Curvularia aeria. *J. Gen. Plant Pathol.* **2018**, *84*, 296–299. [CrossRef]
2. Food and Agricultural Organization of the United Nations (FAOSTAT). Available online: https://www.fao.org/faostat/en/#data/RP (accessed on 15 November 2022).
3. Camejo, D.; Frutos, A.; Mestre, T.C.; del Carmen Piñero, M.; Rivero, R.M.; Martínez, V. Artificial light impacts the physical and nutritional quality of lettuce plants. *Hortic. Environ. Biotechnol.* **2020**, *61*, 69–82. [CrossRef]
4. Shohael, A.; Hrisha, A.A.; Ahamed, T.; Khatun, S. An easy and reproducible field to table technology for the production of hydroponics lettuce in Bangladesh International Journal of Agronomy and Agricultural Research (IJAAR). *Int. J. Agron. Agric. Res.* **2017**, *10*, 37–47.
5. Market Analysis Report. Hydroponics Market Size, Share&Trends Analysis Report by Type (Aggregate Systems, Liquid Systems), by Crops (Tomatoes, Lettuce, Peppers, Cucumbers, Herbs), By Region, And Segment Forecasts, 2021–2028. Available online: https://www.grandviewresearch.com/industry-analysis/hydroponics-market (accessed on 15 November 2022).
6. Siringam, K.; Theerawipa, K.; Hlaihakhot, N. Effect of nutrient solution on growth of lettuce (*Lactuca sativa* L.) cultivated under hydroponic system. *Thai Sci. Technol. J.* **2014**, *22*, 828–836.
7. Hou, C.-Y.; Kong, T.-K.; Lin, C.-M.; Chen, H.-L. The Effects of Plasma-Activated Water on Heavy Metals Accumulation in Water Spinach. *Appl. Sci.* **2021**, *11*, 5304. [CrossRef]
8. Colla, G.; Kim, H.-J.; Kyriacou, M.C.; Rouphael, Y. Nitrate in fruits and vegetables. *Sci. Hortic.* **2018**, *237*, 221–238. [CrossRef]
9. van der Hoeven, M.; Kobayashi, Y.; Diercks, R. Technology roadmap: Energy and GHG reductions in the chemical industry via catalytic processes. *Int. Energy Agency* **2013**, *56*, 12–16.
10. Chen, J.G.; Crooks, R.M.; Seefeldt, L.C.; Bren, K.L.; Bullock, R.M.; Darensbourg, M.Y.; Holland, P.L.; Hoffman, B.; Janik, M.J.; Jones, A.K. Beyond fossil fuel–driven nitrogen transformations. *Science* **2018**, *360*, eaar6611. [CrossRef]
11. Tanabe, Y.; Nishibayashi, Y. Developing more sustainable processes for ammonia synthesis. *Coord. Chem. Rev.* **2013**, *257*, 2551–2564. [CrossRef]
12. Chu, D.-T.; Vu Ngoc, S.-M.; Vu Thi, H.; Nguyen Thi, Y.-V.; Ho, T.-T.; Hoang, V.-T.; Singh, V.; Al-Tawfiq, J.A. COVID-19 in Southeast Asia: Current status and perspectives. *Bioengineered* **2022**, *13*, 3797–3809. [CrossRef]
13. Sapbamrer, R.; Chittrakul, J.; Sirikul, W.; Kitro, A.; Chaiut, W.; Panya, P.; Amput, P.; Chaipin, E.; Sutalangka, C.; Sidthilaw, S. Impact of COVID-19 Pandemic on Daily Lives, Agricultural Working Lives, and Mental Health of Farmers in Northern Thai-land. *Sustainability* **2022**, *14*, 1189. [CrossRef]
14. Randeniya, L.K.; de Groot, G.J. Non-thermal plasma treatment of agricultural seeds for stimulation of germination, removal of surface contamination and other benefits: A review. *Plasma Process. Polym.* **2015**, *12*, 608–623. [CrossRef]
15. Starič, P.; Vogel-Mikuš, K.; Mozetič, M.; Junkar, I. Effects of Nonthermal Plasma on Morphology, Genetics and Physiology of Seeds: A Review. *Plants* **2020**, *9*, 1736. [CrossRef]
16. Perinban, S.; Orsat, V.; Raghavan, V. Nonthermal plasma–liquid interactions in food processing: A review. *Compr. Rev. Food Sci. Food Saf.* **2019**, *18*, 1985–2008. [CrossRef] [PubMed]
17. Park, D.P.; Davis, K.; Gilani, S.; Alonzo, C.-A.; Dobrynin, D.; Friedman, G.; Fridman, A.; Rabinovich, A.; Fridman, G. Reactive nitrogen species produced in water by non-equilibrium plasma increase plant growth rate and nutritional yield. *Curr. Appl. Phys.* **2013**, *13*, S19–S29. [CrossRef]

18. Guo, D.; Liu, H.; Zhou, L.; Xie, J.; He, C. Plasma-activated water production and its application in agriculture. *J. Sci. Food Agric.* **2021**, *101*, 4891–4899. [CrossRef]
19. Zhang, Q.; Ma, R.; Tian, Y.; Su, B.; Wang, K.; Yu, S.; Zhang, J.; Fang, J. Sterilization Efficiency of a Novel Electrochemical Dis-infectant against Staphylococcus aureus. *Environ. Sci. Technol.* **2016**, *50*, 3184–3192. [CrossRef]
20. Shen, J.; Tian, Y.; Li, Y.; Ma, R.; Zhang, Q.; Zhang, J.; Fang, J. Bactericidal Effects against S. aureus and Physicochemical Properties of Plasma Activated Water stored at different temperatures. *Sci. Rep.* **2016**, *6*, 28505. [CrossRef] [PubMed]
21. Su, L.; Lan, Q.; Pritchard, H.W.; Xue, H.; Wang, X. Reactive oxygen species induced by cold stratification promote germination of Hedysarum scoparium seeds. *Plant Physiol. Biochem.* **2016**, *109*, 406–415. [CrossRef]
22. Mitra, A.; Li, Y.-F.; Klämpfl, T.G.; Shimizu, T.; Jeon, J.; Morfill, G.E.; Zimmermann, J.L. Inactivation of surface-borne microorganisms and increased germination of seed specimen by cold atmospheric plasma. *Food Bioprocess Technol.* **2014**, *7*, 645–653. [CrossRef]
23. Maniruzzaman, M.; Sinclair, A.J.; Cahill, D.M.; Wang, X.; Dai, X.J. Nitrate and hydrogen peroxide generated in water by electrical discharges stimulate wheat seedling growth. *Plasma Chem. Plasma Process.* **2017**, *37*, 1393–1404. [CrossRef]
24. Sarinont, T.; Katayama, R.; Wada, Y.; Koga, K.; Shiratani, M. Plant growth enhancement of seeds immersed in plasma activated water. *MRS Adv.* **2017**, *2*, 995–1000. [CrossRef]
25. Graves, D.B.; Bakken, L.B.; Jensen, M.B.; Ingels, R. Plasma Activated Organic Fertilizer. *Plasma Chem. Plasma Process.* **2019**, *39*, 1–19. [CrossRef]
26. Wu, S.; Thapa, B.; Rivera, C.; Yuan, Y. Nitrate and nitrite fertilizer production from air and water by continuous flow liq-uid-phase plasma discharge. *J. Environ. Chem. Eng.* **2021**, *9*, 104761. [CrossRef]
27. Carmassi, G.; Cela, F.; Trivellini, A.; Gambineri, F.; Cursi, L.; Cecchi, A.; Pardossi, A.; Incrocci, L. Effects of Nonthermal Plasma (NTP) on the Growth and Quality of Baby Leaf Lettuce (*Lactuca sativa* var. *acephala* Alef.) Cultivated in an Indoor Hydroponic Growing System. *Horticulturae* **2022**, *8*, 251. [CrossRef]
28. Jamie. Algae in Hydroponics: Types, Causes, Effects, Treating & More. Available online: https://whyfarmit.com/algae-in-hydroponics/ (accessed on 16 November 2022).
29. Stephens, O. Algae in Hydroponics: How to Get Rid of and Prevent Its Growth. Available online: https://thehydroponicsplanet.com/how-to-get-rid-of-algae-in-hydroponics-for-good/ (accessed on 15 November 2022).
30. Date, M.B.; Rivero, W.C.; Tan, J.; Specca, D.; Simon, J.; Salvi, D.; Karwe, M.V. Effect of Plasma-Activated Nutrient Solution (Pans) on Sweet Basil (*O. basilicum* L.) Grown Using an Ebb and Flow Hydroponic System. *Soc. Sci. Res. Netw.* **2022**. [CrossRef]
31. Zhao, L.; Wang, Y. Nitrate Assay for Plant Tissues. *Bio-Protoc. J.* **2017**, *7*, e2029. [CrossRef]
32. Swain, T.; Hillis, W.E. The phenolic constituents of Prunus domestica. I.—The quantitative analysis of phenolic constituents. *J. Sci. Food Agric.* **1959**, *10*, 63–68. [CrossRef]
33. Miller, D.D.; Rutzke, M. Atomic absorption and emission spectroscopy. In *Food Analysis*, 3rd ed.; Kluwer: New York, NY, USA, 2003; pp. 401–421.
34. Bisergaeva, R.A.; Sirieva, Y.N. Determination of calcium and magnesium by atomic absorption spectroscopy and flame photometry. *J. Phys. Conf. Ser.* **2020**, *1691*, 012055. [CrossRef]
35. Wang, P.-Y.; Shuang, F.-F.; Yang, J.-X.; Jv, Y.-X.; Hu, R.-Z.; Chen, T.; Yao, X.-H.; Zhao, W.-G.; Liu, L.; Zhang, D.-Y. A rapid and efficient method of microwave-assisted extraction and hydrolysis and automatic amino acid analyzer determination of 17 amino acids from mulberry leaves. *Ind. Crops Prod.* **2022**, *186*, 115271. [CrossRef]
36. Bradu, C.; Kutasi, K.; Magureanu, M.; Puač, N.; Živković, S. Reactive nitrogen species in plasma-activated water: Generation, chemistry and application in agriculture. *J. Phys. D Appl. Phys.* **2020**, *53*, 223001. [CrossRef]
37. Al-Sharify, Z.T.; Al-Sharify, T.A.; al-Obaidy, B.W.; al-Azawi, A.M. Investigative Study on the Interaction and Applications of Plasma Activated Water(PAW). *IOP Conf. Ser. Mater. Sci. Eng.* **2020**, *870*, 012042. [CrossRef]
38. Thirumdas, R.; Kothakota, A.; Annapure, U.; Siliveru, K.; Blundell, R.; Gatt, R.; Valdramidis, V.P. Plasma activated water (PAW): Chemistry, physico-chemical properties, applications in food and agriculture. *Trends Food Sci. Technol.* **2018**, *77*, 21–31. [CrossRef]
39. Andrews, M.; Raven, J.A. Root or shoot nitrate assimilation in terrestrial vascular plants—Does it matter? *Plant Soil* **2022**, *476*, 31–62. [CrossRef]
40. Jia, Z.; Giehl, R.F.H.; von Wirén, N. The Root Foraging Response under Low Nitrogen Depends on DWARF1-Mediated Brassinosteroid Biosynthesis. *Plant Physiol.* **2020**, *183*, 998–1010. [CrossRef] [PubMed]
41. Noh, S.W.; Park, J.S.; Kim, S.J.; Kim, D.-W.; Kang, W.S. Effect of Plasma-activated Water Process on the Growth and Functional Substance Content of Lettuce during the Cultivation Period in a Deep Flow Technique System. *Prot. Hortic. Plant Fact.* **2020**, *29*, 464–472. [CrossRef]
42. Li, L.; Li, J.; Shen, M.; Hou, J.; Shao, H.; Dong, Y.; Jiang, J. Improving Seed Germination and Peanut Yields by Cold Plasma Treatment. *Plasma Sci. Technol.* **2016**, *18*, 1027–1033. [CrossRef]
43. Ndiffo Yemeli, G.B.; Švubová, R.; Kostolani, D.; Kyzek, S.; Machala, Z. The effect of water activated by nonthermal air plasma on the growth of farm plants: Case of maize and barley. *Plasma Process. Polym.* **2021**, *18*, 2000205. [CrossRef]
44. Than, H.A.Q.; Pham, T.H.; Nguyen, D.K.V.; Pham, T.H.; Khacef, A. Non-thermal Plasma Activated Water for Increasing Ger-mination and Plant Growth of *Lactuca sativa* L. *Plasma Chem. Plasma Process.* **2022**, *42*, 73–89. [CrossRef]
45. Kučerová, K.; Henselová, M.; Slováková, Ľ.; Bačovčinová, M.; Hensel, K. Effect of Plasma Activated Water, Hydrogen Peroxide, and Nitrates on Lettuce Growth and Its Physiological Parameters. *Appl. Sci.* **2021**, *11*, 1985. [CrossRef]

46. Hachiya, T.; Sakakibara, H. Interactions between nitrate and ammonium in their uptake, allocation, assimilation, and signaling in plants. *J. Exp. Bot.* **2016**, *68*, 2501–2512. [CrossRef] [PubMed]
47. Francesconi, S.; Balestra, G.M. The modulation of stomatal conductance and photosynthetic parameters is involved in Fusarium head blight resistance in wheat. *PLoS ONE* **2020**, *15*, e0235482. [CrossRef] [PubMed]
48. Yamori, W.; Kusumi, K.; Iba, K.; Terashima, I. Increased stomatal conductance induces rapid changes to photosynthetic rate in response to naturally fluctuating light conditions in rice. *Plant Cell Environ.* **2020**, *43*, 1230–1240. [CrossRef]
49. Saberi, M.; Modarres Sanavy, M.A.; Zare, R.; Ghomi, H. Improvement of Photosynthesis and Photosynthetic Productivity of Winter Wheat by Cold Plasma Treatment under Haze Condition. *J. Agric. Sci. Technol.* **2019**, *21*, 1889–1904.
50. Rahman, M.M.; Sajib, S.A.; Rahi, M.S.; Tahura, S.; Roy, N.C.; Parvez, S.; Reza, M.A.; Talukder, M.R.; Kabir, A.H. Mechanisms and Signaling Associated with LPDBD Plasma Mediated Growth Improvement in Wheat. *Sci. Rep.* **2018**, *8*, 10498. [CrossRef] [PubMed]
51. Mujahid, Z.; Tounekti, T.; Khemira, H. Cold plasma treatment to release dormancy and improve growth in grape buds: A promising alternative to natural chilling and rest breaking chemicals. *Sci. Rep.* **2020**, *10*, 2667. [CrossRef]
52. Fan, L.; Liu, X.; Ma, Y.; Xiang, Q. Effects of plasma-activated water treatment on seed germination and growth of mung bean sprouts. *J. Taibah Univ. Sci.* **2020**, *14*, 823–830. [CrossRef]
53. Stoleru, V.; Burlica, R.; Mihalache, G.; Dirlau, D.; Padureanu, S.; Teliban, G.-C.; Astanei, D.; Cojocaru, A.; Beniuga, O.; Patras, A. Plant growth promotion effect of plasma activated water on Lactuca sativa L. cultivated in two different volumes of substrate. *Sci. Rep.* **2020**, *10*, 20920. [CrossRef] [PubMed]
54. Liu, C.-W.; Sung, Y.; Chen, B.-C.; Lai, H.-Y. Effects of Nitrogen Fertilizers on the Growth and Nitrate Content of Lettuce (*Lactuca sativa* L.). *Int. J. Environ. Res. Public Health* **2014**, *11*, 4427–4440. [CrossRef]
55. Henry-Kirk, R.A.; Plunkett, B.; Hall, M.; McGhie, T.; Allan, A.C.; Wargent, J.J.; Espley, R.V. Solar UV light regulates flavonoid metabolism in apple (Malus × domestica). *Plant Cell Environ.* **2018**, *41*, 675–688. [CrossRef]
56. Zhou, W.; Liang, X.; Zhang, Y.; Li, K.; Jin, B.; Lu, L.; Jin, C.; Lin, X. Reduced nitrogen supply enhances the cellular antioxidant potential of phenolic extracts through alteration of the phenolic composition in lettuce (*Lactuca sativa* L.). *J. Sci. Food Agric.* **2019**, *99*, 4761–4771. [CrossRef] [PubMed]
57. Tappi, S.; Ramazzina, I.; Rizzi, F.; Sacchetti, G.; Ragni, L.; Rocculi, P. Effect of Plasma Exposure Time on the Polyphenolic Profile and Antioxidant Activity of Fresh-Cut Apples. *Appl. Sci.* **2018**, *8*, 1939. [CrossRef]
58. Xiang, Q.; Liu, X.; Liu, S.; Ma, Y.; Xu, C.; Bai, Y. Effect of plasma-activated water on microbial quality and physicochemical characteristics of mung bean sprouts. *Innov. Food Sci. Emerg. Technol.* **2019**, *52*, 49–56. [CrossRef]
59. Laurita, R.; Gozzi, G.; Tappi, S.; Capelli, F.; Bisag, A.; Laghi, G.; Gherardi, M.; Cellini, B.; Abouelenein, D.; Vittori, S.; et al. Effect of plasma activated water (PAW) on rocket leaves decontamination and nutritional value. *Innov. Food Sci. Emerg. Technol.* **2021**, *73*, 102805. [CrossRef]
60. Liu, C.; Chen, C.; Jiang, A.; Sun, X.; Guan, Q.; Hu, W. Effects of plasma-activated water on microbial growth and storage quality of fresh-cut apple. *Innov. Food Sci. Emerg. Technol.* **2020**, *59*, 102256. [CrossRef]
61. Zhou, R.; Zhou, R.; Wang, P.; Xian, Y.; Mai-Prochnow, A.; Lu, X.; Cullen, P.J.; Ostrikov, K.; Bazaka, K. Plasma-activated water: Generation, origin of reactive species and biological applications. *J. Phys. D Appl. Phys.* **2020**, *53*, 303001. [CrossRef]
62. Han, Y.; Cheng, J.-H.; Sun, D.-W. Activities and conformation changes of food enzymes induced by cold plasma: A review. *Crit. Rev. Food Sci. Nutr.* **2019**, *59*, 794–811. [CrossRef]
63. Wilson, T.L.; Guttieri, M.J.; Nelson, N.O.; Fritz, A.; Tilley, M. Nitrogen and sulfur effects on hard winter wheat quality and as-paragine concentration. *J. Cereal Sci.* **2020**, *93*, 102969. [CrossRef]

MDPI AG
Grosspeteranlage 5
4052 Basel
Switzerland
Tel.: +41 61 683 77 34

Horticulturae Editorial Office
E-mail: horticulturae@mdpi.com
www.mdpi.com/journal/horticulturae

www.ingramcontent.com/pod-product-compliance
Lightning Source LLC
Chambersburg PA
CBHW061513210926
51850CB00008B/1900
9783725853519